W9-CFV-746

THE BLUEPRINT

Withdrawn

Withdrawn

THE BLUEPRINT

AVERTING GLOBAL COLLAPSE

DANIEL RIRDAN

CORINNO PRESS

Published by Corinno Press
P. O. Box 270365 Louisville, CO 80027 USA
For more information, visit www.corinnopress.com
Copyright © 2012 by Daniel Rirdan
All rights reserved.
Printed in the United States of America
First Edition 1.1

Illustration of superconducting cable in chapter seven courtesy of Nexans. The world degradation map in chapter two is adapted and reproduced with permission from Center for International Earth Science Information Network (CIESIN), Columbia University. 2008. Last of the Wild, Global. [Map] From: Last of the Wild Data Version 2, 2005 (LWP-2): Global Human Footprint data set (HF). Wildlife Conservation (WCS) and Center for International Earth Science Information Network (CIESIN). Palisades, NY: NASA Socioeconomic Data and Applications Center (SEDAC), CIESIN, Columbia University. Available at http://sedac.ciesin. columbia.edu/wildareas.

Library of Congress Control Number: 2011963104
The Blueprint: Averting Global Collapse / by Daniel Rirdan
Includes bibliographical references and index.
ISBN 978–1470135881 (paperback)
1. Sustainable development 2. Environmental protection—economic aspects. 3. Global warming—Prevention. 4. Renewable energy sources

❧

for my children,
Jonah and Ellie

CONTENTS

Acknowledgements........ *ix*

PART I   STRESSORS........*1*

   1: Climate Change........5
   what's in store

   2: Land........*11*
   stressors and prospects of further degradation

   3: Sea........*35*
   stressors and prospects of further degradation

   4: Human Habitats........*48*
   stressors and prospects of further degradation

PART II   MITIGATION........*85*

   5: Transportation........*89*
   on land, at sea, and in the air

   6: Buildings........*131*
   cutting their energy consumption by a factor of ten

   7: Energy........*153*
   lower-impact technologies

   8: Nutrient and Water Recycling........*214*
   focusing on living spaces

   9: Land Use........*231*
   *reinventing the food system & bringing a
   new water source online*

   10: Nature Restoration........*279*
   bringing back the wilderness

11: Drawing Down Carbon........*298*
the necessity to go beyond a carbon-neutral economy

12: Consumption........*331*
overpopulation, consumption habits, industry

13: Economic & Political Paradigm........*351*
what exists, what is called for, and getting there politically

Appendix A........*414*
rooftop-mounted PV panels and thin-film PV:
why they don't add up

Appendix B........*417*
carbon sequestration; two maybe's

Notes........*422*

Index........*476*

ACKNOWLEDGEMENTS

FOR PROFESSIONAL HELP, suggestions, and advice I wish to thank in particular to the following individuals: Shai Agassi, CEO of Better Place, Israel; Richard Alley, Evan Pugh Prof. of Geoscience at Pennsylvania State University, USA; Prof. David Archer, Department of the Geophysical Sciences, University of Chicago, USA; Dr. Frederic T. Barrows, Fish Technology Center, USDA, USA; Dr. Richard Betts, Head of the Climate Impacts Strategic Area, Met Office, UK; Dr. Long Cao, Senior Research Associate, Department of Global Ecology, Carnegie Institution, USA; Prof. James Hansen, head of NASA Goddard Institute for Space Studies, USA; Prof. Ove Hoegh-Guldberg, Director of the Global Change Institute, University of Queensland, Australia; Richard Heinberg, author, USA; Prof. Matthew Huber, Purdue Climate Change Research Center, USA; Prof. Hakan Jonsson, Swedish University of Agricultural Sciences, Sweden; Dr. David Keith, Director of the Insitute for Sustainable Energy, Environment, and Economy (ISEEE) Energy and Environmental Systems Group, University of Calgary, Canada; Prof. Fridolin Krausmann, Institute of Social Ecology Vienna (SEC) and Centre for Environmental History, Austria; Dr. Evelyn Krull, The Commonwealth Scientific and Industrial Research Organisation, Australia; Prof. Rattan Lal, director of the Carbon Management and Sequestration Center, USA; Dr. Johannes Lehmann, Department of Crop and Soil Sciences, Cornell University, USA; Dr. Tom Mancini, team leader, Concentrating Solar Power Systems, Sandia National Laboratories, USA; Prof. Gerald A. Meehl, Climate and Global Dynamics Division, National Center for Atmospheric Research, USA; Veerabhadran Ramanathan, a Victor Alderson Prof. of Applied Ocean

Sciences and director of the Center for Atmospheric Sciences at the Scripps Institution of Oceanography, USA; Dr. Constantine Samaras, RAND Corporation, USA; Dr. Allan Savory, founder of Africa Center for Holistic Management, Zimbabwe; Dr. Kevin Schaefer, National Snow & Ice Data Center, USA; Dr. Gavin Schmidt, Goddard Space Flight Center, USA; Dr. Jürgen Schnieders, Passivhaus Institutute, Germany; Michael Soule, Prof. Emeritus of Environmental Studies, University of California, Santa Cruz, USA; Mark Stemmle, Project Manager, Superconducting Cable Systems at Nexans Deutschland GmbH, Germany; Prof. Yoshihide Wada, Department of Physical Geography, Utrecht University, Netherlands; Lars Weimers, Chief Engineer HVDC Systems, ABB, Sweden; Matthew Wright, Executive Director of Beyond Zero Emissions; Prof. Bert van der Zwaan, IPCC lead author, The Netherlands; Prof. Jaap Van Rijn, Hebrew University, Israel; Dr. Sergey Zimov, director of the Northeast Science Station, Russia.

For proofing, editing work, formatting, and research related to the notes, I wish to thank Janaye Gerrelts, Lora Freeman Williams, and Darcy Greenwald.

For his assistance with the generation of maps, I would like to thank Roman Bondarenko. For their assistance with the more complex calculation and projections, I wish to thank Sergey Nikulin and Evgeny Radchenko.

For valuable comments, feedback, and discussions I wish to thank Darcy Greenwald, Nubia Trone, Dana Leman, Irving Frank, Neil R. Biers, Eric Maisel, Phil Ferrante-Roseberry, and Niels and Mark Schonbeck.

Last but not least, I wish to thank my parents for making this book possible, and my wife for putting up with me for those long, grueling months and supporting me throughout the project.

PART I

STRESSORS

WE ARE FACED WITH an impending calamity that threatens to bankrupt the planet. This is one of those times when doing our best is not enough. We must do what is necessary.

Global warming is one of the global stressors afflicting our world. It is not the only one, though. And perhaps it is not even the main one. There are others, such as water shortages and scarcity of fossil fuel. What is on the table is the imminent withering of the planetary ecosystem, and with it much of the manmade world.

Halting the emission of greenhouse gases is easy; it would have taken all of one paragraph to spell it out. Doing so while keeping us minimally comfortable is a different matter. In part, that's what this book is about. And at times, it won't be all that comfortable. But then again, when there is an environmentally-devastating asteroid heading your way, you do what needs to be done. Nature cannot be negotiated with. In part, that's what this book is about, as well.

The blueprint laid out in this book contains the called-for breadth of vision, laying out an utterly new course for our technological and industrial engines. The plan is audacious, casting aside sacred cows and calling for measures that will impinge upon our comfort. It cuts through the haze, with immediately employable solutions. It goes the distance, using myriad computations and models to validate the content.

The intent here is a makeover of the manmade world within fifteen years.

It will cost a lot of money—if you must look at it this way. What it really means is that a lot of people will have to perform. It will take many millions rolling up their sleeves, but the means are at hand.

Neither the deliberate pursuit of profit nor the dynamics of existing governments quite lend themselves to getting us from here to there. Had they been able to do so, they would have—during the last half century, since the alarm bells first rang. Let's put national governments and current economic dynamics aside and consider what actually needs to be done. Function will dictate form; these will be what they need to be.

Get ready to embark on a powerful, demanding journey: with tough choices and tougher numbers, permeated with the grit of concrete. It will be labor-intensive. It will be challenging. It will be taxing. In short, it will be thrilling and meaningful. This is a call to action for all the grownups out there to knuckle down and buckle up. It asks of us to do the job that the adolescents of the past left for the adults of the present to handle.

It is time we steer our civilization on a new course.

1
CLIMATE CHANGE

≈

what's in store

CLIMATE DYNAMICS

THERE WERE TIMES when tropical forests dominated all continents except Antarctica. There were other times when Earth was almost frozen solid from pole to pole. Life has existed in between those two ends of the climatic spectrum.

What has been controlling the climate of the world is a symphony of myriad notes generated by many instruments.

Beyond the annual cycle of seasons, the shortest notes are the minute fluctuations in solar intensity. Minimal sunspot activity is suspected to be one of the instigators in the climate blip that was the Little Ice Age from about 1300 CE to 1800 CE.

Another short-term player is the sulfur haze vented by the occasional volcanic eruption. The haze deflects sunlight back into space. When Mount Pinatubo erupted in 1991, the discharge of aerosols reduced the amount of incoming solar radiation. Consequently, the global mean temperature dipped by 0.6°C for a period of two years.

The occasional changes in warm ocean currents can also impact the climate. Their effect ranges from the relatively mild, as in the case of the El Niño phenomenon, to the relatively significant, as when the Atlantic conveyer belt, circulating warm tropical water northward, got stalled about 12,000 years ago. Climatic changes driven by ocean currents are usually regional rather than global in nature.

Superimposed on these rapid climatic fluctuations are the cyclical changes of the Earth's orbit. These cycles span tens of thousands and hundreds of thousands of years. In some eras, the Earth's orbit is more elliptical, in others less. In some eras, the Earth's axis is tilted slightly more toward the sun, in others it is tilted slightly less. The combined effect of these cycles is to redistribute the heat between the two hemispheres and otherwise widen the gulf between summer and winter temperatures. During a given ice age, when the climate is colder to begin with, these orbital oscillations have a pronounced effect: they are the main instigator in getting the Earth in and out of glacial periods within a given ice age.

A bit of an explanation is in order. A glacial period of an ice age is when North America is under a two-mile-thick ice sheet and when ice cover is widespread. The interglacial period of an ice age is what we have had for about the last 10,000 years: permanent ice sheets that are largely constrained to the polar regions.

The two driving engines that get our planet to swing between glacial and interglacial periods are the changes in atmospheric levels of carbon dioxide (CO_2) and the amplification effect of the reflective ice cover. (The more widespread the ice cover is, the more sunlight is reflected back with consequently less ground warming.) However, the orbital changes of the planet start these two big engines—the extent of ice coverage and the rates of CO_2 emissions—leaning one way at the beginning of a glacial period and the other way at the onset of an interglacial period.

On a scale of tens of millions of years, the thickness of the CO_2 blanket changes markedly. The thicker the atmospheric blanket of CO_2, the warmer it gets. Over the long run, the foremost mechanism controlling the thickness of the blanket is a ponderous interaction between volcanic activities, which emit CO_2, and the weathering process, which locks down the carbon that is in the air.

Over periods of eons, volcanoes belch out CO_2. Everything else being equal, the higher the volcanic activity in a given Age, the more CO_2 released into the air, and the thicker the greenhouse blanket.

Counteracting this mechanism is the weathering process. Rainfall reacts with the CO_2 in the air, creating carbonic acid. The slightly acidic groundwater attacks rocks containing silicate minerals. The ensuing chemical reaction locks into these rocks the carbon contained in the groundwater, taking the carbon out of circulation for a very long time.

The volcano–weathering interplay is probably the greatest climate-engine of them all. When it is all said and done, the ever-shifting balance over millions of years between the rate of weathering and the rate of CO_2 emissions from volcanoes and hot springs accounts for the ponderous oscillations of Earth between an icehouse and hot-house climate through the geological epochs. A Hothouse World is predominantly a tropical world. An Icehouse World is what we have had for the past thirty-four million years.

On a longer time scale yet—that of hundreds of millions of years—is the ever-intensifying radiation of the sun. Four and a half billion years ago the sun output was but 70 to 75 percent of its current level. However, the ever-increasing sun radiation has been compensated by a potent greenhouse blanket in the early period, followed by an overall decrease in greenhouse gas concentrations through the ensuing thousands of millions of years.

Those are the prominent, more obvious instruments controlling climate. There are many ancillary ones, such as the patterns of wind, dust,[1] precipitation, and clouds, which all amplify or mitigate the effects of the key instruments. As the climate changes, so do the patterns of vegetation, soil exposure, and ice coverage—and with them the level of reflectivity of the sun's rays. All of these parameters interact, producing a symphony of dazzling complexity and dynamics.

Then we showed up on the scene.

GLOBAL WARMING

The planet's surface emits the energy from the sun in the form of infra-red radiation, or heat. Some of that makes it to outer space, some is absorbed by the so-called greenhouse gases. Those in turn emit some of the heat downward. The net result is augmented warming of the planet surface. Thanks to this blanket of greenhouse gases, Earth does not have an average temperature of –18°C (–.4°F). The resultant 33°C higher average temperature makes life as we know it possible on Earth.

Carbon dioxide is constantly being cycled through the vegetation, ocean surface, and atmosphere. Most of the landmass, and therefore vegetation, is situated in the northern hemisphere. When it is winter in the northern hemisphere, the bulk of the world's leaves shed and release their CO_2, and consequently the atmospheric concentration goes up a bit. In the summer it goes back down.

At the beginning of the current interglacial period, eleven thousand years ago, the CO_2 concentration in the air hovered around 259–265 parts per million (ppm). This is pretty much how it stayed until about 3600 BCE, when the carbon dioxide (CO_2) levels in the atmosphere started to inch their way up and then plateaued at 276–283 ppm around 480 BCE,[2] where they stayed until the early 1800s.[3]

About that time, we got into the fossil-fuel business and started releasing massive amounts of CO_2 into the air. Some of it was picked up by the ocean, some by the land.[4] However, about half of it remained in the air. And we went from an atmospheric concentration of around 283 parts per million (ppm) in 1807 to 391 ppm as of 2011. This CO_2 concentration is the highest in the last 800,000 years and potentially for the past few million years.[5]

Carbon dioxide accounts for about 77 percent of the effects of our annual greenhouse gas (GHG) emission.[6] Methane and nitrous oxide account for most of the rest. The main source of anthropogenic, or human-induced, greenhouse gases is the combustion of fossil fuel. We use the resultant heat to generate electricity, to warm indoor spaces, to power our motor vehicles and various industrial processes. Other significant sources of anthropogenic GHG emissions are due to carbon outgassing from the soil, from cement production, and from deforestation. Secondary sources of anthropogenic GHG emissions include landfills, rice paddies, the production of steel, and the manufacture of petrochemicals. It is unclear whether livestock emissions should also be added to this tally. Our cattle take in CO_2 from the air and turn it to the far more potent methane at the back end of the process. Hence, no cattle, no extra methane. Yet, in some roundabout way, the domestic cattle of today stand in place of the hordes of bison and musk ox of bygone days—which also contributed their share of converting CO_2 to methane.

At the end of the day, what matters most is the resultant level of warming from it all. Currently, we are at 0.9°C mean global warming, and there is no doubt that human activities are at the root of it.[7] In fact, if not for the offsetting effects of aerosols and minimal solar activity, the warming would have been greater yet.[8]

Under the business-as-usual, fossil-fuel intensive scenario, in which CO_2 concentrations are projected to rise to 872 ppm by the 2090s, an integrated model at the Hadley Centre projects that by that time, the temperature will have increased by 4.4°C to 7.3°C from pre-industrial temperature levels.[9] Under a comparable emission scenario, MIT Integrated Global Systems Model projects between 5.1°C and 6.6°C warming relative to pre-industrial levels by 2100.[10] In accordance, I assume a median figure of 5.5°C global mean temperature increase by the end of the century as a likely outcome under the business-as-usual, fossil-fuel intensive scenario.

As of 2010, the year 2010 was one of the two warmest years on record.[11] In fact, as 2011 came to a close, nine of the ten warmest years in recorded history have been since 2000. During the spring of 2011, fires of epic proportions raged in Texas, which had its driest spring on record. Australia and New Zealand had mega-floods, and the Midwest had record snowfall. This is just the beginning; this is just at 0.9°C warming. These are but first, timid forays of a new weather regimen.

The routine 4°C–7°C oscillations between glacial and interglacial periods[12] take thousands of years to run their course, not one hundred years, as is projected to happen under the current emissions trajectory. Moreover, in the last few million years, the changes have been occurring within the bounds of a certain temperature range. At present, we are *already* at the warm end of the pendulum. Pushing it 5°C farther out may prove to be outside the operational specs of some of the existing species and ecosystems.

In terms of global mean temperature, we are travelling back in time. In our current trajectory, around mid-century we will have gone back in time to the Pliocene epoch, a few million years back. Toward the end of the century, we are likely to have reached the Mid Miocene

Climatic Optimum period, around 15 million years back. And then on to the twenty-second century and further back in time, getting to temperature levels that are likely to have last existed during the Eocene epoch,[13] perhaps around 40 to 50 million years ago—along with ocean acidity not on a comparable time scale.[14]

This is where the similarities may end. It is one thing to have global transitions of climate over millions of years or over many thousands of years, allowing most species to migrate, evolve, or work their way to suitable changing climatic distribution. It is an entirely different ball of wax to turn the dial 5°C–6°C over a one hundred year period for a planetary ecosystem that is largely bankrupt with only isolated, hemmed-in pockets of intact nature.

2
LAND

≫≪

stressors and prospects of further degradation

ANOTHER GLACIAL PERIOD slowly gave way to a spring era starting about 13,000 years ago.

The boundless panorama of the American Great Plains teemed with life. Camels treaded along the Missouri River and giant ground-sloth ambled about, towering over some of the widely dispersed trees of the prairie. Out in the vast open territories, herds of horses grazed the tall grass along with herds of pronghorns, mammoths, and bison. Packs of scimitar cats and majestic American lions closely trailed the grazing masses. Titanic condor-like birds flew under the big blue sky.

About 10,000 years ago, as the most recent spring era gave way to the long summer to come, most of these massive animals were dead.[1]

One thing was different about the recent spring era, setting it apart from the spring era preceding it, about 120,000 years earlier, and from the many prior spring eras: the first settlers had arrived at the Great Plains.

The first signs of humans in the Great Plains were toward the end of the long winter era, about 15,500 years ago.[2] In the centuries and millennia following their arrival, the settlers hunted the camels, the horses, the mammoths, the larger bison species. And one day the big animals were gone.

Until those die-offs, massive herds of millions of herbivores kept the vast steppe grasslands of the world alive, while the predators in turn kept the herbivore populations in check. The giant herbivores fueled plant productivity by eating the steppe grass produced during the rainy season and returning its nutrients through their manure. They trampled down shrubs, preventing these plants from gaining supremacy.[3] They ingested seeds of grassy plants and deposited them afar via their dung.[4] They left little fuel load for possible wildfires, as they kept the vegetation trimmed and the litter grazed away.

With the demise of millions of large animals, eerie silence must have slowly descended throughout.

Until that time, some of the fruited plants had circumvented the trade-off between seed size and dispersal by relying on massive herbivores to disperse heavy seed loads over long distances.[5]

The Kentucky coffee tree produces a large pod containing pulp and seeds meant for dispersal by elephants that are all dead.[6] The distinctive Joshua tree occupies desert grasslands and scrublands of the Mojave Desert. The giant Shasta ground sloth, its companion, used to munch on its fruits and thus help in dispersing its seeds far and wide. People happened, and the Shasta became extinct. Subsequent distribution of the Joshua tree dwindled to present levels.[7]

The population crash of the giant mammals in the Great Plains marked the beginning of the long downward spiral. In the following millennia, it was a lot of stop and go: In some periods and locales, the degradation of the American steppe halted. In most periods and places, it continued.

Future waves of settlers brought with them the pigs and the black rats and the fences. The human juggernaut rolled on. The bison, elk, and antelope—the few large species that survived the impact of the first human settlements—were slaughtered or driven off. Next would come the systematic clearing of the prairie coinciding with the planting of countless fields of corn, soy, and wheat interlaced with cement roadways.

It happened in the Great Plains of North America. It happened in the Pampas of South America. With the surge of human populations, it has happened in one region after another across the planet. We have been inflicting scars on the living tissue of Earth wherever we stake our claim, whether it is the decimation of the Great Lakes beaver population by the Five Nations of the Iroquois in the first part of the 17th Century; whether it is the obliteration of mangrove forests along the gulf coast of Thailand to put in place shrimp farms; or whether it is the truckloads of logged majestic teak trees making their way up north out of Burma's old growth forest.

Mankind is hunting down the last remnants of spectacular, large wildlife species at a brisk pace. Currently, there are an estimated 6,000 elephants left in the wild in eastern Democratic Republic of Congo, down from approximately 22,000 a decade or so ago, due to both the civil war and ivory poaching.[8] With 2011 possibly being a record year, 23 metric tons (henceforth, *tons*) of ivory were seized. This represents about 2,500 slayed elephants.[9] Meanwhile, the ape population in Gabon declined by more than half between 1983 and 2000.[10] Leopard skins make beautiful throw rugs; tiger bones bestow health;[11] and rhinoceros horns reduce delirium in people—alas, only after the rhino is killed for its horn and not beforehand.

At the present time, the remaining massive intact blocks of nature are these whose destruction and conversion for human use has offered the least economic value: the arid deserts, the tundra and taiga, the polar ice sheets, and the Amazon rainforest.

The Amazon rainforest is without a doubt the last grand stronghold of the planetary ecosystem. It covers a region of nearly six million square kilometers. It provides habitat for the largest concentration of species of any land ecosystem. The Amazon is home to one out of every five remaining mammal, bird, and tree species in the world. It is a vast complex of organisms and relationships that has evolved over millions of years.

And it is being decimated, piece by piece.

Under the current emissions trajectory, we are headed toward 5°C–6°C warming by the end of this century.[12] A complex model shows that with 5°C warming, a third of the Amazon will be gone by 2100. However, this warming will have committed the Amazon to a worse fate: an almost complete dieback in subsequent centuries.[13]

Alas, the prospect is more severe than this model suggests. Climate change is neither the only agent nor the most significant one.

FIGURE 2.1. land degradation (heavy degradation in dark hues)

Source: Map data file from SEDAC – Last of the Wild Data Version 2, 2005, Global Human Footprint data set: World, Wildlife Conservation Society (WCS) and Center for International Earth Science Information Network (CIESIN), http://sedac.ciesin.columbia.edu/wildareas/downloads.jsp#Humfoot.

Note: A more nuanced, detailed version in color is found at www.danielrirdan.com

For untold centuries, humans living in the Amazon have been hunting its animals. The scale of hunting has reached crisis proportions in recent decades. Rare and endangered species are extensively sought out, particularly large species with low reproduction rates.[14] In the 1980s, it was estimated that indigenous hunters in Amazonia were killing, or fatally wounding, over fifty million animals each year, the vast majority of which were mammals. This may leave the otherwise undisturbed regions of the forests strangely quiet. Many of these animals act as seed dispersal agents for various species of trees.[15] In their absence, some of the trees may become locally extinct.[16]

Human harvesting of nuts and fruits also degrades the tropical rainforest ecosystem. Those extractive activities remove nutrients from the ecosystem and deprive forest animals of a primary food source. For instance, people pick the fruits of the Moriche, which are a critical food source for the red-bellied macaw and the tapir.

In the pursuit of choice trees, our march to new frontiers is unabated. Crawler tractors and wheeled skidders are pecking away at the forest canopy, working their way through the deepest recesses of the Amazon rainforest. There are over three thousand timber mills scattered about Amazonia, busily priming the choicest trees for conversion into parquet, furniture, chopsticks, and coffins. All in all, the mills take out of the forest about twenty-five million cubic meters of trees each year.[17]

Insofar as the extent of collateral damage goes, this is only the tip of the iceberg. Skid trails, logging roads, and landing areas impose a heavy collateral damage in the form of uprooted stems and deep cuts to the branches. Wounds to plants can provide infection sites for pathogens. Once established, those pathogens can spread to neighboring plants.[18] Most of the damage occurs as people clear vegetation in order to skid the prized trees out of the forest. On average, for every tree being hauled away, over forty are killed or damaged. Logging batters 10 to 40 percent of the living biomass of the forests within the logged area.[19]

Loggers make pathways, and the resultant forest edges are buffeted by winds and drying sunlight. These edges lead to structural changes, including an increased mortality of trees and a reduced living biomass.

Roads fragment the forest. On the heels of fragmentation, canopy closure takes but a few years. Full restoration of the forest ecology is another matter.

Most Amazonian forest mammals are reluctant to cross even small forest gaps. For many forest dwellers, a road cut through the evergreen presents a major habitat discontinuity. The roadway is brighter, hotter, drier, and frequently polluted by noise and chemicals. It offers very little by way of protective cover or habitat. Rents in the forest canopy have created countless tiny islands in which animals are stranded. These forest islands intensify hunting pressures and competition for depleted resources, foremost among the larger mammals, such as the woolly and spider monkeys.

In addition, the population of each island is limited, too limited for a viable minimum to assure survival. The prospect that an increasingly fragmented Amazon can retain its full assemblage of avian and mammalian species is unlikely. Under a conservative estimate, an aggregate area of over 1.2 million sq km of the Brazilian Amazon may have already lost key elements of its large vertebrate fauna due to the combined effects of hunting and fragmentation.[20]

With the loggers come the roads. A 7,500-kilometer-long highway is planned to be cut deep into the nearly pristine heart of the Amazon basin.[21] The road will bring, as it always has, the hunters, the miners, the ranchers, the land speculators, and the agribusinesses. In short, deforestation.

By 2001, a part of the Amazonian forest larger than Turkey had already been cleared.[22] More has been cut down since. In May 2011, two-hundred fifty square kilometers of the Amazon rainforest were destroyed. In the following month, three hundred square kilometers of the Amazon rainforest were destroyed. This rate of demolition is equivalent to razing an area the size of Manhattan each week. In Columbia, over one hundred square kilometers of forest are being razed each year to make room for cocoa plantations.[23] Deforestation in parts of the Peruvian Amazon has increased sixfold in recent years, as rising prices of gold have sent many small-time miners scurrying to the area.[24]

The surging global demands for biofuel and animal feed bring more and more farmers to the edges of the rainforest. As the farmers purchase land from the cattle owners, those in turn press in and obtain land holdings deeper in the forest.[25] Oftentimes, the cattle owners are in truth land speculators, who just retain cattle as nothing more than a cheap way to maintain the rainforest land cleared for the benefit of the next land speculator.[26]

By 2050, under the business-as-usual scenario, a considerably larger area of the Amazon will be razed—as indicated in figure 2.2. This projection accounts only for outright deforestation and does not include areas of selective logging.

Tropical rainforests have rarely if ever burned. This is because rainforests have a dense, nearly continuous canopy cover that helps to maintain a shaded, stable, humid climate. Under these conditions, potential fuels such as litter and woody debris do not accumulate

FIGURE 2.2

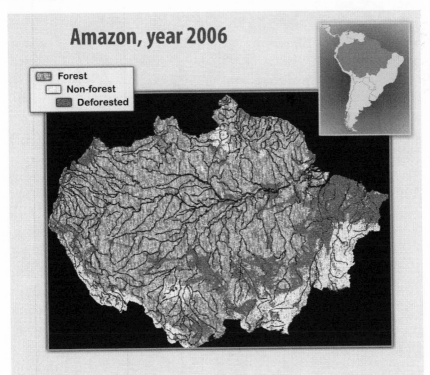

Amazon, year 2006

Forest
Non-forest
Deforested

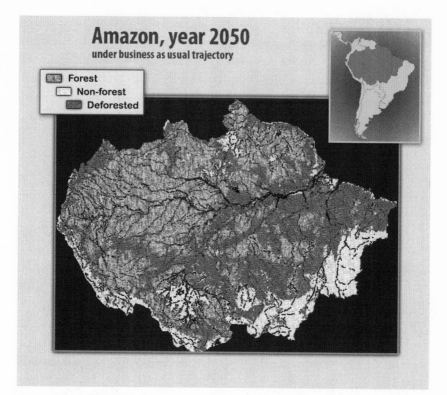

Source of data: Britaldo Silveira Soares-Filho et al., "Modelling Conservation in the Amazon Basin," *Nature* 440 (2006): 520–23.

in the forest understory as they are rapidly decomposed by fungi, microbes, and insects. This is changing. As the large canopy trees are removed or otherwise die, a canopy gap is created, through which sun rays penetrate the forest, both drying the organic debris and heating the forest floor.[27] The new conditions allow grass to grow in the understory, providing fuel and making the forest more susceptible to sustained fire.[28]

Croplands are traditionally cleared via fire. As those abut forest edges, fire often spills over into the forest itself.

Smoke-borne aerosols from fires disrupt normal hydrological processes and reduce rainfall. That is not all. Fires reduce the ability of tropical forests to retain water. They exacerbate flooding, erosion, and seasonal water shortages. Worse yet, forest composition significantly shifts with each consecutive fire event. After a few burns, the forest experiences a drastic proliferation of pioneer species.[29] In other words, under repeated fires, the old growth, closed canopy of the Amazon is gradually being replaced by a low-stature ecosystem dominated by bamboo, grasses, ferns, and smaller, denser wooded trees.[30]

Most global climate-models predict more than 20 percent reduction in rainfall in eastern Amazonia by the end of the century, with the biggest decline occurring during the dry season, when the forest is most vulnerable.[31] A 2012 study found that human land use activities have already begun to change the water cycle in the Amazon.[32]

It was found that if the soil moisture falls 30 percent below its maximum value for a period of two or three years, the large canopy trees swiftly start dying.[33] In the drought of 1997, some parts of the Amazon may have reached this threshold. For a brief period, tree mortality spiked 50 percent above the normal rate.[34] As of this writing, the 2010 drought was the most severe drought on record. During that year, the greenness levels of Amazonian vegetation—a measure of its health—decreased dramatically in most areas and did not return to normal levels, even after the drought ended in the autumn of 2010.[35] Forests are a significant source of moisture that creates rain. Reduced forests mean less rain.[36] As the tropical rainforest system approaches a tipping point, a dry year may push it into a lower state, such as savannah.[37]

The forest woes don't stop there. Prolonged water stress may cause the trees to shut down photosynthesis while respiration continues, depleting their internal carbon reserves. Subsequent carbon starvation reduces the trees' ability to fend off attacking insects or fungi.

Pathogens such as some rusts and powdery mildews are well-suited to drier, warmer, brighter conditions. Extended warmth during droughts can increase their presence, allowing them to overwhelm their already stressed tree-hosts. When an entire forest becomes weakened by environmental stress factors, an onslaught of damaging fungal diseases can result in large-scale tissue death and tree mortality. It is likely that fungal disease was instrumental in the widespread tree mortality that occurred 250 million years ago.[38]

Trees are not the only members of the tropical forest that will be adversely affected by the projected warming. Unlike those in higher latitudes, insects in the tropics can exist only in a relatively-narrow thermal band.[39] Global warming would spell trouble to the native insects of the Amazon.[40] There is nothing trivial about this. Insects play a primary role in decomposition, nutrient cycling, and pollination—all of which contribute to forest regeneration and resilience.

Mammals may not fare any better in extreme heat and humidity. Humans, for one, cannot survive for sustained periods under wet-bulb temperatures that are in excess of 35°C (about 95°F).[41] Similar limits may apply to other mammals.[42] Figure 2.3 indicates projected daily wet-bulb temperatures by 2100 in the city of Manaus and, by implication, the larger Amazon region.[43] The Amazon rainforest may turn from a mammalian wellspring to a mammalian graveyard.

FIGURE 2.3

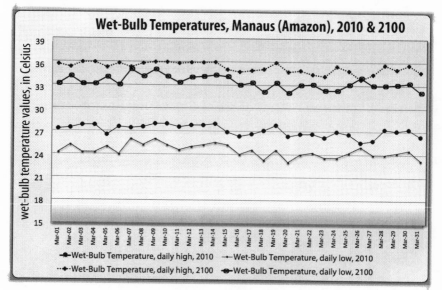

Note: Data extrapolated from Freemeteo.com, Weather History.

It seems probable, if not outright likely, that between now and the end of this century we will have been instrumental in the annihilation of the entire Amazon Rainforest and all smaller tropical forests on the planet. The existing stands of tropical rainforest in the world are likely to be replaced by a mosaic of ranches, crop fields, and patches of open-canopy forests. With them, the climate that gave them rise will also be gone. There would be no way to bring that luxurious world back.

The last vestige of ecologically rich biomes will be gone from Earth for eons to come, leaving behind ghosts of what was and ashes, which shortly thereafter would be scattered by the winds.

———— • ————

Then there is the rest of the biosphere. Or what's left of it, at any rate.

Ecosystems are able to withstand strains to which they are subjected, resisting a change to another state. Most ecosystems have redundant species; they can afford to lose species without actually changing their basic characteristics.

Biodiversity today is being chipped away much in the same way as supporting cables are snipped off from a suspension bridge. Although every species-deletion leads to an overall weakening, nothing observable really happens until one too many species is taken out. When the cumulative changes exceed resiliency limits, they push the ecosystem over the edge—resulting in an irreversible transition of the ecosystem to a new, lower state.[35]

Decades of fire suppression in the western United States have resulted in a dramatic buildup of litter and dead wood. Coupled with a warming climate, this has resulted in many more wildfires than there were thirty years ago, and with a combined burned area six times the size.[44]

Prior to human meddling, the burning of lodgepole pines resulted in random forest clearings, as the distribution of wildfires was haphazard. Fire suppression management put a stop to this. Beetles can now spread throughout the forest, moving from tree to tree without any break in the forest canopy to stop their advance.

Mild temperatures have allowed more insect larvae to survive the Northwest winters. Instead of the two-week window, the beetles fly continually from May until October—attacking trees, burrowing in, and laying their eggs. What makes this worse is that climate-stressed trees produce less pitch, which makes them more susceptible to beetle damage. Historically, the beetles rarely attacked immature trees. This has changed as well; the tree die-offs are on a scale unprecedented in modern times. Due to the freezing temperatures at high elevations, the whitebark pine used to be protected from the beetles. It is now functionally extinct. With its demise, an important food source has winked out from the forest along with the entrapment of snow, which retained some of the water for spring melt. Walnut trees may wither under warmer, drier conditions predicted in the future.[45] And the yellow-cedar in southeastern Alaska is dying in vast numbers as the shallow roots freeze in the absence of protective snow cover in the late winter and early spring.[46]

Hundreds of millions of pine trees in North America have died in recent times. In a vicious cycle, their dried remnants offer readily available fuel for wildfires that may devastate more trees yet. The potential for outbreaks of spruce and mountain pine beetles in western North America's forests is likely to increase significantly in the coming decades.[47] Ecologists now expect that some of these areas will not recover as forests; they will instead return as open grassland.[48] The Greater Yellowstone Ecosystem is a case in point. Within half a century, the projected climate change may result in fires so frequent and severe that burned parts of the Yellowstone forests are likely to come back as open woodland, grass, and shrub vegetation.[49] Our fire suppression practices since the late 1800s have resulted in the build up of a lot of dry wood. It is projected that a dramatic increase in temperature or drought will push the western United States into a fire regime more intense than any in the last three thousand years.[50]

The deserts of the North American Southwest are not faring any better.

Prolonged heat combined with drought delivers a one-two punch to trees, leading at least in one studied case to a breakdown of the water-transportation system of the trees.[51] As trees shut down, biological agents, like bark beetles, can successfully colonize them, increasing the number of dying trees. That is exactly what happened recently across much of the piñon and juniper woodlands covering the Four Corners region, where New Mexico, Arizona, Colorado, and Utah meet.

This region experienced a severe drought from 2000 to 2003, with precipitation levels 20 to 50 percent less than average. This was accompanied by abnormally high temperatures. A large fraction of the weakened piñons was colonized by the pine bark beetle. In total, close to a fifth of all forested area in the Southwest—millions of hectares—has experienced mortality due to severe wildfires and bark beetle outbreaks in the last twenty years. It is projected to get worse.[52]

The decimation of our tree populations in North America has escalated in the past one hundred and fifty years. From Mexico to Alaska, this massive die-off has occurred because of the burgeoning of insects who leave dead trees in their wake. Approximately eighteen million hectares of forest have died since the turn of this century in the Rocky Mountains of Canada and the United States alone, from piñon to pine, juniper to fir.[53]

Reports of extensive tree die-off are coming in from all over the world, especially from semiarid ecosystems where woody plants are already near their physiological limits of water-stress tolerance.

From Africa, there are reports of tree mortality linked to drought and heat in the tropical moist forest in Uganda.[54] In Zimbabwe, the mountain acacia is dying;[55] in Namibia, centuries-old, singular quiver trees are dying;[56] in northern Algeria, the majestic cedar trees are dying;[57] and across the Sahel, significant numbers of trees have perished in recent decades.[58]

From Asia, reports are coming in of tree die-backs in the tropical rainforest of Borneo and the dry tropical forests of western India.[59] The Korean fir is dying in the Korean peninsula;[60] The African juniper is dying in Saudi Arabia; and large swatches of Chinese red pine are dying in east-central China.[61]

Increased mortality of oak, fir, spruce, beech, and pine species was reported in France.[62] In Greece, a severe drought decimated large numbers of Aleppo pine trees, the most drought-tolerant of the Mediterranean pines.[63] Farther north, there was extensive die off of the Siberian spruce in northwest Russia.[64] In the far south, many of the tall *Nothofagus dombeyi* trees have perished in Patagonia, South America.[65] In the sub-humid environment of northeast Australia, multi-year droughts have repeatedly triggered widespread eucalyptus mortality.[66]

Forest die-offs have always taken place, linked to the ever variable climate. However, the magnitude of die-offs and the breadth of these occurrences across the Earth could suggest something beyond the norm. This has been the conclusion of a major study that analyzed many of these tree mortality reports.[67]

There is little doubt that as the temperature needle works its way up, the magnitude of die offs will become more pronounced and calamitous. For one, there is no consistent evidence that trees migrate in response to the warming climate.[68]

If the beetles, under the cover of fire and high temperatures, are the air forces, the invasive plants are the ground troops in an overall trend that undermines the existing ecosystems of the world.

Cheatgrass (*Bromus tectorum*) has taken over sizeable portions of the rangelands, markedly in the Great Basin of North America. Many areas formerly dominated by big sagebrush are today covered with exotic annual grasses, primarily cheatgrass.

Native, perennial bunchgrasses grow in scattered clumps, and thus wildfire doesn't spread readily across a native bunchgrass community. The invasive cheatgrass has provided continuous cover, and the fire spreads easily throughout. Moreover, cheatgrass usually dries out by mid-June, significantly lengthening the fire season and increasing its frequency. Cheatgrass roots grow fast. By the time cold winter temperatures chill the topsoil, cheatgrass roots have thrust down to soil depths where temperatures rarely drop below 2°C (about 36°F). This enables the invasive grass to continue growing throughout the winter. In spring, when soil temperatures rise sufficiently for the native bluebunch roots to resume growing, cheatgrass roots are already well-established. As the young bluebunch roots probe deeper, they find a soil that is devoid of moisture.

Having less sagebrush and bitterbrush spells less forage for wild animals like elk and deer. As most native grasses have disappeared, squirrels have become more and more sparse and with them the falcons and eagles that depend on them.

The displacement of native flora has also been taken up by exotic trees.

The melaleuca tree was introduced to sub-tropical Florida during the early twentieth century from Australia and has slowly displaced the native wetlands vegetation of the region. The melaleuca has hijacked moist habitats and has claimed them as its own. The relentless advance of the melaleuca is facilitated by its profuse seed-generation coupled with its ability to withstand intense fire that kills its native competitors, cypress and pine. With the spread of the trees, there is less forage for deer with subsequent decline in the numbers of the Florida panthers.[69] The native trees of the region are being displaced by the melaleuca while the gigantic Burmese Pythons, whose presence in the area has surged since 2000, may wreak havoc on the native animals. One recent study argues that virtually the entire population of raccoons, opossums, and bobcats has been decimated in south Florida.[70] The giant constricting snakes even prey on American alligators.[71]

Any crack in the American chestnut tree gives the *Cryphonectria parasitica* fungus an entryway. Deep within, the invading fungus grows as it feeds on the tree, depleting the chestnut tree's stores of energy and clogging up its arteries. When the *Cryphonectria parasitica* has spread throughout, the American chestnut succumbs. The fungus *Cryphonectria parasitica* was inadvertently brought to North America at the turn of the twentieth century. The blight rolled from Maine all the way to Alabama, decimating billions of chestnut trees.[72]

Balsam woolly adelgids are small insects that infest and kill firs by sucking on their sap. Since they were transported to the United States decades ago, they have fanned out through Virginia and North Carolina, ravishing fir stands in their wake.[73]

As human-induced activities introduce certain species throughout the world, the biota is becoming more and more alike everywhere. What has kept global biodiversity high through the ages was the separation of biotas regionally. We are now doing away with this, bringing the world closer to dangerous ecological thresholds. In all, invasive species, whether plants or animals, are the leading cause of the extinction of birds and one of the leading causes of the extinction of mammals.[74]

———— • ————

In 2010, Greenland experienced a record-setting surface melt and glacier area loss due to a relatively warm, dry winter followed by an exceptionally warm summer.[75] A few years earlier, a storm surge inundated shorelines in the Mackenzie Delta region of Canada. Most of the alder bushes along the coastline were dead within the year. More bushes died in the years that followed. This surge of lethal saltwater was unprecedented in the last one thousand years.[76]

It is called sea level rise, and it has barely gotten started.

Under the existing trajectory, by mid-century our planet will have reached climatic conditions comparable to that of the Pliocene epoch, three to five million years ago. What is relevant to note is that during the Pliocene epoch, Greenland was partially or completely de-glaciated; the West Antarctic Ice Sheet was also largely de-glaciated; and the East Antarctic Ice Sheet was greatly diminished.[77] During the Pliocene, sea level was between twenty to twenty-five meters higher than present levels.[78]

Based on this and on evidence of sea levels from 0.5 to 3.5 million years ago, a study concluded that within a few millennia we would reach a 20 to 30 meter sea-level rise—provided we stabilize at 387 ppm CO2 levels.[79] Well, as of 2011, we are at 391 ppm and under the existing trajectory projected to reach the 900 ppm mark by 2100 and higher concentrations in subsequent centuries. Dr. Aslak Grinsted of the Centre for Ice and Climate in Copenhagen has inferred that under 6°C warming, the ocean levels will eventually rise between 35 and 40 meters (about 115 to 131 feet), largely courtesy of melted glaciers from Antarctica and Greenland. However, this is not going to happen overnight.

In the short term, recent models show that under a business-as-usual, fossil-fuel intensive scenario, sea level rise may reach 1.5 meters by the end of this century.[80] The projected level varies somewhat from study to study. One model projects around a 1.0–1.5 meter rise.[81] Another model projects a 0.71 centimeter rise once we reach 4°C warming (which the Hadley model projects we would reach between 2058 and 2088 under the existing economic and technological path).[82]

Finally, a paper co-authored by James Hansen, one of the world's lead-ing climatologists, concludes that during this century a multi-meter sea level rise is likely.[83]

Based on these estimates, we might assume that under a business-as-usual trajectory, the ocean will rise between 35 and 40 meters (≈ 115 to 131 ft.) in the next coming millennia and 1 to 1.5 meters (≈ 3 to 5 ft.) by the end of this century.

A meter and a half sea level rise spells the submersion of many of the shorelines, with catastrophic consequences for the salt marshes and the mangroves.

Salt marshes are important habitats for a wide variety of species. They trap pollutants and supply nutrients to surrounding waters. Mangroves are the primary coastal wetland plant community in warm waters and are found along most of the tropical shores around the globe. They are critical habitats for a large number of marine species. Mangroves are also home to thousands of species of birds, insects, mammals, and reptiles. They harbor tigers in Bangladesh and jaguars in Brazil, as these seek their prey at low tide.

About half of the world's mangroves and salt marshes have been lost in recent decades due to dredging for the making of harbors; land changeover to shrimp farms and agricultural use; filling and draining for human settlement; or converting mangrove trees to boats, as is done in parts of East Africa.

With sea level rise, it is believed that marshes and mangroves will be unable to migrate upward. The reasons vary. Human habitats leave little room for expansion of coastal wetlands.[84] Many estuaries are ringed with seawalls and bulkheads, blocking any upward movement of the wetlands. Under the projected sea level rise, most marshes are likely to be submerged by the end of the century.[85]

The Everglades in Florida are vast subtropical wetlands comprised of sedges, mangrove forests, and cypress trees. This ecoregion is home to wading birds, alligators, wood storks, panthers, and manatees. The projected 1.5 meter sea level rise will utterly submerge the Everglades before the end of this century.[86]

Situated in the delta of the Bay of Bengal, the Sundarbans is per-haps the largest tidal mangrove forest in the world. It is home to the endangered Bengal tiger, marine turtles, crocodiles, and freshwater dolphins. Before the end of this century, it is likely to be wiped out by and large due to sea level rise.[87]

The Great Rann of Kutch is the largest wildlife reserve in India. It is home to the last two thousand endangered wild donkeys and to one of the world's largest colonies of flamingos. It is likely to go under in a few decades.[88]

Climate-driven extinctions are already widespread. Once Earth exceeds 4°C warming, major extinctions around the globe are expected—due to climate-related flooding, drought, wildfire, insects, and ocean acidification.[89] A most comprehensive study evaluated 130 reported changes, using them as a basis to analyze and project future extinction rates. One in ten species that has survived to the present could face extinction by 2100 under current emission rates.[90] A more recent study projects that toward the end of the century almost a third of all flora and fauna species worldwide could become extinct if global warming continues. By 2080, the world may lose more than 80 percent of the genetic diversity of some species.[91]

Unusual climatic conditions initiate a flurry of adaptations within the genetic repertoire of each species. Elk are smaller if born on the heels of a warm winter and pocket gophers living in moister areas are bigger than those living in drier areas.[92] Without sufficient mutation, these adaptations are only a temporary fix. Nature runs onto the upper bounds of genetic flexibility and is hitting a dead end.[93] Those with extremely short generation-spans like houseflies, mosquitos, and cockroaches should do well. The remaining 99 percent won't be as fortunate. Using the cold-blooded snake as a case in point, it is assessed that between now and 2100, climate will change 100 times faster than the rate at which animals can adapt.[94] We already know that global warming has forced alpine chipmunks to higher ground. The reduced geographical range has resulted in a severe decline in their genetic diversity.[95]

Beyond the limited physiological adaptability, the best option is to move to areas that offer climatic range for which the animals have evolved—provided the animals can migrate that far that fast.

Some wish to migrate to more hospitable tracts of nature but are unable to traverse past farmlands and fences. Alternatively, what could have been a viable habitat has been claimed by people. In part, this is what happened at the end of the last glacial period; many colder, more suitable habitats had been taken over by humans.[96] Beyond the above, for a significant number of animals there is nowhere to go, even if they could. They are already at the coldest reaches of the mountains. They are already at the farthest reaches of the sea ice biome. Polar bears, walrus, seals, and birds have seen their numbers decline. Lemmings, musk oxen, and reindeer have seen their numbers plummet.[97] Climate change is driving the most massive relocation of species since the beginning of this interglacial period.[98] Those who can, move.[99] Those who cannot will have to face destiny. Then again, a study simulating the projected climate shifts and migration patterns found that relocations are also fraught with difficulties. Whether an animal can make it to a climate-friendly region depends on the temperature swings along the way and throughout the decades of migration. If a simulation of such a migration holds true, more than half would become either extinct or endangered.[100] For one, we already know that butterflies and birds do not keep pace with the changing climate in Europe.[101] Worse yet, new places with hospitable climates may be only half the story; the migrating animal may be out-competed by species already there.[102] Finally, projected massive changes in plant distribution due to climate change may undercut the stability of animal populations.[103]

One species may shift its seasonal behavior pattern, while another, dependent species may not—getting off-sync. Such a change can disrupt the predator-prey interactions, which in turn can cause a drop in the predator population. In Europe, the Pied Flycatcher bird has not changed the time it arrives to its breeding grounds. But the caterpillars, which feed the bird's hatchlings, are emerging earlier. Missing the peak of food availability means fewer chicks survive and consequently the Pied Flycatcher population is falling off.

The climate is one of the drivers. But the blows are coming hard and fast from all sides.

The Tasmanian devil suffered bounty killing in the nineteenth century and poisoning with strychnine in the early twentieth century. Thanks to its high reproductive rate and opportunistic behavior, it recovered nicely—at least in sheer numbers—and by the 1990s, according to one estimate, the wild population stood at about 150,000. However, low genetic diversity in the wake of the excessive hunting means that the animals are extraordinarily similar to one another. This also means their immune system is unable to distinguish between self and non-self. This allows a recent malignant facial tumor to jump from one Tasmanian devil to another without provoking an autoimmune response. For all intents and purposes, they are dying from a transmittable cancer. They are likely to all be dead within a few decades.[104]

From a bat cave in New York, circa 2006, it radiated out to the entire Northeast region: a deadly disease whose main culprit is a fungus named *Geomyces destructans*.[105] This is the first transmittable disease among bats in recorded history. Biologists estimate that more than one million hibernating bats have died up to now, which far exceeds the rate and magnitude of any previously known natural or anthropogenic mortality event in bats, and possibly in any mammalian group.[106] More recently, some rattlesnakes in the wild have also been found afflicted with a deadly fungus.[107]

Then there are the drugs we manufacture.

Each year, the drug industry sells close to one trillion US dollars' worth of drugs worldwide. It is a big party, and *everyone* is invited—whether they want to come or not.

The pervasive herbicide Roundup destroys testicle cells,[108] and messes with hormones.[109] Not to be outdone, herbicides containing atrazine cause reproductive dysfunction and skew the sex ratio of newly born amphibians, fish, reptiles, and mammals.[110] Nasty. The widespread distribution of chemicals has been contributing to the marked rise of freak mutations among frogs. A grotesque spectrum of abnormalities has been recorded, including misshapen eyes and tails, skin lesions, and whole-body deformities.[111] Some newborn frogs have traits associated with both males and females, such as male frogs whose testes contain eggs.[112]

In the early 1990s, the farmers of India started dosing their cows and oxen with an anti-inflammatory drug, Diclofenac. When those cows died later of natural causes, their carcasses were sent to special dumps. Vultures who fed on the treated carcasses accrued lethal doses of Diclofenac. Tens of millions of vultures died. The Oriental white-back vultures used to be one of the most common raptors in the Indian subcontinent. Primarily as a result of the Diclofenac intake, their population has plummeted by about 95 percent in the region.[113]

The results of the antibiotic bonanza have been all too predictable. Rivers in India are flooded with runoff from the big drug factories. As may be expected, these rivers are swarming with bacteria resistant to a wide range of antibiotics.[114] Broader yet, water discharged into lakes and rivers from municipal sewage treatment plants may contain significant concentrations of the genes that make bacteria antibiotic-resistant.[115] There is overwhelming evidence that low-dose antibiotics fed to livestock to promote growth contribute to antibiotic resistance.[116] In the United States, about half of all poultry and meat is contaminated with the drug-resistant strain of *Staphylococcus aureus* bacteria.[117] In 2010, there were about half a million new cases of multidrug-resistant tuberculosis. In 2011, Japan reported its first case of gonorrhea resistant to all single-dose antibiotics.[118] Antibiotic-resistant infections are occurring at a rate that outstrips our ability to fight them with current medications. To make matters worse, antibiotic-resistant genes can jump between bacterial species. Therefore, it matters little in what environment, in what part of the world, or in what bacterial species antibiotic resistance first arises.[119] Drug-use practices represent some of humankind's most significant tampering with evolution, maximizing the evolutionary advantage of pathogens.[120]

A mysterious plague began decimating honeybees. The insects started vanishing from their hives, abandoning their broods and queens to death by starvation. One of the culprits is a new class of synthetic pesticides based on nicotine: neonicotinoids.[121] These chemicals came into vogue in the early 2000s. Older pesticides evaporate shortly after application. The neonicotinoids are more akin to tattoos than to washable markers. Applied to the soil or doused on seeds, neonicotinoids are embedded into the plant, turning it into a tiny poison mill, exuding toxins from its roots, leaves, and pollen. Acting in synergy with fungi and viruses, even low doses of insecticides have proven to be lethal to the honey bees.[122] Likewise, genetically modified maize embedded with the Bt toxins was found to cause mortality among ladybug (ladybird) larvae.[123]

The Earth is subjected to a wide range of toxic chemicals from exhaust pipes as well as from sewage sludge. They come from consumer products, pharmaceutical byproducts, and herbicides. They are emitted from power generation, incineration, and industrial processes. More than eighty thousand compounds are used commercially, and hundreds of new ones are added each year—many of which do not exist in nature. Pesticides, prescription drugs, industrial solvents, mine waste, and plastics have become ubiquitous.

Each year, industrial facilities release billions of pounds of toxic chemicals. There are chemicals for every occasion and taste. There are cancer-causing chemicals, such as benzene and ethylene oxide; there are chemicals linked to developmental problems, such as toluene and nitrous oxide; there are chemicals that cause reproductive disorders, as carbon disulfide; or those that are respiratory toxicants, such as acid aerosols of hydrochloric acid. Many of these toxic chemicals persist in the environment, circulating between air, water, and soil. They are coming in contact with wildlife everywhere. A slew of chemical compounds we generate—from insect repellants to jet fuel to various plastics—not only affect the exposed animals but determine the way genes are turned on or off in the sperm of their descendants. In other words, the toxins affect subsequent generations.[124]

All of these stressors take a punishing toll on the biosphere.

Twenty different species of frogs have been lost at the Monteverde Cloud Forest Preserve in Costa Rica. This includes the spectacular Golden Toad. Historically, species diversity decreased both the prevalence and severity of an infection. While ecosystems are losing species left and right, a fungal infection is hammering the amphibians.[125] The fungus *Batrachochytrium dendrobatidis* has caused dramatic amphibian population declines and extinctions in Australia, Central and North America, and Europe.[126] Over one third of the world's amphibians are at risk of being wiped out. Their decline is likely to accelerate in the coming decades as toxification, climate change, land use and fungal disease deliver multiple blows.[127] Directly or indirectly, human activities are associated with nearly every aspect of this mass-extinction prospect.[128] Frogs have survived glacial periods, the meteor that decimated the dinosaurs, and all the ecological climatic changes in between. But they may not survive us.

The absorbent skin of frogs places them in intimate contact with their environment. When the environment comes unglued, frogs are the first to feel its effects. They are the bellwether of what is to come.

We are currently in the midst of the sixth mass extinction wave in the history of Earth.[129] The current wave has been comprised of a series of extinction pulses across various regions commencing about 45,000 years ago in Australia. First to exit the scene were many of the giant animals and top predators. Extinctions proceeded with the massive clearing of land and its conversion to human habitats and croplands. By the mid-twentieth century, most of the ecosystems of the world were in various levels of degradation.

Since then, there has been a continued, steady decline in the distribution of wild species' populations.[130] Pressures on biodiversity continue to mount. The key stressors driving biodiversity loss have been overexploitation of species, invasive alien species, pollution, climate change, and foremost the degradation, fragmentation, and destruction of natural habitats. Remaining terrestrial biodiversity is increasingly confined to fragmented patches separated by expanding cultivation, infrastructure, and residential and industrial development.[131]

The final solution of the biodiversity question is near. The extinction rate is about one hundred to one thousand times higher than rates typical throughout the planet's history.[132] Over the past fifty years, people have wrecked ecosystems more rapidly and extensively than in any comparable period of time in recorded history. Ten thousand to ten million species now become extinct each decade.[133]

Land ecosystems are unraveling. The fate of biological diversity for the next 10 million years will be determined this century.

We are all responsible due to our actions and inactions. If this has not been a crime, nothing is. If this has not been immoral, nothing is.

3

SEA

～✕～

stressors and prospects of further degradation

FROM THE TROPICAL BEACHES of Tonga to the frigid West Antarctic Peninsula, micro-plastic debris is everywhere,[1] alongside polyester particles from household washing machines.[2]

Indirectly, some plastic we fish out and eat. A lot of it, though, remains at sea. It is estimated that thousands of disintegrated microscopic pieces of plastic are littering every square mile of ocean—perhaps one hundred million tons worth of plastic in all, perhaps more.[3] Many of the tiny particles of plastic are the size of plankton and are being grazed upon by hapless sea animals such as the Pacific krill.[4] Evidently, numerous species of fish, seabirds, turtles, and marine mammals ingest plastic trash. Various samplings were taken. All told, 9 percent of fish were found to have plastic trash in their bodies,[5] 83 percent of lobsters,[6] and 55 percent of the sampled seabird species.[7]

Harmful effects from the ingestion of plastics include diminished feeding stimulus, lowered steroid hormone levels, delayed ovulation, and reproductive failure.[8] The ingestion of plastic debris by small fish and seabirds has been found to reduce food uptake and to cause internal injury and death following blockage of the intestinal tract.[9] For migrating wadepipers, the ingestion of plastic particles hinders the formation of fat deposits on their bodies and thus adversely affects the prospects of a long-distance migration.[10]

Various sea animals have been entangled in plastic: fish, turtles, seabirds, and whales. Young seals are attracted to floating debris and dive in and roll about. Those entangled at a young age with packing bands and such are especially vulnerable. As they grow, the entangling material tightens, eventually strangling the youngsters. Alternatively, lesions from netting or packing bands often get infected. If the infection surpasses the ability of the lymph system to control it, the lungs will often become infected as well, possibly leading to death.[11]

Last but not least, under the ultraviolet rays of the sun, the plastic particles are suspected to leach toxic substances into the water: bisphenol A (BPA), phthalic acid ester (PAE), and PS oligomer.[12]

In addition to plastics, we discharge into the marine environment brominated flame retardants and fluorinated compounds. We flush into the sea antibiotics, anti-depressants, and birth control pills. These chemicals may be laced with mercury that works its way to the waterways from coal power-plants. It is a heady brew, enough to cause one's head to spin—or at least perhaps to cause it to sprout a tumor. And this brew is found throughout the world's oceans.

Noises are another kind of pollutant. Over the last century, human activities have increased noise in the marine environment. Noises emanate from sonar, air guns, offshore drilling excavations and, foremost, from ships.[13] These sounds encumber dolphins and whales, which rely on sound for survival, reproduction, and communication in general. The low frequency noise may inflict whales with chronic stress.[14] Some noises add distraction and may impair foraging activities.[15] Some noises are responsible for a massive, permanent acoustical trauma in various octopuses.[16]

The ocean soaks up a third of all the carbon we emit into the air. This in turn leads to a pH decline and changes in seawater chemistry that collectively are referred to as ocean acidification. Mounting evidence suggests that these changes in ocean chemistry may end up having serious repercussions for marine organisms, potentially altering ecosystem structure, food webs, and biochemical processes.[17]

Foremost, lower seawater pH is believed to affect the ability of corals to form skeletons and of snails to keep their shells from dissolving.[18]

Acidification may end up impacting processes so fundamental to the overall structure and function of marine ecosystems that any significant changes could have far-reaching consequences for the oceans.[19]

We are flooding the atmosphere with carbon dioxide at a rate rarely if ever seen in the history of Earth.[20] It can be said with certainty that the intake of CO_2 by the ocean is outstripping its capacity to neutralize it.[21] The current ocean acidification is more severe and at ten times the rate of the acidification that preceded a mass extinction fifty-five million years ago.[22] This may spell the demise of many marine species, particularly ones that live in the deep ocean.

Ocean acidification is essentially irreversible during any human timeframe. It will take many thousands of years for ocean chemistry to return to conditions similar to those of pre-industrial times.[23] For one, it took one hundred thousand years for the oceans to restore their chemical balance on the heels of the acidification event fifty-five million years ago.

On top of that, the heightened CO_2 levels predicted to occur in the ocean toward the end of this century are likely to sabotage the entire nervous system of some crustaceans and of most fish—affecting their ability to hear, smell, turn, and evade predators. Elevated levels of CO_2 overly stimulate certain nerve signals. And the effect is more pronounced in marine animals, which have lower blood CO_2 levels. This can have serious consequences for their survival.[24]

The marine world was a far different place back in the day before indiscriminate fishing. New Zealand waters once hosted upward of thirty thousand southern right whales. The shores of Queensland once sported countless dugongs, forming miles-long columns along the coasts. San Francisco bay had white sturgeons up to 6 meters ($\approx$ 20 ft.) in length and over 800 kilograms ($\approx$ 1,760 lbs.) in weight. Along the Gulf of Mexico, hundreds of thousands of Caribbean monk seals basked in the sun, and the water was teeming with tens of millions of green turtles.

The first people arrived at Grand Turk around 750 CE and drove iguanas and large fish to local extinction. Around fifty years later, people cleared land in southwestern Jamaica and are likely to have contributed to greater sediment loads, which transformed Bluefields Bay from a clear, free circulating sea grass habitat to a muddy mangrove. A few hundred miles to the west, the first people in Haiti wiped out the local populace of green sea turtles.

About one thousand years later, a comparable thing happened in the far north.

The ponderous Steller's sea cows were 7 to 8 meters long ($\approx$ 23 to 26 ft.). They had gnarled, bark-like black hide on top of which sea gulls perched. As the Steller's sea cows grazed in the shallow water, they moved first one foot and then the other, and thus with a gentle motion, they half swam and half walked, keeping their head underwater. Every few minutes or so, they raised their nose above water to blow out air with a snort. Otherwise, they were quiet. When the sea cows took a nap, they turned on their backs and allowed themselves to drift in the water like massive logs.[25]

Europeans chanced upon a few thousand of these gentle beasts in 1741, in what was apparently their last stronghold: two remote islands in the Bering Sea. Word of the appealing taste of Steller's sea cows spread through the sailing community. Subsequent fur-hunting expeditions relied on their meat for the long months of the journeys.

The spearman would stand in the prow of the boat. As soon as he struck a sea cow with a hook, thirty men jumped in and held onto the animal, and in spite of its frantic efforts they dragged it toward the shore, pummeling the sea cow with constant blows that wore it out. Great pieces were cut from the animal while still alive. The sea cow would sigh when injured but otherwise uttered no sound. As long as he kept his head under water the blood did not flow out, but as soon as he raised his head to breathe the blood discharged like a fountain. Those in the herd near the captured sea cow endeavored to assist him. To this end, some of them tried to upset the boat with their backs, others bore down upon the rope and tried to break it or attempted to extract the hook from the back of their wounded companion with a blow from their tails. When a female was caught, her male companion, after trying with all his strength to free his captured mate, would follow her to the shore, even though he was delivered many blows by the hands of the people. Some were seen to be by their dead mate for days. Twenty-seven years after they became widely known, the last Steller's sea cow was killed.

The pattern has been the same everywhere, regardless of the group of people involved. The Roman Empire grew, and the marine life in the Mediterranean drastically diminished; the native population of Jamaica ballooned, and there was a marked decline in manatees; New Zealand was settled, and the population of millions of seals crashed. Large marine vertebrates such as whales, manatees, turtles, groupers, and sharks have been systematically removed from the ocean by humans and are now largely absent from most coastal ecosystems worldwide.[26] Hunting continues at a brisk pace. Since the early 1990s, the number of whales slain has doubled to about two thousand per year.[27] Reef shark populations have dropped by more than 90 percent in heavily populated areas.[28] In fact, the oceans have lost across the board over 90 percent of their large predatory fish,[29] many of whom have ended up as shark fin soup and the like.

With the top predators went the checks and balances.

The decimation of hammerhead sharks resulted in a population explosion of cownose rays, which in turn ravished the community of scallops.[30] In this manner, the overfishing of sea otters set off population explosions of sea urchins, which in turn have stripped kelp forests down to desolate expanses of bare rock and algae.[31] Approximately half of the coastal areas of the North Atlantic and North Pacific show a decline in predator biomass of more than 90 percent.[32] A major recent study concluded that the loss of the top predators is arguably humankind's most pervasive influence on the natural world.[33]

Widespread devastation occurred when what had previously been regional plagues acquired global dimensions. That was when the oceans began to run down, with dramatic declines across all marine ecosystems: coral reefs, sea grasses, kelp forests, mangroves, estuaries, and the open seas.[34]

The annual world fish catch rose over fourfold between 1950 and 1990.[35]

At present, we annually fish out ninety million tons of marine life.[36] A very crude calculation suggests that had we piled our collective annual sea harvest into a big conical mound, its diameter would be about one thousand meters wide at its base, and it would tower to three times the height of the Empire State Building. This is the biomass we annually take out of circulation, out of the ocean biome. Thus, the biomass of bass fish in Southern California has decreased by 90 percent since 1980—in part due to overfishing,[37] and tuna populations have declined by over 50 percent in the period from 1954 to 2006—to the very limits of sustainability and beyond.[38] Ninety million groupers were killed and captured in 2009, and 20 grouper species are at risk of extinction.[39]

As ocean fishing increased in intensity, stocks began to bottom out. Populations of some species were driven so low that they were unable to recover. In fact, since the 1950s one fishery in four has collapsed.[40] Unlike land species, both small and large fish are vulnerable to collapse.[41] We first depleted stocks of cod, hake, flounder, sole, and halibut in the Northern Hemisphere. As those stocks were exhausted, fishing fleets moved southward until they reached the shores of Antarctica, searching for icefish and rockcods, and, more recently, for the tiny krill.[42] By the turn of the twenty-first century, three-quarters of ocean stocks were overfished, depleted, or exploited up to their maximum sustainable yields.[43] And if fish stock drops below a certain point—which is universally the same across species and habitats—then the number of dependent baby birds born drops precipitously.[44]

The fishing fleet had traversed to the most remote reaches of the oceans. Then their advance halted; they had dropped anchor at the farthest region of the sea. That was a few decades ago. Since that time, we have descended into the abyss, fishing miles deep at what is the last untapped source of fish. Hard at work, we have already caused the collapse of many deep sea fishes around the world such as sharks and orange roughy.[45]

The deep sea is the largest ecosystem on Earth. The deep-sea habitat contains a vast array of unique environments: hydrothermal vents, cold seeps, seamounts, submarine canyons, abyssal plains, oceanic trenches, and recently-discovered asphalt volcanoes. It has been estimated that the number of species inhabiting the deep sea may be as high as ten million.[46] It is suspected that the deep seabed supports more species than all other marine environments.[47]

The fishing fleet has begun availing itself to these diverse deep-sea ecosystems using bottom trawling—a cross between forest clear-cutting and strip mining. Bottom trawling is not something new; it has been extensively used in the continental shelves for many decades. The area of sea floor scraped clean by trawling extends over many hundreds of thousands of square kilometers. In the aftermath of bottom trawling, marine regions have been transformed to barren, flattened expanses.[48] In recent years, a modified bottom-trawling technology has made itself viable for deep water. In the high seas, fishing vessels drag huge nets armed with steel plates and heavy rollers across the seabed. Those heavy, deep-sea rigs do not merely scrape bottom, they raze much of it. In order to catch a few 'target' fish species of commercial value, biologically rich and diverse deep-sea ecosystems are plowed through.[49] Caught by the voracious fishing net, unwanted fish are summarily dumped back into the water, lifeless.

Beyond fishing out the deep sea and scraping its floor with steel wool, we have always used this enthralling biome as a dumping ground for traveling ships: anything and everything from beer cans and medicine bottles to dishes and furniture parts. The most common litter in the deepest regions of the Mediterranean Sea floor is plastic bags, bottles, and aluminum cans. In fact, empty cans were found at the ocean bottom near the shores of Antarctica.

During the steamship era, thousands of tons of coal were burned daily, and the remains were pushed over the rail. These remains formed a toxic substrate currently comprising a part of the seafloor. For decades, defective, obsolete and surplus munitions have been dumped at sea: grenades, mines, smoke devices, and mortar bombs. Some munitions contained sarin, phosgene and mustard gas.[50]

All species have a certain range of adaptation. However, beyond a certain temperature range, acclimation fails, mortality risk increases, fitness is reduced, and populations decline or are driven to local extinction.[51]

With the warming of the oceans, those who can, move elsewhere, to new regions. For tens of millions of years, there have not been bone-crushing predators in the polar regions. This is about to change.[52] There have been sightings of king crabs along the west Antarctic Peninsula,[53] and shark rays (*Rhina ancylostoma*) are anticipated to arrive at the polar seas by the end of the century. Despite the slower warming of the oceans, sea creatures in the tropic regions will have to traverse larger distances than land animals to reach colder water and escape extinction.[54] At times the threat of local extinction is augmented. Mussels and barnacles were able to live high on the shore, well beyond the range of their predators. With the rising water temperature, they are now forced to live at lower shore levels, placing them at the same level as predatory sea stars.[55]

The warming seas may be the undoing of the coral reefs.

A coral reef harbors microscopic algae within the reef tissues. The algae provide photosynthesis services to the coral, while the coral in turn provides a physical reef structure that keeps the algae near the surface where light is abundant. When it is too hot, the coral expels its algae. This is called bleaching, and more often than not, it ends up killing the coral. Mass coral-bleaching started with the onset of global warming, decades ago. At today's greenhouse gas levels, allowing ten years for sea temperatures to catch up, bleaching is likely to be an annual event, committing most reefs to an irreversible decline. This is likely to have triggered the onset of some mass extinctions in the past.[56]

The expected sea level rise is also projected to have adverse effects on coral reefs.

Coral requires light for photosynthesis and typically cannot survive in deep water. Some, like the staghorn coral reefs, may be able to grow fast enough to compensate for the rising sea. Others, like the head coral reefs, grow too slowly and will be claimed by the dusky depths of the rising ocean. Breathtaking in their beauty and riot of color, coral reefs are arguably the most structurally complex and diverse marine ecosystems. They provide habitats for tens of thousands of associated fishes and invertebrates. Yet, remaining coral reefs that have survived to the present are as good as dead; they just don't know it yet.

The seawater acidification level projected for mid-century will push some coral reefs into a negative carbonate balance. That is to say, with the calcification rate falling sharply, some coral's ability to build its reef will not keep up with natural biological and physical erosions. Under these conditions, coral reefs will no longer grow and in many cases will begin to crumble away.[57] The majority of corals appear to have the ability to buffer water acidifcation; however, coralline algae, which hold coral reefs together, seem to be vulnerable.[58]

Severely overfished reefs failed to recover because the diminished number of herbivores failed to control the population explosion of seaweeds on newly bared substrate in the wake of a storm and because of the deposition of silt due to land clearing.[59] Some of the seaweeds used chemicals to kill or inhibit the growth of reef building coral. Bacteria flourished on the heels of seaweed growth and on the heels of organic matter streaming from the land. The corals were soon smothered by algae and bacteria, which consumed the oxygen and suffocated the organisms within the coral.[60]

Reefs are likely to be the first ecosystem to collapse worldwide in the face of the climate changes now in progress.[61] And if the heat of the water won't do them in, other things will: the reduced calcification as a result of rising water acidity, the sea level rise, and the smothering bacteria and algae. It would be a boneyard of coral skeletons. With the death of the coral, seaweeds and blue-green bacterial slime will stake out the reefs. Coral reefs were wiped out once before. It took them over one million years to re-emerge.[62]

Invasive species are another major contributor to ocean-biome degradation.

Ocean-going boats carry seawater as ballast. The water is taken in at the port of departure and is discharged at the next port of call. Along with the water, the ships also release opportunistic species that have tagged along. Accelerated warming of high-latitude environments has increased the chances that species being transported from lower latitudes are able to establish themselves in new marine biomes.[63]

Invasive species have transformed marine habitats around the world. The most harmful of these invaders displace native species, change food webs, and alter fundamental processes such as nutrient-cycling and sedimentation. Only a small fraction of the marine species introduced outside of their native range is able to thrive in new habitats, yet the impact of those few is profound. Once alien species become established in a marine habitat, it can be nearly impossible to dislodge them.[64]

The one-celled parasite *Bonamia ostreae* has spread to all parts of Northern Europe, drastically affecting oyster populations. Slipper snails have had detrimental effects on oyster beds and great-scallop habitats.[65] The lionfish population has rapidly spread throughout the Atlantic Ocean and most of the Caribbean Sea. The proliferating population of lionfish is now working its way around the Gulf of Mexico. A preliminary study indicates that within a short period after the entry of lionfish into the Bahamas, the survival of other reef fish was reduced by about 80 percent.[66]

Initial results from the first comprehensive global assessment of its kind show that only 16 percent of marine ecoregions have no reported marine invasions. In actuality, the number may be smaller yet.[67]

An influx of invasive species can stop the natural process of new species formation and trigger a mass extinction event. This is speculated to have occurred between 375 and 378 million years ago. New species emerge when a local population becomes geographically divided over the long term, such as with the formation of a mountain range. Back in the Devonian Period, as sea level rose, the reverse happened: the continents converged. As a result, some species gained access to environments they hadn't inhabited before. The hardiest of these invasive species, which could thrive on a variety of food sources and in new climates, became dominant. During the Devonian times, the invasive species were so prolific that it became difficult for many new species to emerge. The entire marine ecosystem suffered a major collapse and reef-forming corals were decimated. The number of giant fish, trilobites, sponges, and brachiopods also declined dramatically. In similar fashion, modern ecosystems in which invasive species are rampant may exhibit a similar dearth of new species.[68]

This is a real stew, no doubt about it. But it isn't quite complete without seasoning.

The discharge of organic matter into the oceans has increased in recent decades with devastating effects on coastal ecosystems.[69] Massive amounts of chemical fertilizers, human sewage, and dung from hog and chicken farms stream into the oceans. Joining this barrage are dead livestock that die en route at sea—diseased or otherwise. In 1990, 10,000 sheep died and were unceremoniously dumped from a ship en route between New Zealand and Saudi Arabia.[70] About ten years later, another 270,000 sheep were pitched over while en route to the Middle East.[71] The rotting carcasses have sent large pulses of organic material to the seafloor.[72]

The white pox disease that has been killing elkhorn coral in the Caribbean is caused by a bacteria present in the human sewage that is making its way to sea.[73] The urban run-off may also contain traces of many medicines, such as the hormone levonorgestrel from contraceptive pills, which can lead to infertility in fish.[74] In recent times, pathogens originating in pets and livestock have been the source of many infections afflicting marine mammals.[75] But there is more that needs to be stated.

The torrent of millions of tons of fertilizer and excreta has been fueling an explosive growth of micro algae in the ocean that is far beyond the capacity of marine animals to process—especially since we fished out of existence most of the oysters.[76] When excessive amounts of micro algae die and sink to the sea floor, their subsequent decomposition by bacteria results in two harmful effects: One, it releases carbon dioxide, which reacts with water and makes the water more acidic (i.e., ocean acidification).[77] Two, the decomposition of the algae depletes the oxygen in deep waters. This has resulted in the formation of vast oxygen-starved regions. Many bottom-dwelling animals that cannot swim or scuttle away from those areas suffocate and die. Dead zones in the coastal regions have spread exponentially since the 1960s. As of this writing, there are four hundred such dead zones. Once a region experiences low-oxygen conditions, reoccurrences are highly likely.[78] The low-oxygen phenomenon is largely irreversible, projected to persist for thousands of years.[79]

The explosion of harmful micro algae in nutrient-overloaded regions has caused toxins to move up through the food chain.[80] Countless seals, sea lions, and dolphins have been sickened and washed up along the California shoreline. Sea turtles in Hawaii have been sighted with fist-sized tumors growing out of their eyes and mouths. Some humpback whales died in Cape Cod Bay as a result of eating mackerel infested with algal toxins. Each summer, off the coast of Sweden, blooms of blue-green bacteria turn the Baltic Sea into a reeking, brown colored slush with dead fish floating about. On the southern coast of Maui, the high tides leave foul-smelling piles of green-brown algae. On Florida's Gulf Coast, harmful algae blooms have become more pronounced and longer-lasting. North of Venice, Italy, a sticky mixture of algae and bacteria collects on the Adriatic Sea in the warm seasons. This white mucus washes ashore or congeals into blobs, some bigger than a person. During the summer of 2011, a massive bloom of green algae traveled toward China's east coast. It was a flotilla the size of the Greater Los Angeles Area. Manatees died in large numbers along the southwest coast of Florida from inhaling brevetoxin, a neurotoxin produced by some algal blooms.[81]

In recent years, Stinging Limu has burst forth each spring from spores on the seafloor and has formed large floating mats of fine, dark wool-like strands. It flourishes for months before retreating. This venomous hairy weed (*Lyngbya majuscula*) has appeared in a dozen places around the globe. Upon contact, people's skin broke out in searing welts; their eyes swelled.[82]

The existing ocean ecosystems are waning, while from the recesses of bygone eras other things are emerging: species that were pushed back hundreds of millions years ago and have been kept at bay by fish and sea mammals.[83]

At the turn of this century, the population of the giant Nomura's jellyfish began to multiply in the Sea of Japan. Such occurrences took place every few decades in the twentieth century. However, since 2002, it has become an almost annual event. In 2005, a particularly bad year, the Sea of Japan brimmed with many hundreds of millions of immense quavering bags of blubber.[84]

Jellyfish are on the rise throughout the majority of the world's coastal waters.[85] Evidence suggests that a suite of human activities promote the rise of jellyfish: overfishing, nutrient runoff, warming seas, and species translocations.[86] Overfishing opens up ecological space for jellyfish, makes more zooplankton food available to them, and reduces predation pressure on them. Nutrient runoff both encourages phytoplankton blooms, on which jellyfish feast, and fosters a low-oxygen environment, which jellyfish can tolerate. Warmer seas present favorable conditions for the population growth of jelly species. Translocation is well suited to some jellyfish, which oftentimes flourish in the new environment.[87]

Jellyfish have a suite of successful attributes that enable them to survive in disturbed marine ecosystems and to rebound rapidly as conditions improve. These attributes include a broad diet, fast growth rates, the ability to shrink when starved, and the capacity to fragment and regenerate. These are characteristic of opportunistic 'weed species' and would appear to give jellyfish an edge over fish in environments stressed by climate change, algae bloom, and overfishing. Once they settle in, jellyfish dig in for the long haul.[73]

Slimy jellyfish are taking the place of fish, and toxic blooms are taking the place of phytoplankton. Coupled with rising temperatures, unrestrained runoff of nutrients and toxins will increase the size and abundance of dead zones and toxic blooms. Widening dead zones and pollution will make coastal waters too toxic for aquaculture, and the frequency of disease outbreaks will climb.[88]

We now face losing marine species and entire marine ecosystems within a single generation.[89]

From the coral reefs in the Red Sea to the sea mounds off the coasts of Chile—some of the most complex webs of marine life are unraveling while base life forms are proliferating. The dials turn backward, working their way from the highly evolved ecosystems of the present epoch to the rudimentary seas of hundreds of millions of years ago—a time when fish were absent, microscopic flagellates dominated, and jellyfish may have been the top predators. Human activities may end up transmuting the ocean into a microbial soup teeming with gelatinous, poisonous blobs and a seabed littered with empty aluminum cans.

We are all responsible due to our actions and inactions. If this has not been a crime, nothing is. If this has not been immoral, nothing is.

4

HUMAN HABITATS

≫≪

stressors and prospects of further degradation

GLOBAL WARMING

THE COOLEST SUMMERS of the future may make the hottest summers of the present appear balmy.[1] It is going to get stifling hot in some places.

But then, it is stifling hot in some places right now. Other locations will join them; this is not the end of the world.

But then again, maybe in some places it would be, or at least a good imitation of it.

Two highly populous regions are at the high end of the heat-humidity spectrum: Indo-Gangetic Plain (including Delhi) and the Yangtze River Delta (including Shanghai). It is important to ascertain how these two population centers may fare around 2100 under existing emission trajectories.

For the ability of our bodies to function, Apparent Temperature is a more meaningful indicator than air temperature. Apparent Temperature is the way temperature is perceived, which is a combination of relative humidity and the actual air temperature.

TABLE 4.1. The American National Weather Service, Apparent Temperature ("Heat Index")

Fahrenheit	Celsius	probable results under continued exposure
105°–130°	40.5°–54.5°	sunstroke, heat cramps, and heat exhaustion likely

48

Fahrenheit	Celsius	probable results under continued exposure
above 130°	above 54.5°	heatstroke (life-threatening) highly likely

For the cities of Shanghai and New Delhi, I have obtained the humidity, temperature, and cloud conditions for every hour between 11:30 a.m. and 4:30 p.m. for every day of summer 2010.[2] This data set represents what the situation is at present.

Under existing trends, come 2100, the Indus Valley will experience an average temperature increase of 5.5°C warming and the Shanghai region will experience an average 5°C warming. I kept all parameters of 2010 the same and just tacked on 5.5°C and 5°C respectively to the hourly dry temperature values for each of the two regions. This is a simplistic computation, which in turn is based on a low resolution, global-scale model:[3] a crude set of calculations atop a crude model. Yet, what follows does provide a *sense* of what the weather may be like one lifetime away in those parts of the world.

In 2010, the threshold of 54°C (≈130°F) Apparent Temperature was crossed only during one day in New Delhi. In the 2100 scenario, it was a daily occurrence in July and August, that is to say, heatstroke would be very likely to take place on almost every summer day—absent some mitigation measures, such as air conditioners. In fact, in the simulation, during one particular day the Apparent Temperature surpassed 82°C (≈180°F). This is like being shoved into and then left in an oven set on low heat for a few hours.

All of this is in the shade, not taking into account any possible enhanced effects of the sun rays.

For the significant number of people who work outside in the Indus Valley during the day, the Wet Bulb Globe Temperature (WBGT) is a better measure. This index factors in air temperature, radiant temperature, humidity, and air movement. The WBGT is used to determine the safe limits of working outdoors: how much is too much.

TABLE 4.2

WBGT	NIOSH recommendations
31–32.5	rest required three-quarters of the time

WBGT	NIOSH recommendations
36	no heavy work at all
39 and above	no work of any kind

Source of data: J. Malchaire et al., "Criteria for a Recommended Standard: Occupational Exposure to Hot Environments," *International Archives of Occupational and Environmental Health* 73, no. 4 (2000):215–20.

FIGURE 4.1

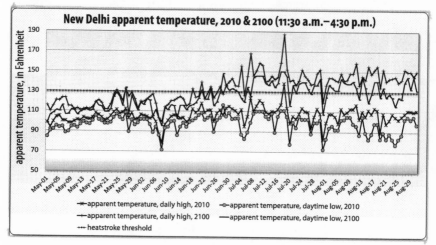

Note: Data extrapolated from Freemeteo.com, Weather History.

FIGURE 4.2

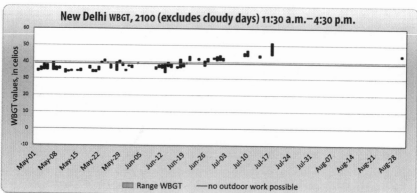

Note: Data extrapolated from Freemeteo.com, Weather History.

I tracked only summer days that were sunny or partially sunny. The Indus Valley is currently experiencing WBGT temperatures that range from 28 to 34, which are still within human tolerance for outdoor labor, at least some of the time. Under the 2100 scenario, WBGT values range from 35 to 44. This means that many days would be too stifling hot to work, with potential for collapse due to heatstroke.

The Shanghai region is also projected to have its share of woes under the existing global warming trajectory.

FIGURE 4.3

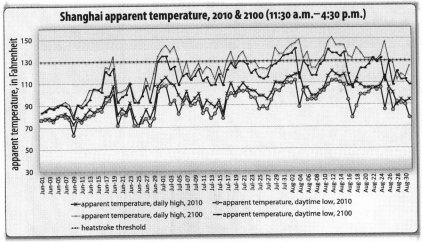

Note: Data extrapolated from Freemeteo.com, Weather History.

FIGURE 4.4

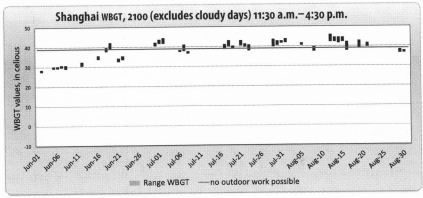

Note: Data extrapolated from Freemeteo.com, Weather History.

Judging by this simulation, Shanghai's Apparent Temperature on most August days would be above 54.5°C (≈130°F) in 2100. These are conditions likely to cause a heatstroke under continuous exposure. Occasionally, the Apparent Temperature would reach past a scorching 60°C (≈140°F) in the shade. There would be multiple days in Shanghai in which the WBGT value would transcend 39, and during those days no outdoor labor would be possible.

Finally, straight wet-bulb temperature (a combination of humidity and temperature) may express the ultimate climatic threshold that humans can withstand, irrespective of clothing, activity, and acclimation. Humans maintain a core body temperature of around 37°C. Skin is always a few degrees colder. Once skin temperature rises to 37°C or higher, the derivative core-body temperature reaches lethal levels. Hence, prolonged periods of wet bulb temperatures above 35°C (≈95°F) are literally intolerable. At the moment, wet-bulb temperatures are usually around 26°C–27°C (≈ 78°F–81°F).[4]

In New Delhi circa 2100, the simulation tracks one day in which the wet-bulb temperature spikes to lethal levels of 37°C (≈99°F). Currently, there is no place on the Earth's surface that even remotely experiences lethal wet-bulb temperature levels. During those hours on that possible future day, there would be air conditioners for those who have them; body ice packs or deep underground shelters for those who could; and possible death for the rest.

Under existing emission levels, is it likely that we would see those temperatures by the end of the century? Yes, and perhaps worse. Coinciding with the overall warming trend, we experience increasingly extreme temperatures; that is to say, the variability is greater.[5] Thus, a more representational projection would indicate the low temperatures as colder and the high temperatures as hotter than assumed here.

SEA LEVEL RISE

As stated in the second chapter ("Land"), under a business-as-usual scenario, average global sea level rise is projected to be around 1 to 1.5 meters (≈ 3 to 5 ft.) by the end of this century. Within a few millennia, sea level rise may reach 35 to 40 meters (≈ 115 to 131 ft.).

FIGURE 4.5. A thirty-five meter sea level rise in populous regions that would be particularly vulnerable.

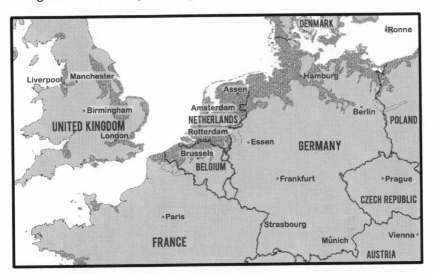

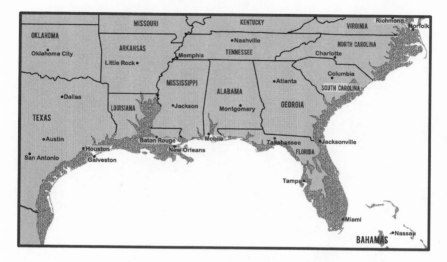

The big impacts of sea level rise are shoreline erosion, flooding, and salinization—coupled with waterborne diseases contaminating drinking water.

In the context of sea level rise, deltas are the most vulnerable areas. By the time the children of today are at the twilight of their lives, there won't be any Mississippi River Delta to speak of—that is, without due levees and walls. However, I will focus on river deltas where the economic impact is projected to be more severe and adaptation measures less likely.

Egypt has a coastal zone more than 3,500 kilometers long and contains 40 percent of the country's total population of 82 million people. At present, a large portion of the 50-kilometer-wide coastal strip lies 2 meters below sea level and is protected from inundation and flooding only by a coastal sand belt a few kilometers wide. The sand belt, which also protects coastal lakes and lagoons, is experiencing rapid erosion associated with the construction of the Aswan dam. Sea-level rise will exacerbate the problem, endangering the fishing industry as one third of the country's fish catch derives from these threatened lakes and lagoons. Inundation and erosion could result in a loss of a significant portion of the northern part of the Nile Delta. With one meter of sea level rise, Egypt may experience the loss of 28,000 sq km of agricultural land and 25,000 sq km of urban area.[6] Under a 5 meter sea level rise, 35 percent of Egypt's agriculture is projected to be impacted.[7]

The rice-growing river deltas in Asia are particularly vulnerable to sea level rise. A World Bank map indicates that a one meter sea level rise would flood half the rice cropland in Bangladesh, home to about 160 million people. This country is the second largest rice exporter in the world. A large portion of Viet Nam's population and economic activity are located in two river deltas, the Mekong Delta in the south and the Red River Delta in the north. The overall effect of sea level rise would be catastrophic.[8] One meter of sea level rise would put an end to the rice harvest in the Mekong Delta, which produces more than half the rice in Viet Nam. A substantial part of the Red River Delta in the north may join it. With the two deltas largely submerged, every tenth person will be a refugee. With a five meter rise, about a third of the population will be looking for a new place to call home; the old one will be gone.

It is projected that between 50 to 200 million people will become climate refugees before the end of this century.

Armoring the coastline can be done, but at a cost. The natural dynamics that occur between water and land would be disrupted. Beaches and wetlands would disappear, and habitat would be lost. However, under existing priorities, when it is the ecology of beaches and wetlands on one side of the scale and the protection of hospitals, wastewater-treatment facilities, power plants, mega-resorts, and coastal roads on the other side of the scale, our collective real estate would tip the scale its own way.

Seawalls are designed to resist the forces of storm waves; bulkheads are meant to retain the fill; and revetments are laid out to protect the shoreline against the erosion associated with light waves. When we talk about a 1 to 2 meter rise, we can stop the rising ocean in its tracks. In that context, a 2010 study determined that a combination of levees and seawalls spanning 100 miles will average about 1.6 billion US dollars.[9] Under those costs, Egypt can put a buffer throughout their entire delta region for what would amount to 0.5%–1% of their annual GDP. (And in fact a much smaller percentage of their GDP, if the investment is spread over a few years.) However, this logic can be pushed only so far, even if we discount the impact such walls would have on trade and fishing.

Would we try to stop sea level rise 35 meters high? One crack, one mega-storm, and a biblical water column could bear down on the coastal population center, possibly killing millions of people. Thus, even if every few decades or centuries we build up the wall ever higher, there will be a point where a disaster will happen or people will finally choose to retreat. The last chapter of the story would be one and the same, regardless: the largest retreat and resettlement campaign in the history of humanity. Or worse, if international borders and immigration policies refuse to yield.

FOSSIL FUEL

Our labor and ingenuity may be the motor of the world, but without fuel this colossal motor won't run. At the present time, this fuel largely means fossil fuel.

As of 2009, fossil fuels comprised 81% of the total energy sustaining our world. Another 10% came from the burning of wood and biofuel. The remaining 9% was mostly nuclear with a bit of hydro power thrown in.[10]

Industrial society is predicated on cheap fossil energy. Our society's very existence is owed to ungodly amounts of fossil fuel injections. In 2010, we went through the equivalent of around 190 million oil barrels worth of fossil fuel a day to keep our industrial society running. Had it all been comprised of liquid oil, the annual volume that we burn through would have been equivalent to the volume of water in Lake Superior.

Cheap fossil energy has made possible the automobile, the aviation industry, mechanized agriculture, and the advent of economic globalization. From airplanes to backhoes to tractors to trucks to ships to scooters to cars: 98 percent of what has fueled motor vehicles has been fossil hydrocarbons, directly or indirectly.

Perhaps 40 percent of the world's dietary protein now comes from synthetic fertilizers derived from carbohydrates, and estimates suggest that if not for that, at least two billion people could not exist.[11]

As of 2008, fossil fuels generated 71 percent of all electricity. Electricity makes possible street lights and residential lighting. It makes possible the ubiquitous media players, computers, telephones, and the Internet Highway. Fossil fuels cool, via air-conditioners or swamp-coolers, many homes during the summer; and they heat, directly or via electricity, most homes in the winter.

Fossil fuels are practically the sole source of petrochemicals: nylon, polyester, formaldehyde, polystyrene, and synthetic rubber. Petrochemicals are in everything. They are in inks and dyes, bottles, packages, food additives, adhesives, and sealants. They are in computers and cell phones, in tires, and in steering wheels. They are in the credit cards we use, trash bags we throw out, and tennis shoes we run in.

As things stand, the dearth of fossil fuels—cheap, expensive, or otherwise—spells the collapse of our society alongside mass mortality whose magnitude goes beyond anything ever experienced before.

Fossil fuel is nonrenewable; it has always been just a question of *when*.

So then, when?

Long before we deplete fossil hydrocarbons, we will stop extracting them from the ground when it takes more energy to extract them than they yield. And long before that, we will stop extracting them when they are simply too resource-intensive to extract. In other words, we will stop when it becomes economically senseless. Thus, the question is not how much resides underground, but how much is economically-recoverable.

One Gtoe (giga-ton of oil equivalent) is the amount of energy embedded in one billion tons of oil. Adjusting for likely political biases and sloppy reporting practices, there were about 127 Gtoe of recoverable reserves of conventional oil as of 2008. This estimate is based on research done by the Smith School of Enterprise and the Environment at Oxford University.[12] Compensating for production since that time, the 2010 reserves may stand at 120 Gtoe.

Beyond conventional oil, naturally-occurring tar can be extracted, processed, and refined to produce the equivalent of crude oil. Natural bitumen, as it is referred to, is reported in 598 deposits in 23 countries.

By far, the largest deposit is located in northern Alberta, Canada. Total recoverable reserves of natural bitumen are estimated at 35.5 Gtoe.[13] For every 3 barrels of oil, it takes the energy equivalent of 1 barrel of oil to extract the nasty tar and refine it.[14] Thus, after deducting the amount of energy required for processing, net energy yield from natural bitumen reserves is about 22.6 Gtoe.

Viscous, extra-heavy oil of a slightly different sort is found predominantly in Venezuela and to a smaller extent in a few other locations. The total recoverable reserve for the so-called extra-heavy oil is estimated at 8.7 Gtoe.[15]After subtracting the energy required for extraction and refining, the available yield would come to 5.8 Gtoe.

Last but not least is oil from oil shale.

Oil shale is a dense rock that has a waxy substance tightly bound within it. When the waxy substance, kerogen, is heated at high temperatures, it liquefies, producing compounds that can eventually be refined into synthetic petroleum products. There are two basic methods to get kerogen out of oil shale. The first is to mine the shale with traditional hard-rock mining methods, then crush the rock, and cook it without the presence of oxygen. The second method is to heat the underground shale rock. The now liquefied kerogen is pumped to the surface. Estonians have been commercially producing shale oil since 1924. China has had industrial production since 1930, and Brazil got into shale oil extraction in 1981.

Oil shale can be found in many parts of the world. Sediments range from small occurrences of little or no economic value to vast deposits covering thousands of square kilometers and containing billions of barrels of potentially-extractable shale oil. In some deposits, the main hurdle for shale oil extraction would appear to be scarcity of water. It takes between one and four barrels of water for each barrel of oil processed.[16] The Green Formation in Colorado is the granddaddy of all oil shale deposits: the Saudi Arabia of shale oil. And if anything, this is an understatement.

Alas, water is hard to come by in Colorado.

TABLE 4.3. Water requirements for shale oil in context

	amount of water used in the Denver, Colorado area, 2005 (population 560,000)[a]	amount of water required for 0.56 billion barrels of shale oil a year[b]	amount of water required for 10 billion barrels of shale oil a year
residential, industrial, agricultural, irrigation, mining usages	265,000 acre feet	...	...
required for related power generation	...	245,000 acre feet	4,375,000 acre feet
required for shale oil processing	...	113,000 acre feet	2,018,000 acre feet

Notes: Figures indicate total withdrawal, not net consumption. In the case of shale oil, net consumption is 76% of indicated total withdrawal. It will take a couple of decades for a major operation to come online. Let's take the year 2035 as a reference point. According to the US Energy Information Administration, world demand for oil in 2035 will be around 40 billion barrels of oil. With declining supplies of conventional oil, it is highly likely that shale oil will be expected to pick up the slack. I have run two figures in the table: production of 0.56 billion barrels (1.4% of expected demand) and production of 10 billion barrels of oil (25% of expected supply).

[a]*Source:* "Estimated Use of Water in the United States County-Level Data for 2005," United States Geological Survey, United States Department of the Interior, last modified August 25, 2010, http://water.usgs.gov/watuse/data/2005.

[b]*Source:* Lawrence J. MacDonnell, *Water on the Rocks: Oil Shale Water Rights in Colorado*, (Boulder, CO: Western Resource Advocates, 2009).

Water demand outstrips water supply in Colorado. The Department of Natural Resources in Colorado projects a 20 percent statewide gap between water supply and demand by 2030. It states that many of the local streams are over-allocated and furthermore, that some climate-change models project as much as a 20 percent reduction in the local water availability in the future. I surmise that no shale oil will be extracted from the Green Formation in Colorado, at least not in any amounts that would matter.

To assess the world's economic reserves of shale oil, I use Mohr's estimates.[17] After excluding the Green Formation and then applying a 25% deduction due to the assumed energy used for the extraction and refining of shale oil,[18] the global recoverable reserves are between 43 Gtoe and 114 Gtoe.

Thus, total reserves of all sources of economically-recoverable oil range between 191 Gtoe and 262 Gtoe.

TABLE 4.4

Oil	economically-recoverable reserves, estimates (in Gtoe)
conventional oil	120
natural bitumen (tar sand)	22.6 (after subtracting energy required for extraction)
extra-heavy oil	5.8 (after subtracting energy required for extraction)
shale oil	43 to 114 (after subtracting energy required for extraction)
Grand Total, rounded	191 to 262

The second of the three forms of fossil fuel whose reserves are to be assessed is natural gas, whose GHG (greenhouse gas) emissions, incidentally, have been shown to be as bad as coal's.[19]

The reserve estimates for natural gas made by *Oil & Gas Journal* and by *BP* are fairly close to each other. I averaged the two estimates and converted the resultant gas volume to its energy equivalent,[20] arriving at 169.4 Gtoe. This estimate includes only what could be regarded as natural gas from traditional sources.

Essentially, natural gas is natural gas is natural gas. However, it is found in different types of deposits. Beyond traditional sources, natural gas is found in a few other types of geological formations.

While gas has been known to exist in coal seams since the beginning of the coal-mining industry, only since 1989 has a significant gas production been realized from that source, with the advent of a new technology. Recoverable reserves of coalbed methane, as it is termed, are estimated to be 21.7 Gtoe.[21]

And we are back to shale. This time, shale gas.

Just as with coalbed methane, it is only recently that we have had the technology to truly tap into this vast natural gas resource. It has been made possible via horizontal drilling coupled with hydraulic fracturing.

Horizontal drilling allows us to drill laterally and thus across much of the horizontal shale strata, coming in contact with far more of the desired, natural-gas-embedded rock. Hydraulic fracturing, or hydro-fracking, cracks the shale rock by injecting at high pressure enormous amounts of water mixed with chemicals and sand. These fissures are then held open by the injected sand particles so that natural gas from the shale can flow up the well.

A shale-gas well can produce over one million gallons of wastewater, which is often laced with highly corrosive salts, carcinogens such as benzene, and radioactive elements like radium. All of these substances may occur naturally thousands of feet underground. Other carcinogenic materials may end up in the wastewater from the chemicals used in the hydrofracking process itself.[22] The wastewater may be hauled to a water treatment plant, purified to the best of our abilities, and then discharged into rivers, sometimes just miles upstream from drinking-water intake plants. There is nothing affable about shale gas. When one accounts for the fugitive emissions of methane as a result of the hydrofracking, the GHG emission of shale gas may be a bit more than that of conventionally drilled natural gas,[23] or a lot more[24] (which is more likely[25]).

Under business-as-usual practices, I assume that in spite of the above, shale gas drilling *will* continue at full steam—in the short term in the United States, and soon thereafter throughout the world. The situation at present seems to support this prognosis. There were more than 493,000 active natural-gas wells in the United States in 2009; this is almost double the number in 1990. And around 90 percent of the wells have used hydrofracking.

We don't know how much shale gas is out there, but we do know that there is a lot of it. An initial assessment in 2011 reckons technically-recoverable shale gas to be 173 Gtoe, spread in fifteen large regions around the world.[26] Yet, this estimate may prove to be overly optimistic.[27]

The last viable source of natural gas is found in reservoirs with low porosity and low permeability. The natural gas in such deposits is referred to as tight gas. It requires multiple fracturing in order for any significant amount of gas to be made available. Drilling for tight gas has been compared to drilling a hole into a concrete driveway: the rock layers that hold the gas are very dense, so the gas doesn't flow easily. Until recently, tight gas was not considered economically viable to produce. Recent technological advances on a number of fronts have made it increasingly possible to extract tight gas. No systematic evaluation of tight gas reserves has been carried out for the world, and for what it is worth, preliminary estimates of recoverable tight gas stand at 30 Gtoe.[28]

In total, about 394 Gtoe of natural gas is estimated to be economically recoverable in the world from all sources.

TABLE 4.5

Natural Gas	economically-recoverable reserves, estimates (in Gtoe)
traditional sources	169.4
coalbed methane	21.7
shale gas	173
tight gas	30
Grand Total, rounded	394

Last but not least is coal.

What we expansively term "coal" ranges from the brown, almost peat like lignite, which has high moisture and low carbon; all the way to the black, carbon-rich, energy-packed anthracite.

The World Energy Council estimates that the world's recoverable reserves are 405 billion tons of bituminous coal, 260 billion tons of sub-bituminous coal, and 195 billion tons of lignite coal.[29] In converting the various types of coal to their energy equivalents, I have arrived at a combined total of 461.6 Gtoe.[30]

TABLE 4.6

Coal	economically-recoverable reserves, estimates (in Gtoe)
All Forms of Coal, rounded	462

Now we can tally all recoverable fossil-fuel reserves: they come to between 1,047 and 1,118 Gtoe. This tally incorporates the whole enchilada. If it is even remotely economically recoverable, it is accounted for in this assessment. It accounts for every scrap of conventional oil, extra-heavy oil from Venezuela, and tight gas from Russia. It includes all black coal and also all available brown, inferior coal. It accounts for all of the viable shale gas and shale oil.

If we have it our way, this will all be consumed within less than a century.

The US Energy Information Administration projects annual global demands for coal, natural gas, and petroleum through the year 2035. From 2036 through 2100, I used the 2035 projected demand figure and adjusted it to reflect anticipated global population for each of the subsequent years, derived from UN's *World Population Prospects: The 2010 Revision*.

I compare these demand figures against the total economically-recoverable reserves that were just chronicled.

TABLE 4.7. Projected global demand for fossil fuels

Year	Coal	Oil	Natural Gas	Total
2010	3,282,497	4,553,631	2,980,152	10,816,279
2015	3,504,505	4,728,955	3,277,116	11,510,576
2020	3,840,416	4,909,608	3,581,964	12,331,987
2025	4,227,481	5,202,170	3,807,972	13,237,623
2030	4,678,301	5,535,765	3,949,884	14,163,951
2035	5,197,665	5,893,874	4,107,564	15,199,103
2040	5,629,104	6,073,303	4,224,682	15,927,089
2045	6,070,904	6,232,068	4,326,999	16,629,971
2050	6,520,795	6,369,019	4,413,800	17,303,614
2055	6,977,720	6,484,532	4,485,432	17,947,685
2060	7,442,231	6,580,537	4,543,312	18,566,080
2065	7,916,235	6,659,936	4,589,515	19,165,685
2070	8,402,044	6,725,576	4,626,065	19,753,685
2075	8,901,043	6,779,202	4,654,215	20,334,460
2080	9,414,651	6,822,366	4,675,073	20,912,090
2085	9,945,590	6,857,320	4,690,222	21,493,131
2090	10,497,230	6,886,392	4,701,281	22,084,903

Year	Coal	Oil	Natural Gas	Total
2095	11,071,081	6,910,352	4,708,800	22,690,233
2100	11,667,884	6,929,396	4,712,930	23,310,210
total demand	645,826,040	562,845,597	390,089,408	1,598,761,044
total supply	461,650,000	227,006,387	394,014,000	1,082,670,387

Notes: Units are in Ktoe (the energy of one thousand tons of oil). Table shows projected demand irrespective of available reserves. The two darkly-framed cells indicate projected supply has run out.

Provided the *rate* of extraction is unbounded, then under projected demands and estimated reserves, we will run out of oil around 2051; we will run out of coal around 2084; we will run out of natural gas around 2101. And if we use the three forms of fossil carbohydrate interchangeably, we will run out of all three around 2077.

This hypothetical scenario can never occur: the rate of extraction is subject to constraints. In fact, the overall rate of extraction is already declining on some fronts.

——— —

The lowest hanging fruits have been picked; the cheap oil, gas, and coal are all about gone.[31] In recent times, we have had to invest ever more resources for extracting and refining fossil fuels in order to arrive at the net energy yield of former years—whether it means drilling under the sea beds or extracting tar from sands and subsequently refining them to a lighter material.

The rate at which we can extract fossil fuel is of paramount importance. Think of it as a pool of water that feeds us through straws. The pool may hold plentiful reserves of water; however, what matters more is the rate of water provided by the straws.

Thousands of individual oilfields have produced ever less quantities of oil. Overall, the oil production of some countries has been declining for some time now, including that of the United States, Norway, Mexico, and the United Kingdom. It is important that we identify the underlying reason. Is it largely because there comes a time when it is more expensive to coax oil from a field while there is a lower hanging fruit elsewhere? Or is there a geological constraint that limits the total output, irrespective of our efforts? If it is predominantly a matter of cheaper prices elsewhere, then fields can be revisited and production revved up again once all the lower hanging fruits have been picked. In such a case, it may get more expensive to drill, but the flow of oil may continue unabated. However, if it is predominantly due to geological constraints, it is likely that there will come a time in which the global extraction rate of oil will begin a path of relentless decline, much as happened in the United States in regards to its conventional oil.

The peak production of conventional oil in the United States was in 1972. It had a smaller peak in 1987 and has declined ever since. As of 2009, the total production is at 56 percent of 1972 production figure. As of 2009, for every barrel extracted domestically, about 1.7 barrels were imported.[32] The fact that the more risky and expensive method of horizontal drilling was widely implemented since 2009 indicates that given half a chance, the United States *will* get its oil domestically. It just couldn't extract more before horizontal drilling came of age.

Perhaps the most compelling piece of evidence for geological constraints is the US oil production during the 1970s. When the price of crude oil was around USD 21 a barrel in 1972, the United States was producing more oil than ever. Production declined in subsequent years. Eight years later, the price of oil climbed to USD 102 (Both the 1973 and the 1980 prices are inflation-adjusted). If America could churn out oil domestically, 1980 was the time. But although they were more than willing to drill and sell for USD 21 a barrel a total of 3.45 billion barrels in 1972, American drilling ventures were evidently unable to do the same for about 5 times the selling price a few years later. Thus, it seems that the decline in production is predominantly a geologically-driven phenomenon. In fact, it appears that a market price of oil plays but a small role, as is evident from figure 4.6.

We also see a comparable lack of correlation in the global arena. Despite a near tripling of world oil prices in the past decade, non-OPEC production hasn't increased since 2004. Geological constraints seem to be both paramount and impervious to market pressures. Some experts think that insofar as conventional oil is concerned, non-OPEC countries have already reached their peak production rates.[33] In other words, this is as good as it gets.

FIGURE 4.6

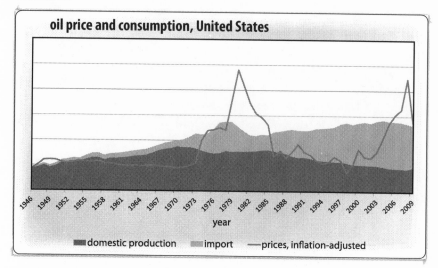

Sooner or later, gas or oil fields start losing pressure and the rate of extraction declines. In order to wring out as much of the oil as we can, we inject high pressure gas, water, steam, or even chemicals to make the oil flow better. However, there is only so much we can do, and more often than not, there is only so much that it makes economic sense to do.

Technology does save the day, if not the day after. There have been significant, ongoing improvements in recent decades in methods of oil extraction. The last two years have seen production rates that buck historical, declining trends. Recent improvements in horizontal drilling made it economical for the United States to coax its depleting fields and raise the total US production rate for the year 2009 and then again in 2010.[34] In the end, these are technologies that expedite the rate of extraction, but as the cumulative amount of oil is the same for a given field, an increased extraction rate comes at the expense of faster declining rates later.[35]

As with conventional oil, natural gas extraction rates peak and then decline. And perhaps in a manner that is even more pronounced.

Pumping natural gas is like poking a hole in a car tire. Production increases rapidly from the time a field comes online. Extraction rates rise as more and more wells are added within a given field. Eventually production plateaus and is followed by a rapid decline.

In the United States, the production of conventional natural-gas plateaued in 1971 through 1973 and since then has broadly declined. In 2009, the United States produced about 56 percent of the natural gas from traditional sources as it produced at its peak in 1972 (which oddly is identical to the performance profile of US oil). Initially, the Americans supplemented their piping gas from Canada. In the early 1990s, more was required to offset the continuing decline in production and increase in demand. Americans started to access non-traditional, more resource-intensive gas sources, which as of 2009 account for roughly half of all natural-gas extraction in the United States.[36]

FIGURE 4.7

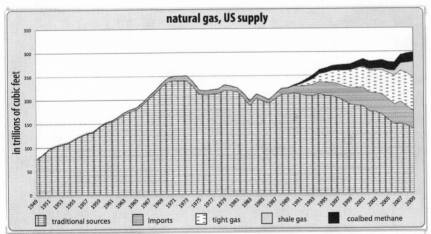

Coal is different from natural gas or oil. It is mined, not drilled.

On one hand, we may find ways to extract what at present is too resource-intensive, and thus the total estimated coal reserves would turn out to be larger than currently assumed. On the other hand, our assessment may show that there is less coal than assumed, and the total reserve amounts would be adjusted down—as has occurred time and again in the past.[37]

The Earth holds considerable amounts of recoverable coal. The question is whether it would be possible to obtain coal at desired rates of extraction until we basically run out of all that can be recovered, or whether there are technical constraints that will see annual coal yields decline long before the end of recoverable coal is in sight—in a couple of decades, as some analysts maintain.[38] I find no evidence, one way or another, for declining rates of extraction from individual coal mines. Combining the production output of all mines is a different story, though.

Imagine coal deposits around the world as thousands of bonfires across a vast plain. Some are large, some are small, some hold a lot of wood, other don't. As the bonfires exhaust their fuel load, one by one they go out. At first, the plain is very bright. Slowly it becomes dimmer as more and more individual bonfires go out. This would occur irrespective of whether individual bonfires snuff out abruptly or die out gradually. In other words, looking at the total picture, it matters little if technology can coax an individual fossil-fuel deposit to give up its energy at a higher rate and deplete faster or do it at a slower rate and have it last longer. Viewing the combined supply, a declining trend prior to the total exhaustion of resources will occur either way.

Now, in reality, some bonfires (that is, coal, oil and gas deposits) die out and new ones are added (as new deposits are being discovered). However, at some stage more bonfires die out than new ones are added—as has been the case with conventional oil and traditional gas fields.

Coal and natural gas are largely used for heating and for power generation. Petroleum is largely used as a transportation fuel. However, it is possible to synthesize most forms of hydrocarbon and make them perform whichever function is called for. Natural gas is a major feedstock in the production of ammonia for use in fertilizer production. Yet, coal can be gasified and made to perform the same function. By the same token, if we run out of oil, we can liquefy coal and use it to run motor vehicles, as they did in Germany during World War II. Alternatively, we can take natural gas and convert it to oil, as they started doing in the large Pearl GTL plant in Qatar.[39] When push comes to shove—as it is likely to—the fossil fuels are interchangeable.

Once the energy required for conversion is accounted for, what matters is the total recoverable energy of those resources stored in the Earth, combined.

A meaningful way to view at things is how much fossil energy is available per capita, year-by-year. As we do that, a few things need to be accounted for.

Through time, we accomplish more with the same amount of energy due to efficiency gains in design. In fifty years, we are likely to get more mileage out of a ton of oil equivalent than we do at the present. For all intents and purposes, it is assumed that 1 ton of oil equivalent in 2099 is the same as 1.35 tons in present-day terms.[40] This was taken into account in the graph in figure 4.8, along with annual population projection.[41]

Earlier, I examined and determined that overall rates of extraction will decline across the board. This is one of the premises that underlies the graph below. The projected schedules of decline were derived from a model created by Steve Mohr.[42] There was some departure from Mohr's model: shale oil, natural bitumen, and Extra Heavy oil take a lot of energy to extract.[43] I made due deductions to arrive at net energy yields. Furthermore, 25 percent of the global shale oil reserves were omitted as I assume that we will not extract oil from the Green River Formation.

The graph in figure 4.8 suggests that in 2099 each person will have, in effect, 29 percent of the energy available in 2010, that is, after factoring in projected efficiency gains and population growth. This does not account for any energy losses due to possible conversions of one form of hydrocarbon to another.

FIGURE 4.8

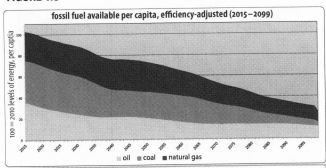

Another meaningful analysis is to compare anticipated demand with the anticipated production rate.[44] Once again, it is adapted from Steve Mohr's model with the adjustments noted earlier. The difference from the previous graph is that possible efficiency gains are not accounted for. The graph in figure 4.9 notes energy availability in absolute terms. The projected demand in this graph is based on data from the US Energy Information Administration.[45]

FIGURE 4.9

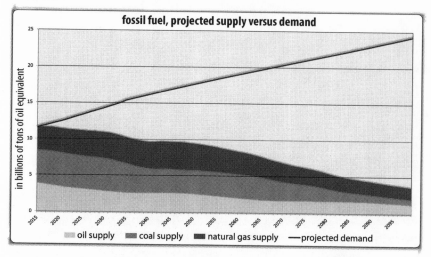

Based on this projection, in the year 2099, the world under a business-as-usual scenario would need over 24 million Ktoe (kilo-ton of oil equivalent) of energy.[46] Combined, all forms of fossil energy would yield about 3.5 million Ktoe during that year. Thus, in the year 2099, fossil-based energy would be able to provide but 15% of the needed raw material for the making of petrochemicals, fertilizers, transport fuel, and energy for the generation of electricity and space heating.

From a broad, planetary perspective, it is debatable whether the main problem is that we are running out of economical fossil-fuel, or whether we are not running out of it fast enough.

Be that as it may, we have gone out on an energy limb that is propped by the gargantuan injection of fossil fuel, allowing the existence of billions of people that arguably would not have lived otherwise. The obvious question is what happens to the many billions of us perched and multiplying on a tree branch, a fossil-fuel branch, that is becoming ever more fragile.

Fossil fuel is not the only vital resource that is projected to shrink and undermine our collective existence.

WATER SCARCITY

Imagine a pool of fresh water that is 1,000 meters wide, 1,000 meters long, and 1,000 meters deep. If you drop the Empire State Building into this pool, it would bob in it like a small Lego brick in a cup. This gigantic pool contains one billion cubic meters of fresh water.

In the year 1900, about 350 such mammoth pools of fresh water were needed to produce food for the people of the world. By 2000, the demand vastly expanded to the equivalent of 3,100 such mega pools across the Earth.[47] The future trajectory is as obvious as it is predictable: more people require more food, which in turn requires more irrigation water.

Our existence is predicated on water by way of drinking, bathing, and cooking. In addition, fresh water is needed for mining, manufacturing processes, and energy production. The largest amount of fresh water, however, goes toward irrigating our crops. Irrigation underpins about 40 percent of all global food production.

The human population in northern Brazil or in Iceland lives amid water abundance. On the other end of the spectrum, those living in western India, northeast China, or the Great Plains live with water scarcity. All told, in 2010 we drew from lakes and rivers the equivalent of 3,500 gargantuan pools of water for use in the agricultural, industrial, and residential sectors. Many of the rivers are sucked dry to irrigate food crops.[48] The Yellow River in China, the Nile in Egypt, the Indus in Pakistan, the Amu Darya in Central Asia, the Rio Grande in the United States and many other mighty rivers are drawn to their last cubic meter. Seventy rivers are close to this stage.[49]

In those regions in which demand exceeds the capacity of surface water to provide, we drill down to obtain more. Thus, the equivalent of 700 mammoth additional pools came into being, for a grand total of 4,500 lake-class pools across the world for our use. By 2030, it is projected that we will need the equivalent of *6,900* such pools of fresh water. This assumption is based on projected future economic activity, population numbers, and agricultural production and does not account for any possible conservation measures beyond those practiced at the present.[50]

The rate of groundwater *depletion* worldwide went from around 126 cubic kilometers in 1960 to around 283 cubic kilometers in 2000.[51] This amount of water depletion can be visualized as a pool of water that is 100 meters deep, 50 kilometers wide and 50 kilometers long. Every year, this hypothetical giant pool is filled to the brim with fresh water. Subsequently it is drained; much of it ends up being flushed in to the salty oceans of the world. Analogy aside, this happens through irrigation followed by evaporation followed by condensation and cloud formation. Given the composition of the planet, most of the rain ends up falling over the ocean.

FIGURE 4.10

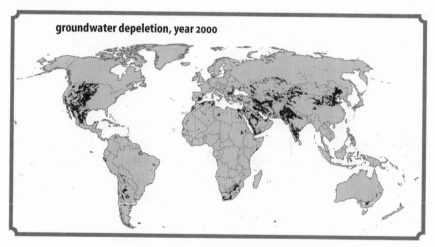

groundwater depletion, year 2000

Source of data: Yoshihide Wada, Department of Physical Geography, Utrecht University.
Note: This map does not indicate the amount of withdrawal but of net depletion, after factoring in the recharge rate. Darker regions indicate considerable net depletion. A more nuanced, detailed version in color is found at www.danielrirdan.com

The regions with the highest depletion rates have been Pakistan, southern India, northeast China, central United States, and western Iran. As aquifers deplete, their respective regions will be saddled with populations that would not have otherwise existed.

The early 1960s saw the en masse arrival of high-yield seeds to India. These are more drought-sensitive breeds that require irrigation to an extent that the traditional breeds do not. During that time, government-subsidized electricity expanded to rural areas together with inexpensive pump technologies. Thus, a Faustian bargain was struck, perhaps the largest in the history of humankind. Fast forward to 2010. India's 100 million farmers have drilled more than 21 million wells, pumping water out of the aquifers faster than they can replenish: squandering the underground reserves of freshwater with thirsty crops like rice, sugarcane, alfalfa, and cotton.

Groundwater meets the domestic needs of more than 80 percent of the rural and 50 percent of the urban population of India. Around 40 percent of India's agricultural output comes from areas irrigated by groundwater.[52] The prodigious injection of water into the food system made possible the accommodation of about 750 million additional people in India.

There will come a time when these augmented rates of water withdrawal will be exhausted. In other words, hundreds of millions of people depend on tube wells that will eventually run dry. Hundreds of millions of people came into the world based on promises of water and food that should have never been made. By 2050, the percentage of overexploited watersheds in India may rise from 15% to 60% and the availability of cultivable land per capita will be cut to half of current levels.[53] The challenge of groundwater availability is compounded by rapidly deteriorating groundwater quality in many areas of India. The prospect of not having sufficient amounts of water to support the existing population-base is all too real in India. Future prospects would compound the problem. By 2030, water demand in India is projected to double, driven by domestic demand for rice, wheat, and sugar for a growing population.[54]

China, next door, is in a long-term predicament just as bleak—as 70 percent of the country's grain is produced with irrigation, to the tune of pumping and using 100 billion cubic meters of groundwater each year.[55]

And then there is Pakistan.

The Indus River has been the lifeline of the country; practically the entire agriculture and populace of Pakistan draw and depend upon this one river. Within the last fifty years, as its population swelled an additional 100 million, the volume of sugar cane produced in Pakistan exploded to four times its historical amount, cattle six times, wheat and rice over five times. Consequently, the once-mighty Indus has shriveled in some areas to little more than puddles.

As of 2008, Pakistan is thought to have a population of 165 million people, of which 41 million live in grinding poverty, 98 million rely on agriculture, about 50 million do not have access to safe drinking water, and 74 million have no access to improved sanitation. This situation is likely to go downhill.[56]

The Indus River Delta is being heavily degraded by salinity, silt, and uncontrolled pollution. Flooding and drainage problems are likely to get worse, especially in the lower Indus Basin, as silt builds up and narrow embankments force the river to flow within relatively narrow beds above the level of the land.

Many rivers that drain glaciated regions are sustained by snowmelt during the summer season. Climate change may cause the glaciers of the Western Himalayas to melt and retreat. When the glacial reservoirs empty, the amount of water in the Indus River will drop.[57] Sooner or later, the snowpack and glaciers are likely to melt away into history—and with them the drinking water, food, energy production, and sanitation that are predicated on them.

Currently, the melt water reaching the Indus Basin feeds 210 million people. This is about to change. Against the prospect of rising population, a study estimates that the Indus would be able to provide water for 23 to 30 million fewer people than are provided for at present.[58] The same plight is projected for the Brahmaputra Basin, currently feeding 62 million people in the Bangladesh area. It is projected that due to climate change the Brahmaputra would be able to provide water for 30 to 40 million fewer people than are provided for at present.[59] This is contrasted against a situation that is projected to see the population of Bangladesh growing from 156 million people in 2010 to 250 million in 2050. More recent studies have ascertained that the extent of icecap melt in the Himalayas has been minimal.[60]

Thus, we can assume that more people in the Indian subcontinent will have access to water than has been assessed.

Come 2020, the Andes glaciers are projected to shrink and affect water availability for 60% of the population of Peru. One of the more affected rivers would be the Mantaro River, where a hydroelectric plant generates 40% of Peru's electricity and provides the energy supply for 70% of the country's industries.[61]

It is projected that there won't be enough water to grow the food needed to meet the population growth and changing dietary demands of the world.[62] Water scarcity will deepen in many parts of the world. Global grain harvests will be threatened; more countries will rely on food imports; and the livelihoods of many people will be threatened. More people will perish from lack of access to fresh water.

DROUGHT

With the prospect of climate change, some regions are expected to experience mega-droughts in the decades to come. The most severe impacts are projected to occur in arid and semi-arid areas, e.g., the Mediterranean basin, western United States, and southern Africa. By the 2090s, the extent of drought-affected areas is likely to grow across the world, while the proportion of the land surface in extreme drought at any one time is predicted to increase ten-fold.[63]

As of this writing, the percentage of land area in the United States experiencing exceptional drought reached its highest recorded levels in July 2011. There is a broad consensus among climate models that the Southwest region will grow drier in the twenty-first century, and that the transition to a more arid climate is already under way. If these projections are correct, the levels of aridity of the Dust Bowl will become the new norm of the American Southwest within a time frame of years to decades.[64] One study projects that the drier conditions in the Southwest will likely reduce perennial vegetation cover and result in increased dust storm activity in the future.[65] Given projected water withdrawal and precipitation patterns, hundreds of counties are likely to experience extreme risk of water shortages in the western regions of the United States.[66]

As it is projected for the Southwest, so it is projected for other regions. Declines in dry-season precipitation in northern Africa, southern Europe, and western Australia are expected to be nearly 20% higher with a mere 2°C mean global warming. By comparison, the American Dust Bowl was associated with average rainfall declines of 10%.[67]

At the National Center for Atmospheric Research, a study by Aiguo Dai indicates that by 2030, subtropical regions will be overtaken by droughts of even greater severity than the one in the Sahel region of western Africa in the 1970s. An increased dryness is projected over the next several decades that eventually lead to severe, long-lasting droughts in most of Africa, southern Europe, the Middle East, most of the Americas, Australia, and Southeast Asia.[68] Projected conditions are shown in figure 4.11 found at www.danielrirdan.com[69].

A projection using a suite of eight climate models shows dramatic declines in soil moisture on a global scale toward the end of the century. Droughts that last a year or longer become three times more common.[70] The results are shown in figure 4.12 found at www.danielrirdan.com

FOOD PRODUCTION

A sizable segment of the world's population is chronically underfed. In recent decades, the number of undernourished people has ranged from an estimated 800 to 1,000 million. For the last fifteen years, about 13% to 14% of the population has been subsisting at various degrees of starvation. This is the baseline, the status quo of where we have been. According to a 2011 study, global food demand could double by 2050.[71]

It appears that in upcoming decades, various trends and practices will collude to affect us all—not just the traditionally undernourished people at the bottom of our social totem pole.

To understand why, we have to go back to the beginning of it all.

The creation of soil begins with an inert and infertile subsoil of clay, sand, and rocks. The first pioneer plants germinate and send their roots downward, into the hard compacted earth. They bring moisture and minerals up from the rocky subsoil to their stems and leaves. Later, as the leaves fall, they cover the ground and shade it from the evaporative and oxidizing effect of direct sunlight. The dead organic material on the surface feeds the soil community, which in turn decomposes it and makes its nutrients available to other plants that start to appear at the scene. Earthworms create tunnels, which are passageways for the infiltration of water and air. As the soil increases its fertility, it becomes more porous. More moisture is retained and the temperature extremes are moderated. As the soil builds up, the richness and diversity of the habitat heighten.

Soil is the substrate for life, the basis for all terrestrial life forms. It is the gut of the earth, the principal digestive organ of the planetary ecosystem. It contains micro-organisms, insects, and worms that live in, contribute to, and feed on various components of the soil. Like the bacterial community in the human gut, the soil is a living community of organisms that produces the necessary conditions for the plant communities to exist.[72]

Soil is also the skin of the earth—the interface between geology and biology. And we are scraping it raw. We see it in brown streams that bleed from construction sites. We see it in sediments deluging rivers downstream from clear-cut forests. We see it where new suburbs and strip malls pave fertile valleys.[73] Worldwide, soil erosion averages approximately 30 to 40 tons per hectare annually. That average is about 30 to 40 times the natural replacement rate.[74]

It is estimated that 24 percent of the global land area, or 35 million square kilometers, has been degrading over the period of 1981–2003, 19 percent of which is cropland.[75] This means that during those 22 years, we lost on average every year 3 million hectares of farmland. To date, close to 20 million hectares would be rquired yearly to make up for the losses and for the increased population.[76] In 1960, when the world population numbered about 3 billion, approximately 0.5 hectare of cropland was available per capita worldwide. It takes half a hectare to provide a diverse, healthy, nutritious diet of plant and animal products for one person. At present, the average per capita world cropland is only 0.22 hectare, or about 40 percent the amount needed according to developed national standards.[77] The difference is made up by a combination of improved yield and the number of undernourished people around the world.

We already commandeer all of the prime land in existence. About the only places left that could still be put to agricultural use are the tropical forests, with their nutrient-poor soils.

The explosive growth of food supply has been funded by extorting fertility from the planetary ecosystem. Our food system short-circuited the natural energy flows of the Earth by decimating significant portions of the biosphere. Every piece of vegetation removed from the planet's surface by roads, housing developments, logging, airports, or farms has been draining the life force of the planet.[78] Nutrients in a functioning ecosystem flow from the physical environment to plants and then to animals. Organic dead matter is decomposed by fungi and microbes and is returned to the Earth. And so it goes around and around. The problem inherent in conventional, modern agriculture is that it is not a closed-loop system. It is a one-way ditch funneling land resources into our gullets and from there on to the landfills and the seas. Conventional modern agriculture practices bleed the earth.

Phosphorus is a nutrient essential to all living organisms, and thus, it is essential in food production. Contemporary agricultural practices accelerate soil erosion and with it the leaching of phosphorus into the waterways. On top of which, much of the crop residue, which contains the mineral, is removed from the field, and much of the excreta, which contains the mineral, is never returned to the field. The end result is a wholesale removal of phosphorus from the cropland, which necessitates constant replenishment from an outside source. As phosphorus is a chemical element that has no substitute, the very existence of our food system hinges on regular input of phosphate fertilizer from a mined rock. This is a recipe for a food crisis.[79]

The world's remaining phosphate rock reserves are under the control of a handful of countries, foremost, China, the United States, and Morocco. At a current consumption trajectory, the global peak in phosphorus production is predicted to occur around the 2030s,[80] after which point demand is projected to exceed supply. Without phosphorus input, it has been estimated that wheat yields could fall from nine tons per hectare in 2000 to four tons per hectare in 2100. The global economically-recoverable reserves of phosphate rock are projected to be depleted within the span of a few lifetimes.[81]

When it comes to the hall of fame for batty land-use practices, burning through our stockpile of phosphorus faces some stiff competition, though. Deforestation and plowing come to mind.

In the first few centuries of the Common Era, potters set up kilns in Southern France and went through the neighboring forests. They started with the nearby trees of the floodplains, and as they cut them down, they started cutting deciduous oaks and, later, holm oaks.[82] Ancient Romans held no monopoly over shortsightedness though. The Chinese cut down trees of the Loess Plateau, notably around 600–900 CE, and the resultant soil erosion earned the Yellow River its name.[83] The first settlers of the Southwest region in North America proceeded to cut thousands of ponderosa pines for fuel and housing. The settlers abandoned the area around 1250–1400 CE, leaving behind a permanent legacy of arid desert. In Haiti, deforestation activities have left a third of its land eroded beyond remedy.[84] Around 1000 BCE, vast tracts of rainforest in the Congo Basin were abruptly replaced by savannas—as Bantu-speaking farmers moved in to the area and likely cut down the rainforest trees.[85] And Solomon Islands, a tropical nation in the Pacific, has decided in recent times to cut down its forests and export the lumber for hard cash. In 2009, its central bank announced that the logging rate had peaked.[86] In some of its islands, as little as 10 percent of the primary forest remains.[87]

Plowing ends up providing the same auspicious aftermath as deforestation. It overturns the soil and thus dries and oxidizes its deep layers by exposing them to the elements. Worse yet, plowing breaks up and collapses the water and air passageways within. By breaking up the soil into ever smaller particles, there is less space in between them. With repeated plowing, the soil becomes ever denser. This causes water to collect at the root zone and potentially drown the plants. The surface becomes more and more impermeable and clay particles begin to predominate.

The plowing technique of the ancient Greeks slowly pushed the soil downhill and created erosion.[88] This was dutifully and unimaginatively duplicated by the Romans—with equally dismal results. Giant cash crop plantations aggravated the problem. Eventually, the fields in central Italy were so degraded that the government had to force people to stay on the land. But to no avail. In the end, grains had to be shipped from North Africa. This bore heavily on the fertile Libyan coast. With the gradual soil erosion, the fertile coastal region of Libya slowly gave way to desert. By the first century CE, the Mediterranean region was stripped bare, bone-white rocks exposed everywhere as they still are to this day. People replicated this practice again and again everywhere. They did it five thousand years ago, and five hundred years ago, and we are still at it today.

In the 1920s, Russell Smith from Columbia University travelled to China. As he stood on the Great Wall of China, he noted the slope below. It was gouged with gullies, some fifty feet deep, stretching as far as the eye could see. The entire countryside was gutted, a desert of sand and gravel, alternatively wet and dry but essentially fruitless. Farmers had cleared the forest and plowed the land. Year after year the rain had washed away the loosened soil. The progression of much of the hilly land under the plow has been forest–field–plow–desert.[89]

In the mid-nineteenth century, Americans started to plow, plucking out the thick mat of perennial grass that held down the soils of the Great Plains, replacing it with drought-sensitive wheat. On the heels of a drought, the crop failed, leaving the soil barren.[90] The wind blew the fine soil particles away. Once those were largely gone, leaving only larger particles, the sand storms began. By 1928, topsoil loss was accounted for in billions of tons annually. A few years later fields throughout Montana and Wyoming were ravaged by high winds. Blowing across the Dakotas, the wind kept picking up dirt until hundreds of millions of tons of topsoil were hurtling east and south, depositing sand over Chicago, Boston, and New York. Sand storms are the final phase in the desertification process.

The failure of rainwater to seep into the soil is the beginning of another process that ends in erosion. Stripping the land of vegetative cover lessens the possibility that rain will be slowed down long enough to seep in. This is a self-reinforcing mechanism. As more soil is carried away, the more impermeable subsoil layers are exposed which causes water to run off faster. The rains continue to come, and the earth continues to erode, but once the plants can no longer get a foothold the process simply continues unabated until it reaches bedrock or other impervious layers.

When the water runs off rather than soaking in, floods are created. Those fill the streambeds with boulders and gravel. The floodwaters eventually carry nothing but inanimate sand and gravel, burying fertile land downstream. Furthermore, once the water is not retained by absorbent topsoil, the springs dry up. As the streams dry up, there is less vegetation left to transpire. It is a vicious circle.

Then there is the issue of salinity due to irrigation. Irrigation water contains some salt, much as rainfall may. This by itself is not the problem. Conventional irrigation applies far more water to the field than crops can handle and the soil can assimilate. Once the water table rises close to the soil surface, water is drawn to the surface by evaporation and the salt stays behind, accumulating. Eventually, a crust of salt builds on the surface and in the long run may render the land barren. Although it is a worldwide problem, salinization is particularly acute in semi-arid areas that are heavily irrigated and are poorly drained and where salt is never completely flushed from the land. Such conditions are found in parts of the Middle East, in China's North Plain, in Central Asia, and in the Colorado River basin in the United States.

By the beginning of the twenty-first century, the soil in many parts of the world as barren, inert, or washed away.

To make up for the lack of soil rotation, modern agriculture practices add a blend of nitrogen, phosphorus and potassium. This ensures that the plants grow full of calories, water, and fiber. But we don't eat fresh produce just for calories, water, and fiber; we eat them also for valuable nutrients. If these minerals aren't present in the soil, they cannot miraculously appear in the plant. Two studies tracked the decline of mineral and vitamin content of fresh produce through the decades. The results speak for themselves.

TABLE 4.8. Nutrition in vegetables through the decades

	study 1 (USA)[a], nutrient decrease (1963 through 1992)	study 2 (UK)[b], nutrient decrease (1940 through 1991)
calcium	29.8%	46%
iron	32%	27%
magnesium	21.1%	24%
phosphorous	11.1%	...
potassium	6.5%	16%
sodium	...	49%
copper	...	76%

[a]*Source:* Paul Bergner, The Healing Power of Minerals, Special Nutrients, and Trace Elements (Prima Publishing, 1997), 46–48.

[b]*Source:* David Thomas, "The Mineral Depletion of Foods Available to Us as a Nation," *Nutrition and Health* 19 (2007): 21–55.

The jury is still out as for the underlying causes. Having said that, we don't employ deep-root plants, which would bring up micro nutrients from the rocky subsoil. Hence, it stands to reason that the soil has no nutrients—except the nitrogen-phosphorus-potassium fertilizers we douse it with.

Modern agriculture is predicated and propped up on prodigious amounts of fossil fuels. These permeate every facet of agriculture: fertilizers and pesticides, operation of field machinery, irrigation, crop drying, long-distance transportation, related processing and packing.

The end result of chemical fertilizers, powered farm machinery, and an increased scope of transportation and trade, was not just an enormous leap in crop yields, but a similar explosion of human population, which has grown over sevenfold since the dawn of the industrial revolution.

The way a nitrogen-based fertilizer is manufactured is by extracting hydrogen from natural gas. Without synthetic fertilizers—and the underlying natural gas—this productivity cannot be sustained. Detailed analysis of nitrogen flows shows that synthetic fertilizers provide close to half of all the nutrients received by crops. In 1900, our fields fed 1.6 billion people with no synthetic fertilizer with 850 million hectares of cropland to work with. At this yield, we can feed about 40% of the world's population with the existing cropland area of 1,500 million hectares. As of 2008, it was estimated that synthetic fertilizer provided for basic means of survival for 48% of everyone currently alive.[91] The obvious question is where does that leave the remaining 52% to 60% of the population?[92]

Last on the list of woes is ground-level ozone.

A stew of chemicals being emitted from gasoline engines reacts with the sunlight to form ozone at low altitude, bringing its concentration to levels harmful to plants. Exposure to elevated ozone typically results in suppressed photosynthesis, accelerated aging, decreased growth, and lower yields.[93] There have been reports of injury to bean crops in Mexico and radishes in Egypt.[94] Reports of visible foliage damage have been coming in from all parts of the world.[95] The most conspicuous damage may have been in the Mediterranean region.[96] For example, an ozone episode in Greece caused such severe reddening of the lettuce that crops could not be sold.[97] The heightened levels of above-ground ozone cause more damage to vegetation than all other pollutants combined.[98]

Within a decade, East Asia is projected to experience yield losses of rice, wheat, maize, and soybeans due to elevated levels of low-altitude ozone.[99] The USDA estimated that ground-level ozone could reduce soybean yields by an average of 23 percent. Overall, global impacts of ozone are likely to be considerable in the next fifty years and may act as a significant constraint on global food production.[100]

Under current practices with the state of the soil being what it is, rising prices and eventual decline in fossil fuel availability will result in anything from significant food stress to a outright global calamity.

We are all responsible due to our actions and inactions. If this has not been a crime, nothing is. If this has not been immoral, nothing is.

PART II

MITIGATION

MADNESS. IT'S MADNESS. ALL OF IT.

There were a few thousand of us in ground zero, Africa. And then, seventy thousand years and a few billion people later, we ended up with *this*? This state of affairs is the best that we can bring about between then and now?

I want to say that the best thing about human history is that it is *history*, but I cannot. We are still at it: the same devil-may-care, self-serving practices of our great-great-grandparents and their irresponsible-great-great-grandparents. Under the existing trajectory, this century is poised to be the grand finale of the last seventy thousand years.

Earth is the only home we are ever likely to have; living under gigantic glass domes in the inert rubble on the plains of Mars is not much of an option. All life forms are interlinked, forming one broad whole. We are all in it together. What's on the table is everything

We have no assurances, but we do have a chance for the redemption of our race from its legacy. If we take on a morally courageous course of action, it may pull this planetary ecosystem from the brink of collapse, and it will change the very identity of what it means to be human. Regardless of the outcome, we won't be the same afterward.

If we stall any further, the window of opportunity may be lost. The natural forces we have unleashed have their own momentum. Furthermore, what is needed requires a level of planetwide coordination and social cohesion never before called for. Once our global human community begins to disintegrate in the wake of food riots and fossil energy shortages, the opportunity to launch any complex, planetary-wide initiatives will be gone.

There are times when it is not enough to do our best. We must do what is necessary.

The proposed plan that follows involves many details. However, three things are big and broad and all-encompassing.

One. The further degradation of wild nature must cease everywhere and by everyone. This includes the constant nibbling at the remaining tropical rainforests and the harvesting of remaining fish in the seas and oceans of the world.

Two. Fossil fuels stay henceforth with the rest of the fossils—in the ground. The blowout party must come to a stop, and we just need to get over it.

Three. No more toxins dumped into the environment.

For every technological solution selected in upcoming chapters, a handful of others did not make the grade. Some did not scale up; some do not deliver; and others may or may not prove possible to implement. I chose what seems to be viable at present. As a general rule, this plan shies away from technologies that are just around the corner. This plan is both unsentimental in its choices and ruthless in its adherence to the facts—as much as these are publicly known. The prospects of profit or loss and the politics of a given situation taints potentially every report and study.

All forms of motorized transportation and energy generation bear a heavy environmental footprint—notably so when you scale them to serve billions of humans on a planet the size of Earth. Some technologies, however, have higher impact than others. The gist of this section is both the shift to lower-impact technologies and the lowering of consumption.

The specifics are but details; they can be endlessly improved upon and amended. And in fact, they should be. What follows in subsequent chapters are not Bible verses but drafts—to be expanded upon and altered over time by countless others.

5
TRANSPORTATION

⚛

on land, at sea, and in the air

FROM THE COLOSSAL bulk carriers traversing the high seas, to the hundreds of millions of cars dashing around on the world's roads, to the never-ending streams of airplanes taking off from the runways—motorized vehicles are running on one form of hydrocarbon or another. They run on light natural-gas and on viscous, tar-like bunker oil, and on everything in between. As of 2008, fossil fuel directly provided 97 percent of all transport energy.

All of this must be done away with. Over with. Kaput.

Within a fifteen-year period all hydrocarbon fuel ought to be phased out, replaced by lower-impact sources of power.

Let's take on one mode of transportation at a time.

ROAD VEHICLES
Internal-combustion engines must join, fittingly enough, the dinosaurs. The din made by thousands of gasoline motors would be gone. And the air of the cities would be fresh and clean—a real break from the cancerous diesel fumes of the twentieth century and from the dung particles and horse manure of the nineteenth century.

In place of internal-combustion cars would come electric cars. In the natural course of events, the bulk of the car fleet is replaced within a fifteen year period. Using the same time frame, internal combustion cars should be replaced with electric vehicles—ones whose engineering should make them last out this century.

One hundred years after the first mass-produced gasoline car, the Nissan Leaf (2010–) is the first mass-produced, highway-capable, everyday car that is purely electric. We have the technology, thus it is simply a matter of ramping up production and working on durability of the cars. Ecologically, we cannot afford to have cars go out of style anymore than we can afford to have houses go out fashion.

In terms of convenience and performance, electric cars are better in some ways and worse in others. Let's take the Nissan Leaf as a case in point.

The Leaf is a compact five-door hatchback car with space for five passengers. Practically in every way, the Nissan Leaf looks like your average economy car, one among many that have been traveling our roads for decades. The Leaf has two modes: "drive," which lets one use the full power of the engine, and "eco," which cuts back some on acceleration, thus providing an extended driving range.

Response in the Leaf is instant and peppy. There are no gears; the engine responds immediately with maximum power. A constant torque is one of the pleasant surprises of an electric car. Those who drove the Leaf report that it handles, well, just like any other car. The Leaf's motor is rated at 107 peak horse power and provides up to 207 pound feet of torque. The car can go from 0 to 60 mph in about ten seconds—which is comparable to the performance of the gasoline-powered Mazda2 (Demio) or Honda CR-V. The Leaf's top speed is about 94 mph ($\approx$ 150 kph). And it is quiet, really quiet. In city driving, the only sound one hears is a faint whir from the motor and the rubber as the car moves over the asphalt.

A nice benefit of an electric car is the ability to pre-heat the car while still reading the morning paper at one's house. Thus, when one enters the Leaf on a freezing morning, the car can be warm and pleasant. It makes practical sense too; as the driver pulls away, the energy from the battery can be applied more toward moving the car and less toward heating its interior space.

The limited driving range due to limited battery capacity is the single biggest drawback of the Leaf. How many miles one can get per charge depends on many things. With climate control turned on, a significant portion of the battery charge is applied toward heating or cooling the car interior and the driving range takes a substantial hit. At high speeds, above 55 mph, the air resistance becomes significant and is a real drain on the battery as well.

Under typical highway driving patterns, with the climate control on, one can expect a range of 60 to 65 miles per charge with the Leaf. Under urban stop-and-go traffic patterns on frigid days, one can expect to average a 63 mile range. Under a typical urban driving pattern (average speed around 20 mph, frequent stops, 57 mph top speed, climate control off) the range is about 100 miles per charge.[1] In summary, most of the time the range of the Leaf is likely to be 70 to 80 miles (≈ 115–130 km) per charge. This is for a new battery. After five years of usage, the battery may lose about 20 percent of its charge capacity.

As a matter of course, the new fleet of electrically-powered vehicles would necessitate an infrastructure of charging spots at places of residence and work. These charging spots, which may be the size of fire hydrants, ought to also be installed in public spaces such as movie theaters, sports arenas, and municipal parking lots. While people shop, work, or stay at home for the night, the vehicle can be plugged in for the six to eight hours required to fully charge the battery, or for thirty minutes with a fast charging option.

Given the fact that the average car is parked 22 hours a day, it is adequate. Given the fact that the average daily commute distance in travel-happy North America is 34 miles, a typical driving-range of 70 to 80 miles with a battery that can be charged at one's destination is acceptable. A network provider, Better Place, calculated that there is a need for 1 charging spot for every 2.5 car on the road. In the United States, this may mean 500–600 million charging spots.[2] In France, it may mean 80–90 million. When one is on the road, the car's navigation system could let the driver know the location of nearby charging points.

Under the Better Place setup, car owners are expected to activate the charging points with RFID cards. The system would identify the driver to be a subscriber and thus the proper account would be charged. In fact, without an identifying RFID card, the socket of the charging point won't even open. This rules out the prospect of someone outside the network pirating power, or of a youngster trying to stick something inside the charger. Another option may be a scanner that would read one's thumb, thus precluding the need to carry—and remember to carry—an RFID card.

Existing technology and software would let the network provider monitor each of the hundreds of millions of charging points and identify those that malfunction. This same software, developed by Better Place, permits the provider to monitor all the batteries in the network, aggregating data on each battery's state-of-charge and anticipated energy demand. The network software can communicate this data to the utility company in real time, allowing the utility company to optimize the allocation of energy based on available supply and drivers' needs.

As stated, the average car is idle twenty-two hours each day. This gives the utility company a lot of wiggle room: different cars can be charged at different times. The network learns the driving habits of each car and makes automated, due adjustments. At the same time, the driver can specify when they need the car charged and ready to go. All of this is quite essential. If the system commenced charging the moment a car is plugged in, the grid would collapse at 9:00 a.m. on any given morning, as millions of cars arrive en masse to the workplace and are all attempting to charge simultaneously. And if the grid didn't collapse, at the very least it would require a much higher generation capacity—needlessly so. The notion of sucking power from the grid the moment every person plugs in his vehicle is environmentally obscene. For that reason, if we do commercialize ultra-fast charging technology, it cannot be used to undermine or otherwise alter the fundamental principle of working off excess energy on the grid.

With ubiquitous charging points, the electric car scheme is viable within city limits. Once we get into the occasional long-distance trip, though, an electric car dependent on battery charges is nothing but trouble. The little solid data we have to go by does not look good. At 55 mph, the Leaf has a 70 mile range. I assume less than a 70 mile range under more typical 60–65 mph highway driving speeds. As no one waits for the battery to drain before getting off the highway in a search of a charging point, we can assume 40–45 miles (≈ 65–72 km) of uninterrupted highway driving before pulling over to charge the battery. Even with the universal 15–20 minute fast-charges, proposed by a consortium of automakers, it is impractical to stop every 40 to 45 miles for half an hour, let alone for eight hours.

Thus, in addition to the diminutive charging points that would dot the cityscape, roadside swapping stations could be installed: places where one can pull in, replace the drained battery with a fresh one, and drive off. This idea harkens back to the time of the horse and buggy, where one would ride a horse, come to a stop station, swap a horse, and keep on riding. This very idea is the brainchild of Shai Agassi, the founder and CEO of the Israeli company Better Place. Given existing driving patterns, it is unlikely that the average driver will have to pull to a switch station more than once a month.

The entire battery swapping process itself takes about one minute, which is less time than we typically spend at a conventional gas station. The process does not require the driver to ever leave the car. As of this writing, there is just such an automated switch-station in Yokohama, Japan.

This switch station somewhat resembles a fast oil change shop, without the staff members. The car comes to a stop at a designated spot and a plate comes up from underneath, mounted atop a hydraulic device. The computer in the switch station requests the computer on board the car to release the battery. Once this is done, the hydraulic plate bearing the drained battery is lowered and gets out of the way. On the heels of the first plate, a second plate carrying a fresh battery comes into position under the car. A latching mechanism within the car adjusts the orientation of the incoming battery as needed, and the service computer asks the car to lock in the battery. A green light tells the driver that the process is complete. The car is now ready to go.

If Renault goes through with its plans, the Fluence Z.E. sedan, planned to be made available in 2012, will be the first mainstream, mass-produced car whose batteries can either be charged in a conventional fashion or be swapped at a switch station.

Better Place reckons that there will be a need for a switch station every twenty-five miles in whichever direction one would go. As a case in point, Israel would need about fifty-five switch stations. It is likely that the existing gas stations, which are already strategically located, will offer battery swapping. Gradually, the gas portion of the business will dwindle to none; with a fifteen-year mandate looming, ever fewer people will buy fossil-fuel cars.

Incidentally, the cost of the infrastructure—switch stations, charging points, and related administration—is about what we pay in a given region for gasoline for one week. According to Better Place, a charging point takes about one hour to install and a switch station takes about two days. In other words, this infrastructure can be set in place in a matter of months or a year in a given geographic region.

Just as it is possible for anyone to go to ATM machines of other banks to withdraw money, so it is essential that any driver can pull in to a switch station of any network to charge their car irrespective of whether they are subscribers of that network. The switch station, in turn, would be expected to stock the various battery types likely to be distinct to each of the handful of car makers.

Thus, if someone is out of charge, it makes more ecological sense to swap a battery with one that was charged at off-peak hours rather than fast-charge on demand one's own battery. With a scheme allowing to charge most batteries in off-peak hours, we won't need to build power plants dedicated to charging our fleet of vehicles.

With a full deployment of switch stations, it would be possible to drive across a continent without ever incurring delays due to the need to charge one's battery. Let's see how it may pan out.

In highway driving, the gasoline-powered Honda CR-V has about a 430 mile range. For most, this range might as well be unlimited, as in effect the driver is likely to take a break before they clock 430 miles of driving. Driving an electric car long-distance will offer a very different highway experience, with a battery swapping every 40 minutes or so. This means that on the 387 mile trip from Los Angeles to San Francisco one would make a quick stop at 9 switch stations. The stopovers may be a snap, but there is no way around the fact that long-distance, highway trips will be less convenient with electric cars:

all ten or fifteen of the long-distance trips the average person takes in a year.

Don't fret over this inconvenience; under the blueprint outlined in this book it is going to be the least of your worries.

Incidentally, a lithium-ion battery *can* have a better range. The Tesla Roadster has a highway range that is in the neighborhood of 225 miles per charge. It can be done. Yet, at this point in time, the Tesla's 53 kWh battery is very expensive, pricing itself out of the mainstream market. Argonne Laboratories has developed a diverse suite of lithium-ion-battery technologies that would enable the next generation of batteries to yield a driving range that is 30% to 70% higher than those available on the market at present. This may translate to an improved highway range of 90 or 100 miles per charge. Meanwhile, folks at Oak Ridge National Laboratories stumbled on an innovative titanium dioxide composite, named mesoporous TiO_2-B microspheres, which further stands to boost battery capacity.[3] Concurrently, an engineering team at Northwestern University created an electrode for lithium ion batteries that with much higher energy density allows the battery to hold ten times the current charge.[4] It is indeed just a matter of time, and perhaps not much at that, until we can produce batteries at the same price point with longer driving range, making long distance trips convenient again.

Yet, as I state in another context, it is by no means clear if regaining convenience in long distance trips is a step in the right direction. If a lower-impact technology offers more convenience than the technology it replaces, it may be good. And if it offers less convenience than the technology it replaces, it may be better yet. After all, we attempt here to reduce our ecological footprint, not to stimulate further consumption. Alas, things that are easy or accessible have a way of doing just that.

Historically, the electric car had an additional major drawback for the consumer. The battery would be packaged as an integral part of the initial purchase. It is a bit like coming to a car dealership and paying USD 37,000 for a new gasoline-powered, economy car: USD 30,000 for the car itself and USD 7,000 down payment toward the USD 15,000 we can expect to pay for gasoline in the first ten years of driving the car.[5] A far more appealing model is to decouple the cost of the electric car from its battery. The customer will buy the car; however, energy usage and its related expense would be paid separately as is done now with gasoline cars. The battery, the energy carrier, would be owned by the network provider.

Much as one buys minutes with a cellphone service-provider, one may either buy kilometers as they drive, or purchase a monthly plan with unlimited miles. The battery itself is licensed out and is not directly paid for by the customer. Furthermore, the initial cost of the vehicle may be subsidized by the network provider, much as a cell phone is sold cheaper by the network provider on the promise of recouping its cost via the expected minutes of usage. When one signs up for the platinum program, with the maximum number of miles, such as for taxi drivers, they could conceivably get the car for free. In the end, just as with cell phones, the customer of course incurs all the costs—batteries, network, and all. However, spreading the cost over the lifetime of the vehicle does make a difference.

It is projected that the price of electric cars without their batteries would be competitive with their gasoline counterparts. It can be assumed that the price will drop further when electric cars are produced in volumes of tens of millions annually.

A concern of someone considering the purchase of a new car has always been their ability to sell the car five or fifteen years down the road. With the rapid, ongoing advances in battery technology, the prospect is all too real that when the electric car is put up for sale, its battery will be antiquated and inefficient by contemporary standards. This is a non-issue with the leasing of the battery. It is simply not part of the equation for the end consumer. The person who may purchase the car years later may subscribe to another, superior battery. Their service is not tied in to that of the original owner of the car.

As of 2011, Better Place has contractual agreements to set up electric car networks in Israel, Denmark, parts of Australia, and Hawaii.

This ought to be carried out for passenger vehicles, motorcycles, pickup trucks, delivery vans, and even buses: all can and should be electrified with batteries to power them. In fact, all of these vehicle categories already sport some electric models.

Electric scooters and motorcycles are ubiquitous in East Asia, nota-bly China. While most electric scooters are capable of a top speed suitable only for city driving, the zev7100 boasts a 78 mph top speed (≈ 130 kph) and is suitable also for highway travel with a range of 30 to 80 miles (≈ 50 to 130 km) on one charge, depending on the driv-ing speed. The company Zero Motorcycles is overhauling its Zero S electric motorcycle; its 2012 model is to have a range in excess of 100 miles (≈ 160 km) and a speed of up to 88 mph (≈ 142 kph).

Smith Electric Vehicles produces an entire array of electric trucks: aerial lift trucks, refrigerated trucks, flatbed trucks, and delivery trucks. Their trucks average a 100–150 mile range and a 54 mph (≈ 87 kph) top speed. They are not going to break any speed records, but they get the job done, delivering merchandise from point A to point B.

Zonda's electric buses in China are a good example of electric buses. They are capable of running between 185 and 310 miles on one charge (≈ 300–500 km). The maximum speed of Zonda's electric bus is 56 mph (≈ 90 kph). Those models are the obvious choice for commuter buses, coupled with switch stations every few hundred kilometers.

When it comes to safety, in truth no one knows if the electromag-netic radiation of electric cars is harmful. For what it's worth, it seems that in the passenger area of the electric car, the electric field generated by the starter batteries may be around 2 mGauss. In comparison, an electric clock may measure 5 to 10 mGauss a few feet away.

The forefront of commercial-battery technology is the lithium-ion battery with iron and phosphate, such as the one developed by A123 Systems. The battery is made from nontoxic materials: lithium, manganese oxide or iron phosphate, and graphite. Today's lithium-ion battery has unprecedented safety. Its cover is designed to sustain shocks, intrusion of fluids, vibration, and crashes. The electric battery also comes with safety disconnects, thermal fuses, and safety vents. It is possible to soak the battery in water with no ill-effects; it is sealed. It is possible to put the battery in the oven, and the only thing that will happen is that the end-caps will melt. The lithium-ion battery has an additional advantage; it does not suffer from the so-called "memory effect" resulting from an incomplete charge. So it is possible to perform a lot of small charges without diminishing the battery's charge capacity.

More than 95 percent of the lithium-ion battery materials can be recovered and reused, and 80 percent could be used directly in the making of a new battery—which is quite imperative as a matter of general environmental policy and more so given the volume of batteries that would be needed. The batteries are expected to last for over 8 years and for 2,000 charges. Thus, if each charge gets 100 miles, the battery is projected to last for 200,000 miles of driving.

Ward's Automotive Group estimates that in 2010 we had 1,000 million motor vehicles on the road. With a vehicle population increasing at about 3 percent per annum in the last decade, we can extrapolate and assume that come 2027 we will have about 1,500 million vehicles on the road. Let's see how we fare on raw material availability.

If there would be a resource bottle-neck in the production of batteries, it is not going to be because of their iron or phosphate content;[6] these minerals are truly abundant. It is going to be because of lithium. Nissan has disclosed that the battery in the Leaf contains 4 kg of lithium. A press release by Renault states that its upcoming Fluence z.e. will come with a battery containing 3 kg of lithium. As the Fluence's swappable battery is more in line with the model of this blueprint, I will use it as a basis for some back-of-the-envelope calculations. Some vehicles, like buses and vans, will require a lot more of the metal, but then again, other vehicles, such as scooters, will require a lot less. So let's say it's a wash and stick with 3 kg as a reference point for the amount of lithium needed per vehicle.

For the year 2027, I assume we need 1.5 billion batteries to equip a comparable number of vehicles. While one may assume that the number of vehicles will keep its steady increase past 2027, one may also assume that the battery efficiency will improve enough to compensate for the growth in demand. So I will stay with the need to provide the equivalent of 1.5 billion batteries with 3 kg of lithium in each—both for the year 2027 and beyond.

Given the average lifecycle of a battery, it comes to 1.5 billion car batteries every eight years. This means that for every 8-year cycle we need 4.5 million metric tons (henceforth, *tons*) of lithium. I assume that the first cycle draws upon 4.5 million tons of virgin material and each subsequent cycle draws upon 900,000 tons (as 80 percent of the needed lithium will be coming from recycled batteries). Given existing usage patterns, let's set aside 21,000 tons of lithium annually for other, existing applications such as lubricating greases, ceramic and glass, casting, and laptop batteries. It seems that for the first 8-year cycle we need about 4.6 million tons of lithium. For every subsequent 8-year cycle we need an additional 1.1 million tons.

At this stage, one may wonder how much of that lithium is to be had.

The ocean is estimated to contain about 230 billion tons of lithium. This might as well be an infinite amount of lithium. However with 30 grams of lithium extracted for every 140,000 liters of water processed, this idea doesn't seem to be going anywhere.[7]

As of 2011, the United States Geological Survey (USGS) estimated global land resources of lithium at 33 million tons, out of which, 25 million tons are likely to be battery grade.[8] When factoring in both the recycling of lithium and other existing lithium usage, this means that with current battery technology all the known lithium resources will give us a bit more than 150 years' worth of car batteries. After that period, which is not that long in the big scheme of things, we better have another way to keep the program going.

I took this analysis as far as I care to and did not sweat the numbers. We only know what we know, and in this case, what we don't know that we don't know is likely to render this analysis moot. In 1972, The Club of Rome warned that the world would run out of gold by 1981. Put another way, we don't really know how much recoverable lithium exists. There could easily be ten times more lithium available to us than is currently assessed. Or maybe not.

HEAVY COMMERCIAL VEHICLES

Hydrogen-powered vehicles exist today. Honda has its FCX Clarity; Mercedes-Benz has brought out the F-Cell; and GM has come out with the HydroGen4. These cars are available for lease in some select locations around the world.

Alas, a hydrogen-powered car requires over two times more energy than a conventional electric car. This is why it makes little sense for it to be the technology of choice.

TABLE 5.1

electricity consumption of hydrogen versus electric cars	
electric car (Nissan Leaf)	hydrogen car (Clarity)
36 kWh per 99 miles ↓	1 kg of hydrogen (requiring 50 kWh) per 60 miles ↓
2.75 miles per 1 kWh	1.2 miles per 1 kWh

Note: To compare apples to apples, I used the published figures of the US Department of Energy, allegedly undergoing the same series of tests. The Leaf clocked in at 106 miles per 33.7 kWh for city driving and 92 miles per 33.7 kWh of highway driving. The Clarity scored 60 miles per kg of hydrogen both for city and highway driving. As this car is lighter than the Clarity (curb weight of 3,354 pounds versus 3,582 pounds), I leveled the field by increasing the Leaf's electrical requirements by 1.07. Thus, the adjusted Leaf consumes 36 kWh for 106 miles of city driving and 92 miles for highway driving. In other words, 1 kWh per 2.94 miles of city driving and 2.55 miles of highway driving. Now, let's look at the Clarity. Using electrolysis, each kg of hydrogen to produce and compress requires 50 kWh. One Kg of hydrogen yields 60 miles with the Clarity. In other words, 50 kWh yield 60 miles, which means 1 kWh will yield 1.2 miles for the hydrogen powered Clarity.

Hydrogen fuel cells are a wasteful way to power a vehicle. So it doesn't make much sense. Except when it does make sense. At this moment in time, buses and mid-size trucks are about the outer practical limit of a conventional electric-battery. Class 6 vehicles can go either way: urban driving may be better off using a conventional electric-battery and long-distance driving may be better off with a hydrogen-based battery. All class 7 and 8 vehicles (12 tons and higher), irrespective of city or intercity travel, are another matter.

Once we get into the heavy trucks, the conventional electric battery technology starts to falter. A 580 kWh hydrogen-powered battery weighs around 1,800 pounds (≈ 800 kg.). A 580 kWh conventionally-powered battery could weigh up to ten times this amount. Thus, to power class 8 vehicles, a hydrogen-fed fuel cell may be the technology of choice.[9]

More specifically, the polymer exchange membrane (PEM) fuel cell may be the most suited for ultra-heavy motor vehicles. To generate electricity, this fuel cell requires pure hydrogen on one side and oxygen on the other. The oxygen is drawn on the fly from good old-fashioned air.

In a PEM fuel cell, pressurized hydrogen is forced through platinum-coated cloth. This causes the hydrogen atoms to break apart into protons and electrons. Oxygen, in effect air, is fed into the fuel cell from the other side. With a membrane divider in the center of the fuel cell, we end up with oxygen molecules on one side of the membrane, and on the other side of the membrane we have the hydrogen protons and electrons. The hydrogen protons pass through the membrane in the center, leaving behind their brethren electrons, which is what we are really after if you think about it. The electrons flow as an electric current to an external motor. This is of course the purpose of the whole exercise. After they do their thing, the electrons are channeled to the other side of the membrane, where they unite with their long lost protons and with the oxygen to generate water and a low amount of heat. The only tailpipe emission is water vapor. A continuous supply of hydrogen and oxygen generates a steady electrical current.

A single fuel cell produces a measly trickle of 0.7 volts. Combined, a long stack of fuel cells bring up the voltage to useful, respectable levels.

Today, platinum is used as a catalyst for the PEM fuel cell. Tomorrow, it may be the abundant molybdenum sulfide, which in 2011 was found to be an efficient catalyst,[10] or a molybdenum-oxo metal complex, which can also work with dirty or sea water,[11] or a combination of iron, carbon, and cobalt.[12]

Heavy trucks are just for starters. The plan calls for using hydrogen to power all the off-road heavy equipment: mining trucks, dump trucks of any kind, and heavy equipment in the utility, forestry, agriculture, and construction sectors.

While no hydrogen-based off-road vehicle exists, the situation with heavy trucks is a bit different: a hydrogen-powered semi-truck is in existence. I refer to the Tyrano from Vision Motor Corporation This is a full-fledged class-8 truck, with 36 tons gross weight.

The Tyrano can store about 40 kilograms of hydrogen, compressed at 5,000 psi. This gives the truck a range of 400 miles of city driving and 240 miles of highway driving. The drive system of this semi-truck is rated at 536 peak horsepower with an impressive 3,200 pounds-feet of torque. Because an electric motor is used, maximum torque is available over the entire rpm range. The Tyrano puts out about three times the torque of an equivalent diesel-powered truck and is capable of holding speeds irrespective of the load or grade. Outwardly, this hydrogen-powered tractor-trailer is exactly the same as any big rig on the road. It just has better performance and is quiet. Refueling the Tyrano takes about the time a conventional semi takes to fuel up.

We have somewhere in the vicinity of 200 hydrogen fuel stations around the world. As it stands, the average hydrogen station has the capacity to fill up perhaps 10 or 15 passenger vehicles in a given day. This is not even in the ballpark. One of those 18-wheelers can gulp down the entire hydrogen supply of one such station. What we need is the equivalent of a full-service truck stop, capable of fully fueling the tanks of 150 semi-trucks a day. Put another way, a full-service truck stop provides on average 1,100,000 gallons of fuel a month. In the context of truck fuel mileage, that is the energy equivalent of 880,000 kilograms of hydrogen a month per truck station. This is a lot of hydrogen. And this is exactly what the plan calls for.

It is possible for a truck stop to generate and store its own hydro-gen. In such an eventuality the truck stop would consume 44,000 MWh, or 44 GWh, of electricity a month for the production and compression of the 880 tons of hydrogen it would likely supply. The amount of needed water for the production of hydrogen would be chump change.[13] The details of a hydrogen-generation operation will be discussed in the seventh chapter ("Energy"). In actuality, though, it is likely that over half of the hydrogen will be generated in central facilities for reasons that will be made obvious in the seventh chapter.

A semi-truck can haul four massive hydrogen cylinders in one trailer and deliver them on a daily basis to truck stops. The largest commercial hydrogen-storage cylinders are those made by Lincoln Composites. Each cylinder is 11.5 meters long ($\approx$ 38 ft.) and 1 meter in diameter ($\approx$ 43 in.) with the capacity to store 200 kilograms worth of hydrogen. Thus, in total a truck can bring in 800 kilograms worth of hydrogen. This averages to 1,100 supply trucks each month per truck stop, which is 8 times the number of required gasoline tanker trucks under the present-day setup.[14] As on some days it will make economic sense to generate the hydrogen on the premises, truck stops will need only a part of their hydrogen provided from the outside. Let's suppose a total of 700 tanker trucks a month rather than 1,100.

Even 700 hydrogen-supply trucks monthly add up to a lot of delivery trips. However, in total it does not seem any worse than the existing delivery setup for gasoline. After all, we have to remember that hydrogen will come from a source a few hundred miles away; this is in contrast to petroleum sources ten thousand miles away, and the necessity for giant tankers and refineries.

As always, safety is a consideration.

Hindenburg is perhaps the most spectacular disaster where hydrogen was erroneously reported as the culprit. While hydrogen did indeed burn in the disaster, a new coating used on the fabric skin of the airship was highly flammable and is likely to have been the primary cause for the major fire engulfing the frame.

The American National Highway Traffic Safety Administration commissioned a review of recently published hydrogen vehicle and safety research from around the world.[15] They concluded that with a proper ventilation system, hydrogen concentrations from a leak into the interior of the vehicle can remain below the point at which it may burst into flames. Hydrogen leakage rate in the case of a collision is within safety norms. Large hydrogen releases in refueling did not generate flames that were likely to spread to flammable materials.

Insofar as the hydrogen tank rupturing, this cannot be permitted to happen—anymore than it can be permitted in vehicles running on compressed natural-gas. During a test, the pressure relief devices were removed to ensure a rupture would be inevitable. The tanks with compressed hydrogen were subject to a propane flame from below. The tank itself ignited in under one minute. Six to twelve minutes later, the tank failed and ruptured under the propane flame. Tank fragments flew 80 to 100 meters (≈ 260 to 330 ft.), and windows would have been broken from the blast wave had they been within 20 meters (≈ 65 ft.) of the ensuing explosion.[16]

A study by Sandia concluded that hydrogen can be handled safely, at least as safely as gasoline. It also recommended that hydrogen be housed in well-ventilated areas.[17] Other voices also caution against parking a hydrogen-fueled vehicle in an enclosed area.[18] While this warning is merited, it would be a non-issue. The application of hydrogen as envisioned in this plan is for heavy trucks in commercial use. Hence, parking facilities for those heavy vehicles can be expressly designed with suitable ventilation to accommodate the particular properties of hydrogen. In the end, it seems that it is as safe to drive a vehicle powered by compressed hydrogen as it is to drive one powered via the combustion of gasoline.

And insofar as tomorrow goes, a team of USC scientists managed to store hydrogen in a safe chemical form that can be stored as a stable solid. It is safe. They also developed a catalyst system that releases the hydrogen, providing the fuel cell with as much gaseous hydrogen as needed to run the vehicle.[19] Another team at the National Institute of Standards and Technology has contrived a plan to inoculate magnesium with iron veins, which collect and discharge hydrogen safely.[20]

As long as we have water and electricity, we can generate all the hydrogen we need. Regardless, electricity generation has a real footprint, and it is sensible and necessary to set some fuel-saving measures. It is within our immediate reach to produce more efficiently designed trucks.

In a highway situation, which is where most of the action is, it takes about 180 horsepower to propel the truck forward at 60 mph along a level road and under no wind conditions. Out of this, 100 horsepower is needed to move the air out of the way, and 80 horsepower is needed to roll the tires of a loaded tractor-trailer. The most straightforward, fuel-saving measure is to simply reduce the cruising speed from 65 to 60 mph. Cummins Engine Company surmises that this would cut back on 8 percent of the fuel.

Moving on to aerodynamics, we can definitely introduce some design features that would reduce wind resistance.

The gap between the tractor rear face and the trailer front face is filled with large vortices at high road speed. It gets worse yet if a wind is coming from the side, causing more air to get in the gap, increasing the drag. The solution is the usage of a deployable seal between the tractor and the trailer. At low speeds, the four-sided "gap seal" would retract against the cab so the vehicle can articulate around corners. Side panels, or skirts, on the sides of the trailer and tractor are another contraption that would significantly reduce air resistance, notably at highway speeds. The side panels, such as the AeroMax skirt, automatically retract to allow better negotiation with the road and avoid related bumps. Some skirts are built to flex when going over a curb rather than snap as solid plastic would do.

The rear face of the trailer can also use a facelift. Improvements would include rounding corners, and otherwise installing a tapered protrusion at the rear to reduce air eddies. Additional fuel savings can be gained by incorporating pneumatic devices that blow air from slots at the rear of the trailer, preventing the separation of air flow.

Tire rolling resistance accounts for nearly 13 percent of the truck's energy use. Instead of the customary dual tires, we can outfit tractor trailers with single wide-base tires. The advantage? Fewer tires under the truck translate to lower overall weight and lower rolling resistance.

We are indeed on a roll here. Let's add one last fuel-saving measure. This one recovers some of the manly character of the semi-trailer that was taken away by the previous aerodynamic additions. Instead of having a tractor hauling a single trailer, it can haul two, three, or more trailers. This saves on fuel. A small detachment of trailers hauled by a single tractor consumes less fuel per given ton of cargo—not to mention resulting in an overall reduction in the number of needed tractors on the road. The American Transportation Research Institute determined that two trailers hitched to a single tractor would consume 15 to 39 percent less energy per ton of cargo than a standard single.[21]

A few regions permit road trains, notably some of the Australian states, where one may see a tractor with five or six trailers barreling down the road in the vast expanses of the Australian Outback. These road trains require bigger turning radii, that is, wider turning lanes. Not all bridges can accommodate them. A study by Rocky Mountain Institute suggests that at least 63 percent of truck trips in United States can be made with combination trucks.[22] It would save not only on fuel but also reduce the total number of tractors on the road. This should give some indication of the worldwide potential.

REDUCING THE NUMBER OF CARS ON THE ROAD

I have sketched out the electrification of all vehicles: from the scooter to the largest of trucks. Running on one form of battery or another, the entire future fleet, about 1.5 billion strong, is to have zero tailpipe emissions. Considering that the source of the electricity will be clean as well, it is a good thing as far as global warming. Considering that there would be no need to construct dedicated power-plants to run them, it is even better yet (see seventh chapter, ["Energy"]). This aside, our transportation infrastructure is a blight on the face of the Earth.

Countless cars congest our streets in Lagos, in Jakarta, and in the thousands of cities in between. Each car leaches toxic zinc and barium into the earth from the wear of tires and brakes.[23] Each car necessitates about one hundred square feet of parking (about 9 m²). And then it sits there more than 90 percent of the time. Our cities are largely built around cars. We have surrounded ourselves in ugly expanses of lifeless concrete and never-ending stretches of dark-gray asphalt. By and large, our metropolises are ugly: having all the charm of dark-gray termite nests that someone stepped on and squashed flat over many square miles.

An-easily accessible motorized personal mode of transport is a problem. It enhances our mobility, our economic activities, and ultimately our environmental footprint. Switching from gasoline to electric cars changes nothing in that regard. Foremost, this is an infrastructure problem.

We could dramatically lower our footprint by living in denser, well-designed urban centers—where walking, for one, can be a mode of transportation. And so we should.

However, in fifteen years, we are not going to radically transform the infrastructures of our cities. This will take decades. In the short term, between now and then, we can introduce a few behavior-changing patterns. I will start with what perhaps is the most obvious one.

Car-sharing is a service that provides its members with access to a fleet of vehicles on an hourly basis. Members reserve a car online or by phone. They are billed at the end of the month either for time or for mileage.

This model is attractive to customers who only use a vehicle occasionally, as well as others who would like access to various vehicle types depending on the needs of the moment. It could be a minivan for going on a family trip or a sports car for a date. Car-sharing is also an attractive alternative to those who wish to avoid the associated costs of car ownership—from outright purchase costs to those related to insurance and maintenance.

Car sharing on a wide scale means the production of fewer cars. It also means more available parking and less congestion.[24] Today, there are more than one thousand cities where people use car-sharing.

While the car-sharing model is most suitable for longer commutes, there are other models that do one better in dense urban settings.

Community bicycle programs are increasingly popular and diverse. These programs are usually organized by nonprofit organizations or municipalities. They offer affordable access to bicycles for short trips within the city as an alternative to motorized public transport or cars. They help reduce traffic congestion, noise, and air pollution. Those programs have proven most successful in Paris, Montreal, and Hangzhou—with many thousands of bikes in each of the respective bicycle-sharing programs. Numerous cities have introduced bike rentals. A cumulative learning process has taken place. Some things worked, some failed. Out of this, a viable model has emerged. In one variation or another, bicycle-sharing ought to be introduced in all of the larger cities.

The bikes are equipped with anti-theft and bike-maintenance sensors. If the bike is not returned within the subscription period or returned with significant damage, the bike-sharing operator withdraws money from the user's credit card account. As a painted bike has little resale value, the rental bicycles are painted with distinct colors top to bottom. Furthermore, the bikes may possess singular designs with unconventional dimensions that offer little value in any attempt at disassembly and resale of their parts. The bikes themselves may be secured using an electric lock, requiring either a smart card to unlock them or a code provided by the operating company.

A RFID tag allows tracking of the bike. Those tags help to note the pattern of usage. They make it possible to match demand and stock, with an ongoing redistribution of the bikes according to the time of day.

Rental bicycles would provide a seamless connection from subway stations to major parking lots, letting people travel in the dense hub of the city without the need to bring in the car. Another advantage of bike-sharing systems is that the smart cards allow the bikes to be returned to any station, facilitating one-way rides. Instituting cycling lanes in all cities is an integral part of the plan.

Not everyone has the leg strength to ride a bike in hilly terrain, or cares to go through the experience. An electric bike effectively flattens hills, increasing the average speed, and eliminating the 'groan' factor. Just as buses have added wheelchair features to cater to passengers with disabilities, the inclusion of electric bicycles will bring bike-sharing to a wider audience. With the inclusion of electric bicycles, some people who otherwise wouldn't consider utilizing a bike-sharing program may give it a try.

Electric bicycles are very much in existence; China has over one hundred million of them.

Both e-bikes and conventional bikes can have different models. All may come with baskets to carry things in. However, some may also sport child seats. Other optional models may be trailers that accommodate luggage. We can definitely expand the offering and go beyond bicycles. Recumbent trikes lend themselves even better to trailers. They cater to the elderly or otherwise to those who want more stability and a more relaxed ride. Trikes also lend themselves better to mounted canopies that offer partial protection from either the sun or rain.

Such a vehicle-sharing plan for the urban center ought to also include fully motorized vehicles. This is where the CityCar comes into place. In 2003, the designers at the MediaLab's Smart Cities Group of MIT set out to design a personal transport vehicle for the dense urban environment.

In a conventional automobile, the engine, suspension system, axles, and steering column take considerable space. Intent on reducing a car's footprint, the MIT team changed basic automotive design principles and invested all of these components in the wheels of the car. Each of the vehicle's four-wheel assemblies is a self-contained mobile unit, supplying its own motor, suspension, brakes, and steering. The implications are considerable. Each of the CityCar's wheels can fully spin on its own axis, allowing the car to rotate 360 degrees. This provides an astounding degree of maneuverability, such as performing lane changes and parallel parking by moving sideways. When a CityCar isn't in use, it collapses into a more compact form and stacks against other CityCars, much like a shopping cart. In this manner, three to four vehicles fill an area traditionally used for one standard-sized, conventional automobile. Stacked one onto the other, the CityCars form a rack, which is to be connected to the city's power grid. Ideally, a vehicle would be fully charged by the time it cycles through to the front of the stack. The front compartment of a CityCar accommodates two passengers and the rear compartment provides ample storage space, such as for baggage or groceries.

The CityCar can be checked out at one location, such as a subway station or a sports arena, and be returned at another. The vehicles wouldn't replace busses and trains, but they would fill in their geographic and logistic gaps. It is easy enough to imagine the CityCar being integrated into a bike-sharing type of program. With the motorized CityCars as part of a vehicle-sharing program, there would be little need to bring full-sized, private cars into dense urban areas. Those commuting from the suburbs may park their cars in large parking lots at the outskirts of the downtown area. From those lots people would be able to check out any of the personal transport vehicles offered by the municipality.

Hiriko Driving Mobility has acquired the license for the technology, and it is hoped this ultra-compact car will become available in Europe in 2013.

Another way to improve our transportation pattern may come from an entirely different direction.

Driver-assistance systems can prevent many road collisions. Radar sensors can scan surrounding traffic conditions, monitoring the vehicle's blind spots or maintaining a safe distance from the vehicle in front. Infrared detectors improve night vision. And fatigue sensors sound an alarm if there is a risk of momentary driver drowsiness.[25]

The Google Driverless car has taken this technology to its culmination. Google has produced a driverless car incorporating artificial-intelligence software coupled with a GPS device, a laser scanning system, and a bank of sensors. The souped-up car not only drives itself, but with more precision and fewer errors than that of human drivers. As of 2010, the diminutive Google driverless fleet of six cars has put in more than 200,000 miles in real-life road conditions.

The advantages of self-driving cars are significant. Most of the million or so fatalities each year from car collisions worldwide are due to bad judgment calls. Hence, the question is not whether the robotic software is error-free, but whether it is at least as good as human software. As it turns out, it is *much* better, markedly so when compared to drunk, tired, distracted, agitated, or texting humans. Needless to say, such robotic cars could be equipped with an override, which permits the human passenger to assume control at any time. Due to higher-precision driving and faster reaction times, robotic cars can drive in closer proximity which would benefit the overburdened road systems. For driverless cars in a platoon formation, air resistance would be minimized (except for those in the front) and thus reduce their energy consumption. But it goes beyond energy savings.

Notably in commercial vehicles, a significant portion of vehicle trips are for deliveries. In other words, some trips by their nature don't require human presence at all if the driving is done by a robotic system. This would free people from spending hours playing chauffeur to vegetables delivered in a semi-trailer or playing the role of delivery boy for cross-country apartment moving; a local team can wait at the destination.

Blind people, elderly, or those otherwise unable or unwilling to drive can finally enjoy the benefits of having a car. Driverless vehicles would lend themselves to having a more optimized design without the driver's seat. People who occupy the cars can engage in activities other than driving. The list of advantages seems boundless. For one, it is estimated that up to 30 percent of people driving in downtown areas are looking for parking.[26] Driverless cars can just drop off their occupants next to an office building in a downtown area and then spend their time hunting for parking and be called upon later when the person is ready to be picked up. As passenger vehicles are parked on average twenty-two hours a day, a self-driving car can serve more than one person. A driverless car can drop off a member of the family and immediately head out and serve another member. In fact, there would seem to be little point for privately owned driverless vehicles. You just hail one when you need it—all of two hours a day on average. There goes the huge maintenance expense and enormous environmental burden of producing more cars than are truly needed.

Everything considered, private ownership of cars and trucks in cities needs to come to end; it is too resource-wasteful to retain—notably so if scaled to seven billion people. The technological promise we have to make is to manifest a car within five minutes from the moment it is requested—whether a driverless car, car sharing model, or just using something less resource-wasteful.

AVIATION

Each month, about three million flights emit disagreeable gases from the burning of about 20 million tons of oil,[27] mostly high up in the atmosphere, where it hurts the most.

The impact of aviation on climate is complex, and our knowledge of it is filled with a fair amount of uncertainty. However, when we blow at the fog of uncertainty, three things are firm enough to remain standing: old-fashioned CO_2, formation of cloud trails, and ozone-enhancing gases.

CO_2 emission is the easiest obstacle to overcome. We can already produce fuel that is carbon-neutral: biogas. In 2010, Sasol made synthetic fuel via the Coal-to-Liquid process. This fuel was tested and found to meet all demands and standards of jet fuel. The first flight was made that year with this fully synthetic fuel. In producing this jet fuel, the raw material was coal. However, natural gas or pure methane would have done just as well. Thus, jet fuel can be made from methane; methane can be distilled from biogas; biogas can be made from various organic waste products. And it is a good thing. For starters, biogas is carbon neutral: the carbon stored in the fuel came from the atmosphere in the first place. Better yet, unlike camelina, jatropha, and algae, which are regarded as candidates for a carbon-neutral jet fuel of the future, biogas does not require land—vast tracts, at that—for its production. Biogas is discussed in greater depth in the eighth chapter of this book ("Nutrient and Water Recycling").

In the context of the effect of aviation on global warming, this still leaves the formation of cloud trails and ozone-enhancing gases.

As the airplane emits water vapor, in some altitudes and in some regions, those can condense into cirrus clouds. These in turn trap reflected infrared radiation and thus contribute to global warming. Fortunately, there is a fair chance that we may able to significantly reduce aircraft-induced cirrus clouds. For the cirrus clouds to form, the airplane needs to fly in an area that has moist air. By avoiding those moist zones, the effect won't take place. The detection of these super-saturated zones from measurements of ambient temperature and moisture is a technical challenge, but one that can be partially overcome. In conjunction with weather detection instruments, it should be possible to avoid about half of the cirrus cloud formations by adjusting flight levels one to two thousand feet either up or down.[28]

NOx (nitric and nitrogen dioxide) emissions are the biggie. These emissions are an inevitable outcome of burning fuel. At high altitudes, NOx emissions have a relatively strong effect on the production of ozone, a greenhouse gas.

In summary, we cannot annul the warming effects of flight: we can eliminate the CO_2 effect; we can reduce the water-vapor effect; but we can do nothing about the NOx effect. There is no technofix that will take us all the way. We cannot hydrogen our way out of this one. There is no fission, fusion, battery, or algae that can give us commercial flights that do not have an adverse effect on global warming.

Aviation as we know it needs to come to a close. And we just need to get over it.

Until and if proven to have negligible ill-effects, only a few routes would be offered: the über-long distance variety. The plan is to retain air routes facilitating travel between far-off destinations, where travel by sea and rail may take weeks. Some of those routes will be busier than others. All, however, will have a greatly diminished number of flights. Given existing traffic patterns and the Boeing market forecast for 2029, we can make some educated guesses as to how it may look:

London–Singapore and London–Hong Kong routes are likely to be the busiest and may have a few daily flights. London–Mumbai and Tokyo–Washington DC will be less busy and may have one daily flight; and routes such as Johannesburg–Atlanta may go once a week. The total monthly cap on aviation fuel consumption is set at 2 percent of 2008 levels. This comes to about 1.6 billion gallons a year.[29]

Let's determine how many roundtrip flights that would buy us.

Considering the air routes to be retained, the average roundtrip would be 19,000 km (≈ 11,800 miles). As a basis for calculation, I used the most likely candidate for the really long flights, the Boeing 777-200LR. In consulting the *Boeing 777 Flight Crew Operating Manual 2008*, it seems that a 19,000 km roundtrip burns off 131,850 kg worth of fuel, which is the equivalent of around 43,950 gallons. The self-imposed cap of about 1.6 billion gallons of fuel per year will thus yield about 36,000 long distance roundtrip flights per year. I assume a seat capacity of 301 passengers and a 95% seat occupancy rate. (All considered, this seems more likely than the 80% occupancy rate of the present.) This means that about 10 million unique passengers will fly each year.[30] It does sound like quite a few people will get to fly long-distance. This isn't so.

To put it in perspective, I will compare it to existing and anticipated demand under a business-as-usual scenario. Inferring from Boeing's report for existing and foreseen passenger miles for the various routes,[31] it appears that in 2009 we had in total 160,000 roundtrip flights in current air corridors, and that in 2029 under a business-as-usual scenario, we will need to beef it up to 471,000 roundtrip flights to fulfill demand. Assuming we would experience the higher seat-occupancy rate of 95% under this plan, we would fulfill only 22.5% of existing travel of those routes and only 7.6% of anticipated demand for 2029. As stated, it will provide zero flights on all other, shorter existing air routes.

Under the broad plan laid out in this book, little fuel is allocated to go around. It is worthwhile to consider whether we can save on some of the allotted fuel with layovers.

A nonstop ultra-long haul flight takes off with a payload of tens of thousands of gallons. The takeoff is a fuel-draining, herculean task. It consumes much less fuel to do the same maneuver carrying a smaller fuel payload, which a shorter flight obviously makes possible. And then there is the flight itself. Much of the fuel in a long, nonstop flight is just dead weight, at least until it is used much later on. As it turns out, a flight with a layover has its own problems. One needs to take off twice, and push against the relatively thick air of the lower atmosphere twice before reaching cruising altitude in the rarified air high on up. The verdict? It seems that up to 13,000 km it makes more fuel-sense to fly nonstop. Beyond that, it is fuel-advantageous to fly with one layover. In other words, Newark to Singapore can achieve some fuel savings with a layover. The bulk of the routes, though, would be better off sticking to the nonstop routine. So no news-breaking story here. It seems that with our need to avert further global warming, we would be stuck with Austerity Airlines till further notice.

Come hell or high water, some functions of aviation will need to be retained: aerial firefighting, emergency evacuation, and search and rescue. Let's also throw in news and traffic-related choppers, a memento to the Roaring Twentieth Century.

As for the period past 2027, there is room for cautious optimism. MIT submitted to NASA a few designs of commercial airplanes, such as the D *double bubble* that are estimated to consume 70 percent less fuel than current planes—along with reduction of NOx emissions. If so, then in a few decades we would replace the existing fleet with designs that would permit us to triple the allotted flight mileage. Having said that, I regard it as throwing good money after bad. There is something potentially superior in every way to aviation, which I will discuss at the end of this chapter.

So if not air, then what would mostly ferry us to far-off destinations? Trains and ocean liners, of course.

MARINE TRANSPORT

With aviation largely out of the picture, marine transport will assume, once again, a cardinal role in ferrying people between continents.

Ships are slow. If they average 55 kph ($\approx$ 35 mph), they are doing well. This happens to be the service speed of the *RMS Queen Mary 2*, the grandest ocean liner, and one of the few in existence. One has to differentiate an ocean liner, which ferries people across the ocean, from a cruise ship, which is a floating leisure hotel. What ocean liners don't make in speed they make up for in volume, that is, the number of people they can pack in.

In reckoning the true potential capacity of an ocean liner, we have to go back about one hundred years, to the golden era of ocean liners: to the time before swimming-pools, movie-theatres, and other space-consuming, cruise-like facilities were incorporated.

Carolina, at 19,500 gross tons, had a passenger capacity of 1,550 people. Her sister, *RMS Carmania,* initially had a passenger capacity of 2,650 people, also at 19,500 gross tons. *SS Kaiser Wilhelm der Große* had the capacity for 2,000 people at 14,000 gross tons. If we extrapolate from those figures, a modern ocean liner the size of *Queen Mary 2*, at 151,400 gross tons, may have a passenger capacity of 16,000 people. If we assume 1,000 personnel, a large modern ocean liner would be able to ferry 15,000 passengers.

I used the timetables of cargo ships as a basis and amended them to reflect the cruising speed of the *Queen Mary 2*. This yielded possible journey times between various key destinations: Hong Kong to Los Angeles would take seven days. It would be just over four days from New York to Hamburg, Germany. And it would be nine days from Rotterdam, the Netherlands, to Cape Town, South Africa. These journeys would take longer or considerably longer with stops on the way.

In 2009, about eleven million people came from Western Europe to the United States. About twelve million Americans went during the same time to Western Europe. Travel by ship will change all of this, of course, and in ways we cannot really predict. Just the roundtrip itself would take about ten days.

It is highly likely that in the absence of many air routes, people will be less inclined to take long-distance trips. Which is wonderful, ecologically speaking.

But let's suppose that the number of unique travelers will stay the same, 23 million in all. If we further assume 80% capacity (compared to existing flight occupancy rates), then it translates to 1,917 round trips across the Atlantic. It will average just over four ocean liners starting on that trans-Atlantic route each day. It also means that if a round trip, with restocking, disembarking and all, takes an ocean liner 2 weeks, there would be a need for about 80 ocean liners servicing the US–Western Europe route.[32] Here I assume that the ocean liner would ferry people across the Atlantic to the nearest point in Europe—a four or five day trip. Everyone would disembark in the United Kingdom or France and from there zip their way in a few hours on a high-speed rail to their final destination in Western Europe. This makes more sense than to spend a few additional days at sea to get to a potentially closer port city in the Mediterranean or the North Sea.

Ocean liners of the future aside, let's take a look at the existing commercial fleet.

At the moment, we have a few tens of thousands of commercial ships. These are comprised of container ships, bulk carriers, tankers, passenger ships, service ships, and tug ships. The 15,000 or so bulk carriers and oil and chemical tankers account for about 60 percent of the gross tonnage of all ships. They ferry across the oceans petrol, grain, steam coal, iron ore, and cement. These are the heavyweights of the fleet. The largest of the bulk carriers boasts a 350,000 ton carrying capacity and is over 300 meters in length. Not to be outdone, the largest of the tankers is close to a 450,000 ton carrying capacity and could accommodate over 3 football fields.

Passenger ships range from ferries, for day trips, to extravagant gigantic cruise ships, in effect floating mega resort hotels that have tropical gardens, upscale restaurants, and shops. Tug boats are the tow-trucks of the sea, for either barges that cannot move themselves or boats that are disabled or otherwise need help in crowded harbors. Then there are the cargo ships each potentially carrying thousands of containers as they haul millions of TV sets and jeans and apples and countless other products in between. Last but not least are the numerous fishing boats of various configurations, which catch, process, and freeze fish.

Completing the picture are a few tens of millions of recreational boats: kayaks, racing boats, yachts, dinghies, ski boats, houseboats, jet boats, and inflatable boats.

At the moment, large ships are running on the foulest of the foul, bottom-of-the-barrel, high-sulfur, tar-like substance, bunker fuel, which the only thing green about it is the greenhouse gas it emits. As of 2007, sixty thousand of us died prematurely due to smokestack emissions from ocean-going ships.[33] There may be only a few tens of thousands of these polluting behemoths, but they emit almost half the amount of particulate matter as the entire fleet of cars.[34] The question is how to run a commercial fleet with a lower-impact technology.

Nuclear. What else?

SOURCE OF POWER TO RUN THE LARGE SHIPS

Currently, about one hundred forty marine vessels are nuclear-powered. These range from aircraft carriers to submarines to icebreakers. Only a handful of civilian nuclear-powered vessels have been constructed. These include the *USS Savannah*, which was a showboat and was laid to rest a few years back. They also include the *NS 50 Let Pobedy*, an Arkita-class nuclear-powered icebreaker that is currently in operation in the North Sea.

Using the existing inventory of ships, this means that about thirty thousand vessels will go nuclear within a fifteen year period: cargo ships, container ships, bulk carriers, tankers, passenger ships, and service ships.

There are significant advantages to powering ships with nuclear energy. This source of energy requires no oxygen and produces no exhaust gases. The nuclear reactor aboard a ship is a reliable, compact source of continuous heat that can last for years without new fuel. Upgrading ships to nuclear will also free up some storage space aboard that is currently taken up by massive amounts of liquid fuel. The US Navy has accumulated 230 million kilometers of accident-free experience over the course of 6,200 reactor-years. Nuclear-based marine propulsion is a mature technology with an exemplary record.

Hyperion Power Generation has designed a 70 MW thermal / 25 MW electrical mini reactor. It is based on a technology developed in Los Alamos National Laboratory. Think of this mini reactor as an oversized, 2.5-meter-high battery. It is factory-sealed and can be shipped to the propulsion factory to be fitted alongside the other propulsion components. And that about sums it up.

Under this scenario, about ten years down the line, a ship will dock in a special factory facility; the vault surrounding the reactor will be pried open; the module will be cut out of the ship; and a fresh module will be fitted in. It does echo something of the switch station described earlier in the context of electric cars. Back at the factory, the battery will get a second lifetime: the wasted fuel pins will be taken out and new ones inserted. The shell holding the fuel will be good for one or two refuelings before it will have to retire due to thorough contamination. It is worthwhile to note that the electric generation components are partitioned off from the nuclear reactor itself.

Two of these mini reactors working in tandem would be enough to power the largest of the existing tankers or mega-cruise ships.

In contrast with past practices in the nuclear industry, the mini reactor is designed for mass production with the potential for thousands to come off the assembly line each year. Nonetheless, it is expensive, likely to be in the tens of millions of dollars a pop. For this reason and for security reasons, the most viable business model would mimic that of Better Place. Namely, the nuclear battery and its content are strictly owned by the manufacturer and leased out. More specifically, the manufacturer could sell electricity on a monthly basis. This would make sense, whereas tacking on a 50 or 100 million dollar power generator atop a 35 million dollar bulk carrier wouldn't make economic sense.

Further down the line, Hyperion plans to introduce smaller modules, which would be suitable for more modest vessels, such as ferries. Unlike the highly-enriched uranium found in existing military nuclear vessels, the Hyperion module will be under 20 percent enrichment. Instead of using pressurized water as a coolant, the Hyperion module will use liquid metal (lead-bismuth). No pressure means that in case of rupture, the nuclear fuel inside will not spew far and wide.

The smallest marine reactor in operation seems to be a 26 MW capacity reactor, powering a ship with a 10,000 ton displacement. However, it is conceivable to have a nuclear reactor power a ship as small as one with a 1,000 ton displacement.

Although our hypothetical manufacturing company will continuously track each of its nuclear batteries wherever they may be on the planet, it is essentially a hands-off mode after the ship is outfitted with its reactor. The nuclear battery has passive and active shutdown mechanisms with two redundant shutdown systems in case of failure.

Nuclear ships may spell the potential for terrorist attacks, right? That's right.

Nuclear terrorism aside, if some people want to make a living hell for the rest of us, there is precious little we can do to prevent it—as the Iraqi citizens found out in 2005, with a bomb detonating seemingly every few days in various public places.

Conventional bombs can be placed on school buses or at day care centers. Typhoid bearing salmonella can be added to ice cream or salad dressing mixes. A fleet of trucks can drive down city streets, inconspicuously spraying anthrax spores. Liquid-chlorine tanks can be made to rupture and the gasified poison quickly spread far and wide. A dust cropper can spray a blistering agent over a sports arena or Agent Orange over a broccoli field. Sarin gas can be released into the ventilation systems of large office buildings. Pneumonic plague can be released in a public bathroom. Blocks can be placed on a railroad track and derail a train full of passengers.

And stealing some radioactive material and fashioning a dirty bomb is another threat, of course, although the execution of some of the previous scenarios seems more straightforward.

Maybe one of the reasons why no one has cared to do this also has to do with the fact that the stuff is both lethal to those around it and is not enriched enough to make much use out of, beyond a conventional bomb that is tainted. I fancy that if someone wanted to hijack highly toxic nuclear material, one would have done so by now, as for decades we have had many thousands of trucks hauling radioactive material in large casks from various nuclear facilities. I reckon it would be easier to hijack a truck and take possession of the cask, than to take over a ship, hack away around the reactor, and sneak off with the fifty ton hulk. By the time the massive casing is cold enough to attempt prying it open, the lead bismuth inside would have frozen solid, fused with the fuel pins.

What if an elite team of terrorists used a rotary cannon with armor-piercing, depleted-uranium ordnance, specifically targeting the section where the reactor is housed—rupturing both its vault and the case itself? While there is nothing within the reactor that can generate a nuclear explosion, it may create some sort of a meltdown. Fancy this happening aboard an ocean liner packed full of people. But then again, such a terrorist team can just torpedo an existing, fossil-fuel powered cruise ship and get the same results.

In Cleveland, in October 1944, an LNG tank cracked. Its content spewed out and formed a vapor cloud that filled the surrounding streets and the storm sewer system. The natural gas ignited, resulting in an explosion and fire over a fifty-city-block area. Thousands of homes were destroyed and 128 people died in all. If someone wants to bomb an existing nuclear ice-breaker ship, we can assume it can be done—and the resultant meltdown will be horrific. If someone wants to bomb one of the LNG carriers that routinely dock in various terminals around the world, we can assume it can be done—and the results are likely to be just as horrific.

The threat of nuclear terrorism has been with us for decades: right along with any of the other mentioned threats. I don't think that making our civilian fleet nuclear would change the status quo. If someone wants to hit where it hurts, they will do it one way or another, nuclear-powered ships or no nuclear-powered ships. We cannot proof the world, short of resolving the underlying conflicts, that is. In conclusion, I don't know whether fifty or two hundred fifty people would die as a result of a nuclear meltdown of a vessel on the heels of a hypothetical terrorist attack. But I do suspect that if we continue to burn billions of gallons of noxious bunker fuel alongside other petroleum products, we will be faced with a worldwide calamity whose magnitude goes beyond anything we have ever faced.

Regardless of the cause, let's consider the repercussions of a nuclear ship that has sunk. What is the potential for contamination? This is of course a problem, one that has been plaguing the oceans of the world with the occasional leaking of various oil tankers and the occasional dumping of barrels containing low-level nuclear waste. At any rate, in the case of a sinking nuclear ship, perhaps the answer is that not much will happen, although in truth it is unknown. For one, the encasing lead would solidify in the water, providing inert shielding. And chances are that the protective shell won't rupture in the first place.

———————

Coal pollutants affect all major body organ systems and contribute to heart disease, cancer, stroke, and chronic lower respiratory diseases.[35] The burning of fossil fuel does not occasionally emit noxious fumes; it does so as a matter of absolute fidelity. Many hundreds of tons of toxic mercury are being released into the air we all breath, and the rate has been on the rise.[36] Premature deaths are not accidental; they are an integral part of the energy system we have put together. As we inhale the cocktail of gases and particulate matter from the burning of fossil fuel, two million of us die prematurely.[37] I say, when it comes to nuclear energy's waste products, we can and should do better.

In Sweden, after years of meticulous research and careful consideration, authorities have elected a site five hundred meters underground in crystalline bedrock near Östhammar. Construction is planned to begin in 2016 and be completed six to eight years later. A comparable type of repository is under consideration next door in Okiluoto, Finland. If approved, nuclear-waste disposal is planned to commence in 2020. As part of the due-diligence process in choosing this site, 5,000 meters of tunnels and shafts were created. More than 100,000 cubic meters of rock were excavated, and rock samples from 53 deep boreholes were collected. This is just for starters.

Long-term storage of radioactive waste requires the stabilization of the waste into a form which will neither react nor degrade for extended periods of time. The fuel pellet is contained within a highly corrosion-resistant fuel rod made of zirconium-based alloy. A bundle of these rods is placed in a canister whose interior is made of graphite cast-iron, which assures mechanical strength. The exterior of the canister is copper, which is highly corrosion-resistant in oxygen-free conditions. In fact, we have found copper that is millions of years old that has retained good protective characteristics.

The twenty-nine ton copper canister is welded shut using a special technique assuring that it will hold out through the next glacial period. To make sure the welds are perfect, high-velocity electrons bombard the seal between the lid and the canister, fusing the cask with a strong weld without weakening the surrounding material. Next, X-ray beams are employed in searching for any flaws in the welding, and ultrasonic and eddy-current inspection systems analyze the surface for any tiny aberrations. In eddy-current testing, a coil carrying an alternating current generates a magnetic field when brought close to the copper. The field induces eddies of current whose flows change in the presence of surface imperfections. When the welding process is completed to satisfaction, the canister is ready to be shipped to its permanent home.

The canister is transported and deposited in a tunnel within a repository deep below the surface—as far down as the Empire State Building is high. The tunnel contains a series of pre-drilled holes: one for each canister. A canister is lowered into a hole, and then the hole is filled to the top with bentonite clay, which when mixed with water swells to provide a watertight barrier and protection against earthquakes. The immediate area of the tunnel above the hole is backfilled with clay blocks and pellets. The filling of the tunnel is to prevent water from flowing into the tunnel. Once the entire disposal tunnel has been filled, a thick concrete plug is cast at its end. When the facility is filled to capacity, it will be shut down permanently and the building above it dismantled. Once again the surrounding pine trees would grow on top.

In the extremely unlikely case of a major earthquake, one that would also damage the canisters and tear them open and somehow cause the radioactive material to come in contact with groundwater, the radiation dosage is estimated to be one thousandth of the total annual average radiation we are exposed to in our daily life—as the nuclear fuel is in solid form, and the release would be extremely slow.

Never in our illustrious, colorful history has the concern for life been as extreme as in the Scandinavian procedure outlined above. One cannot help but be touched by the acute concern for life. However, some find this scheme to be inadequate. Some are concerned that far in the future, some people, digging hundreds of meters down, will stumble on a canister, somehow tear it open, and be exposed to a lethal dose of radiation. The concern is commendable, and one can only wish that it inspired a comparable level of deliberation in regard to the hundreds of thousands of children that die every month from hunger.

Be that as it may, we definitely seem to have another technical option that will take the disposal of nuclear waste to the next level.

Subduction is a process whereby one tectonic plate slides beneath another and is eventually reabsorbed into the mantle. The subducting waste disposal method forms a high-level radioactive waste repository in a subducting plate, so that the waste will be carried beneath the Earth's crust where it will be diluted and dispersed through the mantle. A subducting plate is naturally predestined for consumption in the Earth's mantle, and with it, the canister passenger, which would work its way ever deeper into the bowels of the earth. RadWaste has devised a steel-based canister that would melt under heat and pressure once it reaches three hundred miles underneath the surface of the Earth. The heavy radioactive material will slowly sink deeper toward the planet core. The burial in a Subduction zone is truly a one-way, permanent solution.

It is worthwhile to remember that high-level nuclear waste is not contingent upon this plan. It already exists, whether we want it to or not. We have decades worth of waste sitting aboveground in dry cask storage facilities across the world, waiting for a permanent home.

———

Established nuclear technology is very wasteful; only about 5 percent of the fuel material being fed to the reactor is actually being put to use. Before we are handed a ledger to determine how much uranium we may have available to work with, it behooves us to shave off some of the fuel requirements with better designed ships. Luckily, there are a few relatively inexpensive measures we can introduce to that end.

A Hamburg-based company offers sails to outfit existing and new ships. These are not your great-great-great granddaddy's sails, though.

Instead of a traditional sail fitted to a mast, SkySails use large towing kites to contribute to the ship's propulsion. Shaped like paragliders, they are tethered to the vessel by ropes and respond to wind conditions and the ship's heading. The kites operate at altitudes of a few hundred meters, where stronger and more stable winds prevail. By means of dynamic flight patterns, the SkySails are able to generate five times more power per square meter of sail area than conventional sails. Their double-wall profile gives the SkySail towing kites aerodynamic properties similar to the wings of an aircraft. Hence, the SkySails system can operate not just downwind but also at winds coming from the sides.

Depending on the prevailing winds, the company reports that a ship's average annual fuel costs can be reduced by 10% to 35%. Under optimal wind conditions, fuel consumption can temporarily be cut by up to 50%.

Currently, the company offers towing kite propulsion systems for cargo vessels with an effective load of between 8 and 16 tons. In the future, the company plans to offer towing-kite-propulsion systems to ships with an effective load of up to 130 tons. For the larger ships, we can put to good use the work done by the Wind Challenger Project at the University of Tokyo, whose engineers designed a large, collapsible sail system that is reckoned, once built, to cut fuel use by up to 30%.

Another fuel-saving technology suitable for larger vessels is the air cavity system, such as the one offered by the DK Group, based out of Rotterdam.

When air is pumped rapidly out of small holes in a ship's hull, the swarming bubbles quickly join together and coat the hull with a layer of air a centimeter or two thick. This reduces drag, because air offers far less resistance than water. Air has less than 1 percent the viscosity of water, so for all intents and purposes it lubricates the ship as it moves forward. Tests have shown that the air pump for the bubbles would use about 1 percent of the ship's available power, which is more than made up for in the increased engine efficiency the air cavity hull provides. After accounting for the fuel required for the air-compressor pump, the net savings are around 7% to 9%. It is possible to retrofit existing ships with this technology in just fourteen days.

More fuel saving measures can and therefore ought to be implemented.

Anti-fouling coating would prevent barnacles and weeds from clinging and adding drag.

Last but not least, we can have better propulsion. Propellers used by today's ships do little to cut down on drag, causing air bubbles that reduce efficiency. By contrast, an innovative propeller by Pax Scientific has succeeded in channeling water in the spiraling pattern in which it naturally travels. A full 10 percent saving in energy is possible with this propeller.

Back to nuclear fuel and its reserves.

The OECD Nuclear Energy Agency estimates the total global identified uranium resources to be 6.3 million tons. We go through 68,000 tons of that stuff every year. At this rate, we will use it all up in about 90 years. This also accounts for mining what are currently uneconomical, hard-to-extract resources. At that stage, we either discover more recoverable uranium, or we don't. Chances are that the first round of discoveries done decades ago is just that: a first round. The overwhelming probability is that known reserves may prove to be but the tip of the iceberg. But one cannot base an entire industry on a fuel source that may or may not prove viable beyond the next ninety years.

This point may prove moot. In fact, to some extent I am counting on that. Existing nuclear technology is extremely wasteful. I don't regard it as any more than an interim step. I will discuss this at length in the seventh chapter ("Energy").

For the short term, if we have fifty years' worth of uranium that we can count on, it will suffice. Fifty years of known reserves will yield 126,000 tons annually to work with. If 68,000 tons of it is powering our existing stationary nuclear reactors, that leaves 58,000 tons per year.

No one knows for certain how much fuel our existing marine fleet consumes a year. Estimates are in the hundreds of millions of tons. I will assume that 300 million tons of fossil fuel are used by the larger ships, those that are to be outfitted with nuclear power.

I assume 7% fuel savings due to the air cavity (air bubbles) system. Furthermore, I assume a quarter of the fleet would trim an additional 20% of its fuel consumption due to SkySails. Call it a wash between the introduction of ocean liners and the phasing out of coal and oil freight, which largely will go away under the proposed plan. Let's shave an additional 10% due to the revolutionary new propeller.

Eight thousand tons of petroleum-based fuel gets about the same mileage as one ton of uranium.[38] When everything is factored in, this all comes to about 30,000 tons of uranium a year to run the entire fleet.

So we are good, at least for the first fifty years. In fact, that still leaves about 28,000 tons a year we can play with. And I do have designs on it, to be described in the seventh chapter ("Energy").

SOURCES OF POWER FOR SMALLER BOATS

We are not going to equip a yacht with a nuclear reactor. Nuclear propulsion is for the big, commercial vessels—whose owning entities would be under the same type of regulatory oversight currently governing existing nuclear power-plants.

Conventional sails aside, the only environmentally viable fuel would seem to be hydrogen for the small to mid-size ships. Not much information nor many working models exist. However, what we have is enough to deduce that it is a viable route.

About 1,000 gallons of diesel gets the same mileage as about 1,345 pounds ($\approx$ 610 kg) of hydrogen, given fuel properties and efficiency rates of both diesel-based ship motors and hydrogen PEM fuel cells. This conversion ratio is largely derived from the performance and specs of a hydrogen-powered tourist ship, the ZemShip.[39]

The 11.5-meter fuel tanks of Lincoln Composites, discussed earlier, can hold up to 200 kg ($\approx$ 440 lbs.) of hydrogen at 5,000 psi. These tanks would be suitable containers for the hydrogen fuel of larger vessels. Having 3 of these cylinders below deck would be the equivalent of having 1,000 gallons of diesel on board.

This is a problem, actually. At 5,000 psi compression, hydrogen requires 6.8 times more volume than diesel in order to provide for the same mileage. That can translate to up to 6.8 times less cruising range for a given ship.[40] Or, alternatively, less space for the payload or passengers in exchange for more room the fuel would take. Or more likely a combination of the two: somewhat less space for cargo and a somewhat more frequent need to refuel. A hydrogen-powered boat does not represent progress of convenience and range. It is simply what we need to do, all things considered.

For the larger, commercial boats we could store hydrogen in cryo-compressed cylinders in a liquid form. In other words, liquid hydrogen would be stored under very cold temperatures. After accounting for the extra thickness of the tanks, the liquid hydrogen takes about 3 times more volume to get the same performance as diesel. Technologically at least, this is distinctly better than 6.8 times more volume of compressed hydrogen in a gas form.

A cryo-compressed cylinder has an aluminum liner wrapped with carbon-fiber composite. It has a stainless steel outer shell with a vacuum gap. It would take about 48 kWh to generate 1 kg of hydrogen plus 20 kWh more to liquefy it. This comes to a total of 68 kWh per 1 kg of hydrogen. However, the cryo-compressed tank would be able to safely hold hydrogen up to 5,000 psi ($\approx$ 350 bars) before it would need to vent some pressure that would inevitably build up. Routinely venting hydrogen is not really an option; it depletes the ozone layer. However, it would take many weeks of storage before it would come to that. Hence, for larger, commercial ships, which have predictable travel patterns, the cryo-compressed liquid hydrogen seems most suitable. For the boats with occasional or unpredictable use, such as yachts, compressed gaseous hydrogen is more suitable.[41]

FREIGHT AND PASSENGER TRAINS

Some long-distance travel would be picked up by ocean liners. Much of the rest will be diverted to private vehicles, buses, and rail.

Some trains are presently pulled by diesel locomotives, some by electric locomotives. Under the plan drawn up in this book, the mandate is straightforward: electrify the entire rail network.

Beyond the gradual replacement of diesel locomotives, this would mean the installation of overhead lines above the tracks. It may mean tunnels have to be altered for due clearance. It may also mean a new set of railway signals. What is called for is the electrification of tracks spanning hundreds of thousands of kilometers. This would require a great deal of engineering work, but it is not landing-a-manned-missions-on-Mars kind of stuff. There is not much to say beyond that. The tracks are already where they need to be. If there is a dense network of passenger trains in Europe or India, it is for a reason. And if rail is sparse or non-existent in the heart of Australia or Northern Canada, it is for a reason too. Having said that, the future dearth of aviation will increase passenger rail traffic.

With the phasing out of regional aviation, some new railroad tracks will have to be laid down: perhaps Los Angeles–New York and key major cities in between; perhaps Montreal–Florida and the key cities in between; perhaps Beijing–Hong Kong and key cities in between (which is being planned anyway). In the absence of flights, one may entertain the notion of a train going from Abu Dhabi to London; however, it is questionable whether the traffic would be there to support this train route.

The Beijing–Shanghai High-Speed Railway is a 1,318 km route whose electric bullet trains average 329 kph (≈ 204 mph). The Wuhan–Guangzhou High-Speed Railway is a 968 km line with bullet trains that average 313 kph (≈ 194 mph). Currently, these are the fastest commercial train routes in the world. As a basis for any calculation, I will use an average speed of 310 kph (≈ 193 mph).

In a world largely devoid of commercial aviation, the advantage of high speed rail over cars comes to the forefront when it comes to long trips. It is easy enough to envision how it may look. People may board a train in Los Angeles in the late afternoon and may sit down to type away on their laptops or get comfortable with a book. Later, they may have dinner on the train. When they wake up in the morning, they may proceed to have their breakfast just before they arrive at their destination, New York. This is an eighteen-hour, leisure trip. In contrast, such a car trip is fairly miserable. It entails driving oneself for four laborious days.[42] The assumed eighteen hour train journey from Los Angeles is 5,200 km long. It assumes and factors in both the extra distance and time for stops in Las Vegas, Salt Lake City, Denver, Kansas City, Chicago, Detroit, and Cleveland.[43] For those who board in Denver, the trip to New York would be only eleven hours. This is good. Or maybe not, as longer, one-week train rides will discourage many from venturing to far-off destinations for juvenile reasons. Hence, we would end up conserving energy.

With a long-distance car trip averaging 2.4 passengers, an electric car would consume 0.08 kWh per passenger kilometer. That is to say, if one divides the energy consumption of a car among its average 2.4 passengers over a distance of one kilometer, it would come to 0.08 kWh. This corresponds to the train's electric consumption, averaged among all of its passengers. Naturally, a train's consumption per passenger kilometer can be a bit better or a lot worse, depending on the number of people in a given day.

High-speed electric trains and electric cars may end up with comparable levels of electric consumption per passenger kilometer, but in a way this is comparing apples and oranges. The train works off prime-time electricity. In contrast, the batteries of an electric car demand fewer resources, as they can work off excess power. They can charge themselves at off-peak hours, when the power-generation plants are just spinning their wheels. Thus, using trains may have a bigger ecological footprint than private modes of transportation.

As the time travelled between two destinations gets shorter, the time to get to the railway station and from there to the final destination looms ever larger. Thus, the appeal of the commute from San Jose to Los Angeles by a bullet train will also depend on the convenience of the public transportation available at the train stations.

The World Health Organization reports that the EMF radiation of trains in passenger coaches is between 5 and 50 uT. This is less than being exposed to one's TV or bedside clock. Experience has shown high-speed rail to be the safest mode of travel, hands down. In fact, with the exception of one accident in 2011,[44] no fatality has ever been recorded on high-speed rail. So perhaps there is more to high-speed rail than I have suggested here.

VACTRAINS

One mode of transportation holds the potential to replace to a significant degree all other modes of long-distance travel: the vacuum train, or vactrain. This idea has been making the rounds for about a century. However, as there are no working models, I am reluctant to make any definitive assertions; we first need to do some pilot tests and also conduct extensive environmental and anthropological impact studies. At first glance, the idea is exciting.

In a vactrain system, a magnetically levitated train moves through vacuum-sealed tubes. The key is the lack of resistance—from air and from tracks. Removing these two sources of drag spells all the difference. First, it means that the train can go incredibly fast, reaching a top speed of 4,000 miles per hour (≈ 6,500 kmh). Second, it means that the amount of energy required to move the train is very small. There are new energy costs to maintain the vacuum and to cool and power the magnets, but these energy costs are but fraction of that of aviation—while the train is being a lot faster and potentially with zero pollution. A high speed is not a source of discomfort; acceleration is. Yet, with keeping oneself to 1 G acceleration, it is possible to reach 5,000 miles per hour in under four minutes.[45]

The idea is to construct a backbone: a tunnel or a series of parallel tunnels that traverse on land and connect the population hubs of the world with hundreds if not thousands of secondary vacuum tunnels with sharper turns facilitating lower speeds for more localized travel.

A train route through such a possible backbone will work its way from New York down to Mexico, arriving at Los Angeles within about 45 minutes. From there, it will work its way up through Washington state, Alaska, and over a fifty mile bridge through the Bering Strait, into Siberia, and down to the populous region of eastern China. From San Francisco to Beijing the trip would take an hour and a half. Forming a gentle arc, the backbone will go through the populous regions of Northern India, reaching Europe within one hour or so . One could live in India and commute to London, and then catch the afternoon train home—back for supper. And a 47 mile train commute from New Jersey to Manhattan can be done in two minutes at a lower speed, or ten minutes at a far lower speed.

6

BUILDINGS

〰️

cutting their energy consumption by a factor of ten

THIS CHAPTER FOCUSES ON slashing the energy needs of our homes. This boils down to three things: cooling, heating, and lighting. We can bring down radically the energy usage of each without sacrificing comfort.

REDUCING HEATING AND COOLING LOADS

The sun is beating down on rooftops; we might as well put this to good use. It behooves us to install passive solar water systems on our roofs. The building code should mandate solar water heaters in climates where it makes sense and where a viable roof orientation is available. The same thing goes for solar water-heaters for swimming pools.

Solar heaters can never provide all of the hot-water needs at all times, as cloudy days make such a scheme impossible. It also makes little sense to cover an entire roof with an overbuilt unit to accommodate days with sparse sun. A household can get between 30% and 90% of its hot water from the solar water heater units, depending on the amount of solar radiation in a given region.

The energy required in manufacturing the solar water heater and its basic components is offset after using the solar heater for half a year, on average, and then it starts to have net energy savings.[1] Installing water heaters on all viable rooftops is estimated to reduce primary energy consumption in the United States by 1 quad.[2] The *entire* primary energy generated in the United States per year is 94.6 quads (2009). That includes transportation, residential, commercial, industrial sectors, and all heat byproducts of electrical generation.[3]

Then there is the matter of the roof itself. If a roof is white, most of the sunlight striking it is reflected back into space. But if a roof is of dark color, it absorbs more heat.

The one hundred largest metropolitan areas in the world take up about 0.26% of the Earth land area and contain 670 million people.[4] If the ratio holds true, then as of 2011, with 50% of the urban population, all the metropolitan areas of the world take about 1.4% of the Earth's land area, that is, 2,040,000 sq km. It seems to be in the ballpark of other assessments.[5] If we assume roofs to be 25% of the urban area and pavements and roads to make up 35% of the urban area,[6] this comes to about 510,000 sq km worth of roofs and another 715,000 sq km of paved area.

A study assessed that light, cooling colors applied to all the roof and paved surfaces of the world would result in the equivalent of eliminating 25 to 150 billion tons of CO_2 from the atmosphere.[7] To put it in context, it is estimated that in 2005 global GHG anthropogenic emissions were equivalent to 44 billion tons of CO_2. So light-colored roofs can offset anything from half a year to over three years' worth of anthropogenic GHG emissions, based on 2005 levels. Another study disputed this. It concluded that white roofs will cause a slight increase in net global warming—as they reduce cloudiness, allowing more sunshine to reach the surface.[8]

Having roofs in light colors may or may not make a difference in climate, but they reduce the cooling load on a house, hence reducing its energy demands. The air above a bright-white roof is markedly cooler than over a dark shingled roof. On summer days, peak temperatures of black roofs in New York City were 24°C (about 43°F) higher than white roofs.[9]

A study estimates the energy balance for buildings in the United States.[10] It shows that a light-colored roof almost always reduces the cooling load more than it increases the heating demand. Hence, a cool roof is a net energy saver. Applied atop commercial buildings, light-colored roofs in sunny Arizona saved 6 kWh per square meter once the reduction in cooling and increase in heating demand were reckoned. California and Florida averaged 5 kWh/m² of net saving, and Texas averaged 4 kWh/m². Much colder and cloudier New York averaged a net saving of 2 kWh/m². Illinois, Michigan, and Maine had a bit over 1 kWh/m² in net savings. If one were to average the net savings of all commercial buildings in the United States across all climatic zones and population centers, the net savings would be 2.52 kWh/m².[11]

In accordance, a building code could be instituted mandating light-colored roofs in all tropical and temperate regions of the world. The subarctic and arctic regions don't have enough rooftop or solar exposure to matter, one way or another.

Let's get concrete. *Solar reflectance* is a measure of the ability of a surface material to reflect sunlight. *Emittance* is the ability of a material to release absorbed heat. Solar Reflectance Index (SRI) incorporates both solar reflectance and emittance into a single composite index that measures a material's temperature in the sun. SRI quantifies how hot a surface would get relative to a standard black surface (assigned value, 0) and relative to a standard white surface (assigned value, 100).

Flat roofs and low steep roofs[12] would have a minimum required SRI value of 100.[13] Some existing products meet this standard. National Coatings Corporation has an acrylic top coating, the A590, that is bright-white and reaches an SRI of 110. While many products with SRI values of 100 and above come only in white, at least one brand, Cerama-Flex by Endurance Building Systems, offers an SRI of 102 in a range of light colors such as antique-white, vanilla, and sand.

It is a bit much to expect this performance to exist on tile or metallic steep-slope roofs. Hence, the mandated minimum SRI value for those kinds of roofs is to be 65. The clay roof tiles of Ludowici Roof Tile make the grade with desert-like yellow and tan hues, with an SRI OF 65. The concrete tiles by Eagle Roofing Products with their natural colors have SRI values in the 70s. And the Pac-Clad metal roof by Petersen Aluminum Corporation that comes in an almond hue achieves an SRI OF 66.

A typical asphalt shingle-roof, so popular in the United States, may have an SRI value that is in the 20s. That's low. Too low.

Paved areas—sidewalks, roads, and parking lots—take up about 35% of the total urban area. This is almost one and a half times the rooftop area. They create urban heat effect, which in turn adds to the heat loads on houses in the city. Asphalt roads reflect a meager 5% of the light when they are black, and they age their way to 15% reflectance when they mellow and are light gray. Aged cement-based sidewalks and roads reflect about 20%–35%. Time to lighten up the situation.

We don't know yet what percentage an aged white-cement sidewalk would reflect, but new white cement has 70%–80% reflectance. Adding slag to cement will make roads and pavements reflect 70%–75%. Cementitious coating, such as that offered by Emerald Cities, results in surfaces that reflect 50%–55%. All paved areas in tropical and temperate regions could get a makeover with one high-reflectance surface or another. The tinted spray cement of Emerald Cities may hold the greatest potential. The company applies reflective pigment to white cement in a variety of light green and earth tones. The product is either sprayed or rolled on. It can be applied to any paved surface. The result is not only a surface with higher reflectance but one that glows softly at night. The tinted surface can be easily seen by drivers on country roads without street lights. Using these tints, we can have bike lanes and pedestrian crossings that are color-coded for safety, and by and large we can transform a city's ambience from that of dominant lifeless-gray hues to that of gentle earth tones.

Let's get back to buildings.

There are other ways to reduce the cooling loads of buildings. First, though, something needs to be stated in the context of commercial buildings. We don't need to define thermally comfortable interiors in such rigid terms as those used nowadays. We can sharply cut down on energy needed to cool or heat a commercial building if we simply dispense with the fixed, year-round narrow band of temperature and replace it with a weather-sensitive range of acceptable indoor temperatures.

Numerous studies by ASHREA have shown that the human body adapts to the seasonal and local climate. Occupants will consider different indoor temperatures comfortable based on the season and location. The graph below shows correlations between people's subjective impression of comfort and indoor temperature in hundreds of real buildings.

FIGURE 6.1

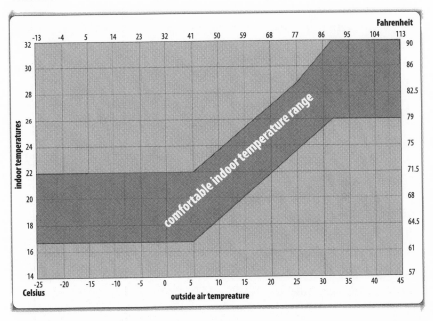

Source: Adaptation of the graph in Gail Schiller Brager and Richard de Dear, "A Standard for Natural Ventilation," *ASHRAE* 42, no. 10 (2000): 21–28.

Note: This chart shows the thermal comfort range for indoor conditions, given outside temperature and assuming some indoor breeze at the hotter end of the spectrum. Thus, if it is –20°C (–4°F) outside, the comfort range for the indoor temperature is somewhere between 17°C and 22°C (63°F and 71.5°F). However, if it is 40°C (104°F) outside, the comfort range is around 26°C to 32°C (79°F to 90°F).

This is significant. Increasing the thermostat setting from 24°C to 28°C (about 75°F–82°F) in the summer will reduce cooling loads by more than a factor of three for a typical office building in Zurich and by more than a factor of two in Rome.[14] Each 1°C of broadening the acceptable temperature threshold in buildings with HVAC may translate to 7%–15% in energy savings.[15]

A noticeable airflow in the interior spaces shaves off an additional 1°C. Thus, the code will mandate buildings also to be either naturally ventilated or coupled with mechanical ventilation, such as fans or updraft towers.[16]

Shade saves on energy, too. It may be obtained by having nearby trees; clustering buildings next to each other; having moveable external shades and overhangs; and texturing the outside walls with deeply carved patterns, such as ridges that are inches deep.

In temperate regions, the placement of broadleaf deciduous trees near a house is wise. Such an arrangement provides shade in the summer, while still letting large amounts of heat-carrying solar radiation strike the house in the winter, as the leaves shed. In colder climate regions, a dense array of coniferous evergreen trees is a suitable choice. They would soak up much of the cold wind and therefore reduce the need to heat the house.

Effective house orientation can also reduce the energy needed to keep a house cool. To the extent possible, windows should be facing north (in the northern hemisphere, that is). There should be as few windows facing east or west as is practical; overhanging shades cannot quite block the more horizontal rays of the rising and setting sun. And whenever applicable, smart windows can be installed. During summer days, these windows turn completely opaque when the sun is out in force, reflecting sunlight away. During winter days, they switch to full transparency. Researchers Ho Sun Lim, Jeong Ho Cho, Jooyong Kim and Chang Hwan Lee have developed smart windows such as this that are stable and inexpensive to make.[17]

A house constructed of materials that absorb and hold heat can be very useful in places where nighttime temperatures drop to sufficient levels to provide satisfactory natural ventilation. This absorbing thermal mass can be masonry walls, adobe bricks, stones, or phase-change materials.[18] During the day, the walls absorb the heat from the sun, keeping the indoor space relatively cool. At night, the walls release the heat. If a given region has frigid nights, the released heat serves well the interior of the house. If a given region has mild nights, there is a need for some night ventilation that would flush out the heat stored in the walls, priming them for the next-day duty.

Each of these measures saves a considerable amount of energy. Combined, they serve to dramatically reduce the cooling load on a building. In mild climates, such as that of Los Angeles, shading, orientation, thermal mass and exterior colors take it most of the way.

The remaining balance of summer heat can be brought to comfortable levels via the use of natural ventilation—utilizing the cold-hot air differential and cross winds.

Here is how it works: Warm indoor air rises and escapes the building at the top. The escaping warm air reduces the air pressure of the remaining indoor-air. The reduced air pressure indoors sucks in cold air from outside at the lower levels. And so there is a constant cycling of incoming colder air from the vents at the lower levels and outgoing warm air at the top of a building. This principle can be utilized via solar chimney or atrium. In mild climates, the entire cooling load of a well-designed building can be handled with natural ventilation schemes such as these.

Wind-driven ventilation packs substantially more oomph than hot-air dissipation, described above. With wind-driven ventilation, air flows over an opening at the top of a domed roof and creates a suction effect that removes hot air that accumulates inside the top of the dome. Cashing in on that principle is the Windcatcher X-Air made by the British company Monodraught. It is a roof-mounted unit designed for commercial buildings. Using compartmentalized vertical vents, fresh air is brought inside through the vents in the Windcatcher, and stale warm air is expelled using the natural effects of the wind. For temperate climates like that in London or San Francisco, it eliminates the need for an air conditioner altogether, while in hotter climates it will eliminate the need for air conditioning during some parts of the day.

It may be hot outside, but a few meters underground it may be a different story. Wherever and whenever the temperature below ground is significantly colder than the daytime air temperature, earth tubes may be employed to great effect. In a highly-insulated house, before ventilated fresh air is brought in, it can pass through a series of deeply buried flexible tubes that circulate liquid in a closed loop.[19] The earth keeps the tubes relatively cold, which in turn cool the incoming air before it enters the building. In this very manner, 70 percent of the cooling load during the summer was reduced in a house in Illinois.[20] Near New Delhi, India, earth tubes made possible indoor temperatures around 24°C–31°C (≈ 75°F–88°F) while outside temperatures were 31°C–48°C (≈ 88°F–118°F).[21]

Provided there is a marked difference between day and night time temperatures, cool night air can be sucked into a cavity in the ceiling or walls. Later, during the day, the incoming hot air can be cooled by passing it through the ceiling that has been chilled at night in the manner described. This scheme lowered the energy bill by 30%–40% in a building in London.[22] In Beijing, such nighttime ventilation eliminated the need for an air-conditioner altogether.[23]

Lafayette, Louisiana has summers with daily average high temperatures around 32°C (≈90°F) with a corresponding relative humidity that is in the 60s. The recently built Le Boise House in Lafayette has an optimal orientation, light exterior colors, overhangs providing shade, and windows that are mostly facing north with glass that lets in a minimal amount of heat. The house is also heavily-insulated—to an extent not attempted before in humid-subtropical climates. During spring and autumn, neither active heating nor active cooling is needed in the Le Boise House. It is enough to occasionally open windows on both sides of the house. For the hot, humid summer months, a 1-ton ductless mini-split air conditioner is used to both cool the house and dehumidify the air. The thermal mass of the walls absorbs some of the indoor cooled air and releases it at night. The end result is a house that has less than 15 kWh cooling and 15 kWh heating load per square meter annually. The Le Boise House was the first passivhaus-certified house constructed in a humid subtropical climate.

This brings us to passivhaus construction and the cardinal role it is to play in future constructions in climates ranging from mild Mediterranean to cold continental.

It could be snowing outside and our homes would be warm and toasty—mainly working off our body heat, appliances, and daylight. Pity it took us so long to connect the dots. Pity that once we did, in recent decades, so little has been done to change our cavalier construction practices. Houses built true to the best engineering and construction practices in temperate climates require 15 kWh or less per square meter per year.[24] In other words, they require between 15% to 20% of the space heating requirements of new, conventionally-constructed buildings that fulfill local European building codes.[25] Most of the trifling 15 kWh worth of heat that is required comes from the body heat of occupants, appliances, and the incoming heat from windows. Whenever needed, minimal electrical heating of the incoming air makes up the difference.

Passivhaus is not an architectural style but a construction standard, which can be applied to most housing styles and achieved with myriad construction materials. Houses built to this standard are known as passivhauses. To construct a passivhaus, a few technologies and principles come together.

An advanced window-technology takes full advantage of sunlight. The windows in a passivhaus let the heat of the sun warm the house in the winter, while assuring that the indoor warmth does not leak outside. These windows are not your grandparents' windows—nor probably your windows, for that matter. Typically, the windows have three glass panes filled with krypton,[26] have low-e coating (reducing heat transfer), and have frames filled with foam insulation. These windows feel warm even when the outside temperature is well below freezing. They may lose a bit of heat to the outside during the heart of winter, but they more than make up for it with the incoming warmth of the sun during the day. Thus, windows built to such standards provide a net gain of energy.[27]

Superinsulation is what makes the incredibly low heating and cooling load possible. Along with ultra-high performance doors and windows, passivhaus buildings in temperate climates have walls, roofs, and foundations with insulation R-values ranging from 40 to 60.[28] This superinsulation may translate to wood-framed buildings with 16-inch-thick double-stud walls or to masonry buildings with over 10 inches of exterior rigid foam. In between the ground and the concrete slab may be foot-thick rigid foam,[29] isolating the foundation from seasonal ground freezing.

An important side note: Until now, beefing up the insulation of an existing building may have required a 20 centimeter thick ($\approx$ 8 in.) outer layer of polyurethane foam, which in turn may have called for structural changes, such as deeper window sills. In Germany, researchers at the Fraunhofer Institute have developed vacuum insulation panels that can do the same job at 2 centimeters thickness ($\approx$ ¾ in.).[30]

It will do little good to have a super-insulated house if there are crevices through which the heat would leak. It is imperative to have insulation applied *continuously* around the building envelope. This requires insulation about ten times tighter than conventional, existing homes.[31] If you were to turn the passivhaus upside down, it would likely bob on the sea a day or so before it would sink. Careful sealing of every construction joint in the building envelope is used to achieve this. Getting the house to be airtight is as much a matter of design as a matter of attention to details during the construction process. Then there is the matter of thermal bridges.

When a stud or a junction box connects the outer sheath of the house with the inner dry-wall, this forms a bridge, a thermal bridge. The walls may be otherwise well-insulated, but that matters little if the heat leaks through the numerous bridges crisscrossing all the walls of the house. Hence, the need for a construction that offers no thermal bridges and keeps an intact, unbroken insulating envelope. Double stud walls are one way to eliminate thermal bridges.

With an ultra-tight envelope, air exchange with the outside is nil. Hence, there is a need for a ventilation system that continuously brings fresh air into the house and vents the stale air. Trapped, stale air would quickly generate dangerous levels of carbon dioxide emitted by the occupants. Furthermore, it would accumulate moisture, which is a breeding ground for mold, fungi, and dust mites. And for the first few years, carpets and wood panels would outgas chemicals we would not care to breathe. Without anywhere for the gasses to go, this is asking for trouble.

The way to keep a steady stream of fresh air coming into a passivhaus is via an energy recovery ventilator. With such a unit, the air in the passivhaus changes every couple of hours. The result is a living space bearing a startlingly fresh air that is more associated with the outside than with the indoor environment. In a conventionally constructed house, one would have to open all of its windows three hours a day to achieve a comparable level of ventilation.[32]

At the heart of the energy recovery ventilator is a heat exchanger. Before the relatively warm, stale air leaves the house, it goes through a heat exchanger and warms up the incoming cold fresh air. The hot outgoing air and the cold incoming air do not mix. They go through a series of ducts, or passageways that share a common wall. The common duct-walls are heated by the outgoing air and in turn heat the incoming, cold air. This kind of system can transfer 80% to 90% of the outgoing heat to the incoming air. An energy recovery ventilator also provides an additional service. The unit exchanges moisture between the two air streams, which is important when the incoming cold air is very dry.

The energy recovery ventilator is an essential part of the central European passivhaus and makes it possible to build homes without conventional heating systems. In regions where the winters are bitterly cold, heat recovery may be insufficient to bring the incoming outside air up to the required temperature. Consequently, the heat recovery system is often supplemented with a small powered heat pump.

All of these elements—insulation, thermal bridges, air tightness, ventilation—are modeled in a dedicated software that enables designers to accurately predict what the energy performance of the house would be. The modeling software takes into account almost every aspect of a house, including the site's weather patterns and solar orientation; the type of construction and the materials used; the window designs and locations; and the ventilation-system design. This simulation is quite essential. One cannot achieve the exacting standards of a passivhaus without putting all specs into the modeling software and seeing if the given configuration results in sufficiently low heating and cooling requirements.

A passivhaus upfront cost is about 5% to 8% more than that of its conventional counterpart if we factor in the cost of a central heating system, or rather the lack of one in the case of a passivhaus. One calculation shows that after eleven years, a passivhaus' s reduced energy bill makes up for the initial higher construction cost—and then it pulls ahead.[33] As of 2010, there are an estimated 15,000–20,000 passivhauses worldwide, the vast majority of which have been built in the marine climate of northern and central Europe.

Two theoretical studies investigated the potential energy savings of well-constructed houses in the mild Mediterranean climate,[34] such as is found in parts of Spain, Italy, Greece, and coastal Southern California.

The results are just as dramatic as those of passivhauses in the mild marine climate of Europe.

A house of conventional, lax construction in Seville, southern Spain, without mechanical cooling has indoor summer temperatures that widely fluctuate between a toasty 26°C–29°C (≈ 79°F–84°F) during the night to an unbearable 32°C–37°C (≈ 90°F–99°F) in the middle of the day. In contrast, with white exterior surfaces and night ventilation, we obtain indoor temperatures that are 25°C–26°C (≈ 77°F–79°F) during the night and 28°C–29°C (≈ 82°F–84°F) during the day, which are more or less within an acceptable comfort range.

Another example is in Granada, southern Spain. A simulated passivhaus construction employs but 17% of the heating and 47% of the cooling energy of a house of the same size that complies with the local building code. During summer nights, the warm, indoor air is ventilated via the stairs, which act like a chimney. In Portugal, a house conforming broadly to passivhaus principals has managed 8% of the heating-energy needs and 12% of the cooling-energy needs of a newly constructed conventional house that meets local energy standards.[35]

The Isabella Eco Home is a passivhaus built in the far colder region of the Superior National Forest, Minnesota. In the coldest months, average nighttime temperatures dip from –18°C to –23°C (≈ 0°F to –10°F), while average daytime temperatures hover at about –10°C to –7°C (≈ 14°F to 20°F). The Isabella Home has 12-inch staggered double-wall-stud framing (around 50 R-value) and 30-inch heavy-timber framing on the roof (around 90 R-value). More so than in a milder climate, south-facing windows are paramount for a house in that region. With its excellent thermal envelope, the Isabella passivhaus has fewer annual heating requirements in North Minnesota than a conventional house situated in a mild Mediterranean climate.

Te Studio has built a passivhaus in Wisconsin, which has a comparable climate to Minnesota. The house achieved equally satisfactory results. It has become evident that in passivhauses in the Great Lakes, Eastern Europe, and the populous regions of Canada and Scandinavia, it would no longer be enough to heat the incoming air with the outgoing air. An additional modest source of heat has to be placed in the house: the energy equivalent of two to four air dryers, to be used during the coldest portions of the year. It matters little where the heater is placed; the house is so airtight that the heat will spread throughout the house.[36]

A conventional door can lose as much heat as 20 square meters of well-insulated wall.[37] The walls in a passivhaus may have R-50 to R-75 insulation values but the door may have an R value between 1 and 4. Ouch. Optiwin comes to the rescue. The company makes the Frostkorken entry doors that have a middle layer made of cork and achieve an R value of 10. They have another door model that embeds 2-inch vacuum-insulated panels and has R value of 50, suitable for a subarctic climate.

Once we get to the subarctic climate, the challenge to build homes with low-energy requirements becomes more demanding yet. In the polar regions, it is not just ice-cold; there is a lot less sun to work with. In the big scheme of things, the fate of housing in the subarctic regions matters little, as practically no one lives there. However, for the hell of it, let's take the passivhaus approach for a spin.

In the frigid climates of Siberia, Northern Scandinavia, and Canada, it is straightforward enough to bump the wall insulation up: getting the walls to R-75 and roof to R-115, as is the case in the SunRise Home in Fairbanks, Alaska. The windows become the bottle neck. Even the best windows lose more heat than they admit in the subarctic and arctic climate. There simply aren't windows that adequately retain heat for those ultra-frigid regions. The solution is to attach R-20 thermal shutters on the outside, to be hermetically shut at night and thus prevent heat leaks. On top of this, one would hyper-insulate window-frames on both sides.

To summarize, the better the thermal envelope around the house, the less we have to make up for it by way of heating and cooling. This means highly-rated insulation, airtight and without thermal bridges. Once we have a sound thermal insulation around the house, the cooling and heating loads are relatively miniscule. In many situations, they can be handled through passive means, that is, wielding the sun and wind to our ends. More than anything else, a passivhaus is an engineering tour-de-force.

LIGHTING

We can also cut way down on our lighting-related energy needs.[38] Let indirect sunlight provide interior illumination—and of utmost quality, at that.

In many cases, atria, skylights, and windows do just that. Beyond these passive measures, large commercial buildings can benefit from mirrors that track the sun throughout the day, diffusing and then admitting in light that otherwise would be lost during the morning and early evening times.

A variation on that idea is a mirror on the roof of a commercial building tracking the sun's daily movement and refracting the sunrays to a tube that extends all the way to ground level. The light pipe, as it is called, comes with a highly reflective inner coating that enables the pipe to deliver the sunlight into the bowels of the building. The light passes through a film that serves to diffuse the light before it illuminates the interior of the building.

The Core Sunlighting System by SunCentral, Inc. has taken this principle to the next level. The system has a canopy that collects and concentrates sunlight and then feeds it through ceiling duct pathways throughout the building. It provides illumination deep into the core of a building, eliminating much of the need for artificial lighting.

The canopy is placed along the length of the building façade that has the best exposure. Within the canopy there is a moveable array of mirrors. Powered by motors, the battery of small square mirrors tracks in unison the movement of the sun throughout the day and redirects the rays in a fixed direction toward the building. Parabolic mirrors concentrate the light before it exits the canopy and is piped into the building interior. Within the interior, the concentrated sunbeams pass through ducts that are coated with a highly reflective film and are housed in the ceiling cavities. The bottom of those ducts is made of prismatic films, which admit the light through to the rooms below, assuring even, diffused light. Even when the sunlight is at its brightest, the lighting is most agreeable for the occupants, without a discomforting glare. The end result is an excellent color rendering that has shown to create a buoyant work environment. The delivery ducts of the Core System come with electric lamps of variable brightness that are programmed to supplement the sunlight to maintain the desired level of light. If a cloud shades the sun or if the sun is setting, the artificial light will automatically brighten—compensating. The net result is a continuous, even supply of satisfactory light to the office space.

With the Core System, lighting-related energy savings range from 25% to 75% on days when the sun shines for at least six hours. The system integrates easily into existing commercial buildings. SunCentral is patenting additional technological improvements that would permit the placement of the collectors on any and all sides of a given building. This would practically assure that the entire span of each floor would be illuminated with piped, diffused sunlight. And this is exactly what the new building code would require. Whereas sunlight cannot adequately illuminate all interior commercial spaces via skylight or windows, the new building code would mandate the incorporation of something along the lines of the Core Sunlighting System. This system may find a place in hospitals, office buildings, universities, hotels, high schools, and even some restaurants.

We can also cut back on electricity by moving away from the on-off light switch. A dimmer not only can create nuanced ambiance but can also save on energy: by as much as 50 percent.[39] Light sensors attached to a dimmer may automatically adjust the level of artificial light, conforming to the variable amounts of incoming sunlight. And occupancy sensors may make sure that in some spaces the light turns off when the room is unoccupied for a certain length of time.

In a residential context, it is technologically straightforward to have lighting settings such as "TV time," "dinner time," "romantic," and "low key." Instead of pressing the on-off switch, one presses a button corresponding to the desired lighting setting. This will both provide ambiance and save energy on many occasions.

All of the above translate to substantial energy savings.

Then there is the matter of the light bulbs themselves. The antiquated, energy-guzzling incandescents, along with the mercury-laden fluorescents, are to be phased out. LED light bulbs are to be the new standard.

Light Emitting Diode (LED) bulbs are far more energy efficient than incandescent bulbs and come without the toxic baggage of the fluorescents. LEDs are available in fluorescent-like tubes as well as in conventional type A light bulbs. They come in various colors; they are very sturdy; and they do well in dimmers. They are found in flashlights, street lights, and under kitchen cabinets.

Delft University of Technology in the Netherlands is currently testing an intelligent street lighting system, which reduces the use of electricity by up to five times. The street lights are equipped with LED lighting and motion sensors. This enables the installation to dim the lights when there are no cars, cyclists, or pedestrians in the vicinity.

LED light bulbs are rated for 50,000 hours. While they have a hefty upfront cost at the store, with over 50,000 hours of service, they more than make up for it.

TABLE 6.1

for 50,000 hours	light bulb cost	electric bill	total cost
incandescent (60W)	$14.5 (=25 bulbs)	$360	$374.5
LED (15W)	$45 (=1 bulb)	$90	$135

Note: The price of a quality LED bulb (48 lumens per watt) is assumed to be $45 once they are mass produced at a very high volume. The price of a single incandescent 60W lasting for 2,000 hours is $0.57. The price of electricity is 12 cents per kilowatt-hour. All prices are in US dollars.

As people may not be able to stomach the upfront cost, utility companies can sell the bulbs at a pittance, and the electric bill would reflect the higher cost assumed for LED bulbs until they are paid in full. Put another way, at no stage will the customer make a higher payment than he does now. The upfront cost of the bulb will be only a fraction of its true manufacturing cost. Each monthly electric bill will include a fraction of the bulb cost. The total bill, though, won't be any higher than it is at present. While there is a bulb surcharge from one side, the electric consumption is lower due to the LED technology, and the net result is the same monthly payment—at least until the bulb is paid in full, at which time the bill will go down.

Right out of the gate, we slash energy usage with LEDs down to 25% of what it is with incandescents. Today, off-the-shelf, middle-of-the-road LED bulbs generate about 50 lumens of light per watt. Tomorrow, they will be capable of putting out 200 lumens per watt,[40] bringing the energy consumption from 25% of that required by incandescents down to 6%.

LED TVs—or the upcoming OLED TVs on flexible plastic,[41] for that matter—consume on average 80% less energy than their plasma counterparts. Much as with light bulbs, this one is a no brainer. Having TVs in everyone's home may be a top national priority, but at least we can cut loose those power-hungry plasma television sets.

To bring it all together, the most effective strategy to lower energy in the building sector is comprised of few things. First, it starts off with reducing heating and cooling loads through a high performance envelope. Second, it provides as much of the reduced load as possible through passive solar heating, ventilation, and cooling while optimizing the use of daylight. And third, it uses efficient electric lighting to compliment the sunlight.[42]

All of this is accounted for and reflected in a proposed comprehensive code that would cap the environmental footprint of a building.

BUILDING CODE

Out of the total energy consumption used by all end-sectors (e.g., industry, transport, agriculture), currently about 32 percent is used in homes, offices, and businesses.[43] This is only the operating energy of existing buildings, at that. It does not include the energy in the construction process and fabrication of raw materials prior to that. Operating energy mostly comes down to heating, cooling, and lighting.

The potential exists to reduce the energy use of new buildings by a factor of four to ten compared to current practices.[44] A new building code needs to be instituted to mandate that. It will set a cap on the total energy needs of a house—and a lot more besides. This building code would curb the environmental footprint of the house throughout its life: raw materials and related transportation, construction, operating-related energy, and post-demolition impact of its materials. One day, we will have benign building materials and technologies, and the focus of the code will shift from a less harmful footprint to a footprint that is actually benign to the environment. But that day is not today.

The building code would establish three distinct caps over the lifecycle of the house: land footprint cap, scarcity cap, and energy cap. The building code is to be strictly performance-based. That is to say, the designer and the owner have full discretion as to how they will remain within the mandated environmental footprint. Building code would be context specific. In the hyper-crowded, highly-stressed region of Bangladesh, the mandated cap over ecological footprint of a house would be different than that mandated in the sparsely populated regions of Canada.

The land footprint cap would limit the disturbed area that it takes to extract (e.g., metal, limestone) or to grow (e.g., tree plantation, bamboo) the raw material. Added to it would be the amount of earth disturbed by the house itself and the footprint of demolition at the end of its lifecycle.

We will assess commodity abundance and set a cap on the amount of non-renewable, relatively scarce materials. Iron and limestone are not renewable, but they are abundant enough that they might as well be. On the other hand, it is profoundly immoral to use up something that may run out in few hundred years, that is, in a handful of lifetimes. Billions of tons of raw materials are used in construction each year. Sustainability is to be a paramount consideration. If wood is used, it will come from wood plantations. Moreover, it will not be harvested at rates higher than the trees can grow. Renewable or otherwise abundant building materials include straw-bale, mud (such as adobe or cob), glass, and steel. Beyond that, the goal is to have building materials that can be recycled; to have construction that lends itself to deconstruction when the house reaches the end of its useful life; and to have components that are decoupled from the building for easy replacement. It is a given that all materials and their manufacturing will have minimal to zero global warming potential, which is to say their manufacture or existence will not have a net warming gain.

Then comes the last of the three caps, the energy cap set on a building.

The embodied energy is location-sensitive and project-sensitive. Within reason, it depends on how far the building materials have to be ferried, and what the local construction processes are like, such as whether a given supply is derived of virgin material or recycled material. Having said that, some construction materials are inherently more energy-intensive than others. Generally speaking, the more highly processed a material is, the higher its embodied energy. For instance, compressed mud blocks take less energy to produce than bricks. At times, though, it is not as straightforward as that. It takes two or three times less energy to produce Glued Laminated Timber than a comparable steel beam. Yet, steel is likely to be recycled many times over.

Energy embodied in building materials is but a small portion of the energy consumed by a house throughout its lifecycle. A study found that the initial embodied energy of a typical office building is 1.3 MWh per square meter. In contrast, the operating energy overshadows it with a cumulative value of 19.5 MWh per square meter at the end of a fifty-year period.[45] By implication, it may be worthwhile to select energy-intensive building materials if those result in energy savings during the operation of the house.

Unsustainable hunting aside, until about ten thousand years ago, we did not have to worry about our energy footprint. Since then we should have—but didn't.

The thing is this: we cannot pull energy out of thin air. It has to come from somewhere and at the expense of something else on Earth. Beyond the money people pay for other people's labor, there is the issue of the land that the energy generation will appropriate at the expense of natural habitats and the life they sustain. Money cannot buy its way out of this one. We cannot offer a given ecosystem money in exchange for its existence in order for us to set down a power plant. Hence, the energy requirements for a building would be capped for each member of the merry men: for each homeowner, irrespective of how much money some of them may be in a position to fork out. The broader economic principle is touched upon in the last two chapters ("Consumption" and "Economic and Political Paradigm").

No one would monitor or dictate actual energy usage once the house is occupied. The cap would be for the projected energy usage of a house over a seventy-five year period.

This projection would be comprised of the energy used in the acquisition of raw materials, their processing, manufacturing, transportation to site, and construction. To this would be added the projected energy requirements for lighting, heating, and cooling. Factored in would be anticipated heat leakage and passive solar gain of the given house design in a given site and climate. To the energy tally would be further added recurring energy expenses, through routine maintenance and ongoing operation of the house. Finally, the tally would also count the projected energy related to demolition and the recycling of materials. The total estimated required energy for a seventy-five year period would be compared to the cap set by the building code.

While the energy requirement is per square meter of livable space, there is to be a sliding scale depending on the house size. Very small housing units suffer from something akin to economy-of-scale; their per-square-meter energy usage is inherently higher.[46] The sliding scale would account for that fact, thus, it won't penalize the construction of ultra-small units. On the other end of the scale, people may elect to have 5,000 square foot mansions (about 465 m²). In this case, the burden for the large home size is on the homeowners. In other words, houses ranging from the extra-large to the obscene would have an ever-decreasing allotted energy consumption per square meter.

Different houses in different climate zones may end up having different mandated energy caps. In the case of a commercial or industrial building, the specific cap would also depend on its intended function. In the case of a residential house, it would also depend on the type of dwelling: a detached house, duplex, or an apartment complex.

The energy standard set for the various climatic zones ought to be as stringent as contemporary technology and fabrication methods would make possible. However, the building code cannot require construction standards that in a roundabout way would result in construction costs exceeding a certain limit.

A low-energy house will drastically reduce the monthly energy bills. The difference between the energy bills of a conventionally-constructed house and a passivhaus would be the upper cap on how much and how far the building code may rely on more expensive technology in setting the energy cap. The upfront cost could be higher but the energy bills would be lower, thus the total sum would be no more than it is now.

In fact, not only is there no expectation that the combined, total amount would exceed current rates but the payment structure would also be comparable to what is paid for at the present. It would have the same payment scheme as explained earlier with the LED lightbulb.

The home owner will pay the same cost upfront that they would have for a conventional counterpart house. The energy bills they get would also be comparable to that of a conventional house. However, here is what will be going on behind the scenes: the actual energy usage under the new construction standards will be a lot lower than that of today's houses. Beyond paying for the actual energy bill, the remainder of the payment would be used to defray the initial higher construction cost for which the owner was not billed upfront. This would go on until the government loan—which in effect it is—would be repaid, at which time the homeowners would pay only for the actual, reduced energy bills. From the consumer end of things, it means that the house, with its possibly more expensive building materials and appliances, would have the same cost as a conventional house. And a year or twenty years later, their energy bill would drastically come down, when the cost of the house is paid in full.

The code would not mandate a technology so expensive in relation to a conventional one that it would require more than twenty-five years of repaying the added cost. Having said that, the projected cost of a new technology would be assessed under the assumption that the technology is to be widely used, hence, mass-produced. At the moment, high-end triple glazed windows are a niche item and are very costly. If mandated by code, they would be ubiquitous, and their price would tumble down. When assuming what technology is economically viable, the building code would assume the given technology to be mass produced and priced relatively low.

In order to get total energy consumption to make the grade and meet the code, one way to do it is by producing energy on the premises.

The mandated energy cap will not depend on or assume that. However, a wind turbine on top of a proposed skyscraper, akin to that at the Pearl River Tower, may offset some of the energy needs, that is, after deducting the seventy-five years' worth of embodied energy of the combined replacement parts of the wind turbine.

As technology and engineering continuously improve, the construction standard bar will rise ever higher.

The vast majority of houses are by definition in existence. Hence, any dent made in the existing market segment would have a significant impact on energy consumption. The code would require people to upgrade their house insulation or appliances. The homeowner would be mandated to introduce improvements, which would pay themselves off. In the long run, they will cost less than the existing energy bills. And the government will make the loan as described earlier.

7
ENERGY

≫≍

lower-impact technologies

TRANSCONTINENTAL GRIDS

NOT EVERY POPULATION CENTER in the world has the needed energy
resources in its vicinity. Accounting for this, the plan calls for the
interconnection of myriad sources of energy with population centers
that are hundreds or thousands of kilometers apart.

Specifically, fourteen transcontinental energy regions would be
established and cover most of the land area of the world, and more
to the point, would service over 95 percent of the world's population.

The backbone of each self-contained, transcontinental region is to
be a grid of ultra-high-power lines. Hence, the sunny Syrian Desert
will supply solar power that would serve people throughout the immediate vicinity and also all the way up to the northern reaches of its
designated region, at the Baltic Sea. Wind turbine installations in the
windy Great Plains of North America will provide power reaching all
the way to the population centers of the East Coast.

Every one of the transcontinental regions has its own specific needs
and its own distribution pattern of natural resources. As a case in point,
I will cover in detail one region: the North American region, which
encompasses the contiguous United States and the population centers
of Canada. Henceforth, the term "North America" refers not to the
continent of North America but to the energy region as delineated

153

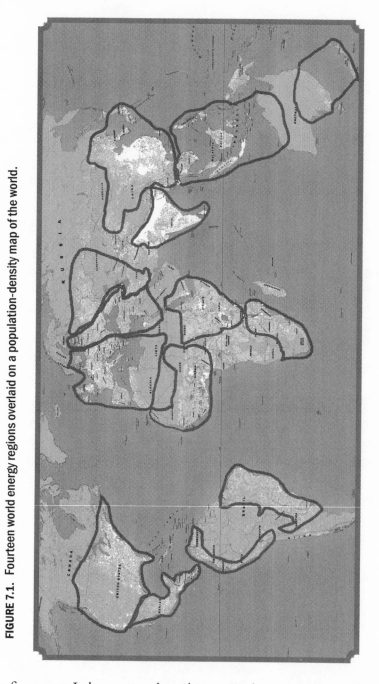

FIGURE 7.1. Fourteen world energy regions overlaid on a population-density map of the world.

in figure 7.1. I chose to analyze this particular energy region because it has the most comprehensive statistics available.

The delineation of the planet to fourteen regions reflects population distribution on one hand and the distribution of wind and sun resources on the other. The delineations also take into consideration the existence of large mountain ranges, which may bar long-distance transmission lines from passing through them.

The small population segments outside those energy regions would be off the grid—as most of them are at present. More on that later.

Before we get down to the specifics, a few basic things about electricity need to be explained.

A power-generation plant has two figures useful for our purposes: its capacity and its yield ("capacity factor"). The capacity is like the diameter of a pipeline. The larger the diameter is, the more water—that is, electrical current—can potentially pass through it at any given time. How often and how much water is flowing in the pipeline is another matter. A 100% yield ("capacity factor") means that the pipe provides water at its full capacity, day-in and day-out. No electricity-generation technology is quite there. Nuclear is the closest, with an average yield of 90%. In contrast, with solar photovoltaic power, the metaphorical water flows only during daytime, when the sun is overhead. It dries up to none during the evening and throughout the night. Hence, the power output of a photovoltaic array is intermittent, and its yield is low, around 30%.

Capacity, the rate at which energy can be delivered, is measured in watts. The total, actual energy produced or consumed is measured with watt-hours—the very same thing that joules, Btu, and calories measure. I chose to use the watt-hour unit when discussing electrical energy.

TABLE 7.1

order of magnitudes, energy amounts
1,000 watt-hour (Wh) = 1 kilowatt-hour (kWh)
1,000 kilowatt-hour (kWh) = 1 megawatt-hour (MWh)
1,000 megawatt-hour (MWh) = 1 gigawatt-hour (GWh)
1,000 gigawatt-hour (GWh) = 1 terawatt-hour (TWh)

What matters is the pattern of the yield: how often, how predictable, how much. Ergo, what matters is the rate and frequency of power, and how dependable its yield pattern is in a given twenty-four-hour period and throughout the year.

Now we are ready to get down to business.

As stated, the plan calls for fourteen transcontinental electric grids. Each energy grid would cover thousands of square miles, connecting remote power plants from various sources to the myriad population centers within a given region. This is very different from the existing setup, where the power sources are on average no farther than 25–50 miles from every household and destination.[1]

Under this plan, a grid of underground superconducting direct-current (DC) cables would connect the power stations with the consumers. Superconducting cables are not your garden-variety transmission lines. They have an attribute that makes the whole scheme of transcontinental grids possible.

The passage of an electric-current in conventional conductors, made of aluminum or copper, incurs heat and energy-loss, as the electrons responsible for the current continuously collide with the atoms of the conductive metal and thus lose energy. Now, something interesting happens when you cool a conductive material: it begins to lose its resistance. Better yet, once the material is cooled sufficiently, the resistance abruptly drops to zero, as the flowing electrons move in an orderly fashion and do not collide with the atoms of the conductive metal.

This is a game changer. This makes it viable to generate power in the Chihuahua Desert in southern Texas to serve customers in Toronto, two thousand miles away—provided of course that the power line is kept duly chilled along the way.

Most conductive materials need to get close to absolute zero[2] before they manifest zero electrical-resistance. Alas, it is virtually impossible to have a power line kept at close to absolute zero temperatures. However, some conductive materials have zero-resistance at the relatively warmer temperature of 70 Kelvin.[3] The foremost conductive materials to have this property are two chemical compounds: yttrium barium copper oxide (YBCO) and bismuth strontium calcium copper oxide (Bi-2223).[4]

This is very good news, as at 70 Kelvin, nitrogen is liquid under a normal atmospheric pressure. Liquid nitrogen is a mature, readily available technology, which can keep duly chilled the superconductive cable made of either of these two compounds. A refrigerator unit is to be placed every 5 to 10 kilometers to keep the liquid nitrogen at temperatures between 66 to 70 Kelvin. Along the refrigerant units, a series of pumps will keep the nitrogen flowing. Cryogenic, vacuum, and refrigeration systems capable of meeting the capacity requirements of long DC cables are in existence.

The technology of superconducting DC cables makes possible a low-voltage transmission, as there is no need to compensate for transmission losses. This in turn makes it possible to incorporate a second important piece of technology: voltage source converters (VSC). If superconductor cables in this plan are the highways, the voltage source converters are the on-ramps and off-ramps. They allow some power to come off the main line and feed local, AC (alternating current) transmission lines. Furthermore, with the VSCs this can be done in precisely controlled amounts. These converters would be placed along the cable route and let us feed the line from myriad power plants along the way.

Another important attribute of the VSC technology is its ability to reverse the flow of power. If the superconductor power line is cut, the grid operators can reverse flow from that point to service the entire circuit minus the very junction that is faulty. The loop design of the grid (shown below) will assure that even if a point in the loop is cut or otherwise malfunctions, the current will just flow the other way and provide power for the entire length of the loop minus the very point the power is severed. The VSC technology has been in operation since 1999 and is considered mature enough for a full-scale deployment.[5]

A report by the Electric Power Research Institute (EPRI) concluded that a large grid of superconducting DC cables is practical and ready using today's technology.[6] Another study finds that such cables, at 10–15 GW capacity, are indeed achievable but do require significant design-optimization and equipment development.[7] At present, our manufacturing and engineering facilities of superconducting cables are Mickey Mouse class. Our claim to fame is a measly 660-meter-long superconducting AC cable at the Holbrook substation in Long Island, New York, which has been in service since 2008. In addition to the 660 meters we already have, we will need 182,000,000 meters of superconducting cable for North America.

FIGURE 7.2. North America with the called-for network of primary transmission lines.

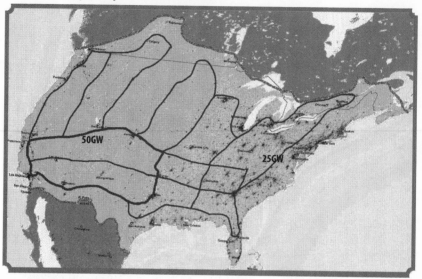

It is not anything I would have put on the table if there was another, more mature technology that fits the bill. There isn't.

Well, there is an option B: high-voltage overhead lines carrying direct current (HVDC). However, it is not exactly a mature technology, either. Furthermore, it has an ugly and a considerable footprint in the form of giant pylons that, with projected power needs, would require over one hundred-meter-wide corridors. Even at their highest commercially-available voltage, the direct-current overhead lines lose considerably more of the transmitted power along the way.[8] Furthermore, having superconducting DC cables would also eliminate the accompanying risks of possible damage due to ice, snow, lightning, and tornadoes. When the power is all underground, one doesn't have to worry that a tornado will knock out the power line supplying power for tens of millions of people. With underground superconducting DC cables there is no electromagnetic radiation to speak of, either.

There is some toll-road fee for getting the power on the main DC power line of such a grid and transferring it back to AC as it is siphoned off at some locality. It also takes some power to keep the nitrogen chilled and moving. Not too much, though. Whenever some electric current leaves the highway via a VSC off-ramp to connect with a local AC grid, there is a 1.5 percent toll on that energy. In addition, we would lose 20 MWh per 1,000 km of cable, due to the power required for refrigeration along the way. However, this is 0.4 percent of the anticipated power per 1,000 km of cable (i.e., 20 MWh out of the transmitted 5,000 MWh).[9]

Fancy a superconducting tape. The heart and soul of the tape is a superconducting material, 1 micron thick, sandwiched by layers of steel, silver, and copper. For one meter of superconducting cable, there is a need to wind within it 115 meters worth of this superconducting tape. The tape is wound around a hollow core in which super-chilled nitrogen is flowing. Around the tape is a shielding copper, more cooling nitrogen, a layer of insulating vacuum, and a layer of insulation. In all, about a 20-cm-thick cable to support 1-micron-thick superconducting material that makes the whole thing work.[10] (See figure 7.3.)

FIGURE 7.3. A cross section of a superconducting cable.

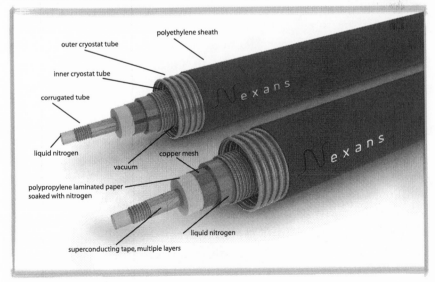

Source: courtesy of Nexans, Inc.

The cable would come in 100–500 meter long sleeves, in effect self-contained segments. Thus, if a cable is punctured and there is a vacuum breach, only a given segment would be affected—not a 3,000-kilometer cable stretch. In such an eventuality, the segment can be replaced with another or be repaired on site. There is no need for a trench; it is possible to drill horizontally underground and pull the cable through, leaving the surface undisturbed. It is to be a bipolar system made of a pair of cables laying side by side, about 0.5 meters apart. In some nodes of the grid, two bipolar pairs would be required. In other nodes, as many as five pairs would be needed. Each pair of cables would be set 10 to 50 meters apart to reduce the chance that more than one set of cables would be damaged in any given instance. All considered, I will assume each cable within the pair has a 5 GW capacity, which is currently feasible.[11]

We have the basic technological building blocks for the above scenario engineered out in recent years; we have plenty of people that can be trained; we have all the needed raw materials; and we have about fifteen years to lay down the most sophisticated and far-reaching grid ever devised.

ENERGY NEEDS OF NORTH AMERICA

Come 2027, how much juice would North America need every hour of the year?

For arriving at the electricity needs of 2027, I used as a basis the energy projection for that year done by the US Energy Information Administration in their Annual Energy Outlook, using the National Energy Modeling System. I assume that, come 2027, technology and products offered to the public will only be the energy-efficient variety. Therefore, I used the agency's Best Technology Case projection, which assumes exactly that. This projection also accounts for the anticipated population growth. While this plan assumes a decline in population numbers (see the twelfth chapter ["Consumption"]), I went along with the population increase assumed by the US Energy Information Administration all the same.

The projected energy needs for the year 2027 in North America depart in various ways from the Best Technology Case to better reflect the assumed improvement in housing technology and overall energy restructuring indicated in other chapters.

What follows is a series of tables.

TABLE 7.2

energy needs, residential, USA	2008	2027
space heating & cooling	4,465 trillion Btu	3,000 trillion Btu
water heating[a]	1,440 trillion Btu	525 trillion Btu
cooking	250 trillion Btu	296 trillion Btu
drying clothes	70 trillion Btu	83 trillion Btu
other needs	427 trillion Btu	506 trillion Btu
electricity	1,380 TWh	1,222 TWh
Total	...	1,779 TWh

[a]Reduction of 25% due to solar water heaters on rooftops.

TABLE 7.3

energy needs, commercial, USA	2008	2027
space heating	1,797 trillion Btu	1,400 trillion Btu
water heating	460 trillion Btu	409 trillion Btu
cooling	30 trillion Btu	36 trillion Btu
cooking	170 trillion Btu	201 trillion Btu
other uses	1,291 trillion Btu	1,530 trillion Btu
other fuels	290 trillion Btu	344 trillion Btu
electricity	1,442 TWh	1,454 TWh
Total	...	2,027 TWh

In both residential and commercial sectors, the big change is the electrification of all the heating devices. First I extrapolated future heating needs, and then I computed their anticipated electrical equivalents. One watt-hour is the same as 3.41 Btu.

TABLE 7.4.

energy needs, industrial, USA	2008	2027
electricity	1,114 TWh	1,159 TWh
fossil fuel (excluding nonfuel feedstock)	8,339 trillion Btu	8,650 trillion Btu
Total	...	3,186 TWh

Note: assuming gas heating process would be in 2026 at 80% efficency average, while resistence heating is at 100%. And so it is a 20% reduction of 2026 projected Btu figures for non electrical usage.

From all the sectors, the projection of the industrial sector is on the most shaky ground. The called-for technological and economic changes are so sweeping that it is hard to project the resultant power needs. Entire industries, such as those related to fossil fuel, would go by the wayside, while new industries, such as those related to new forms of energy and to recycling, would come online. In the end, I suspect that the actual power consumption would be less than suggested here due to a marked decrease in consumption in the private sector. More on this in the twelfth chapter ("Consumption").

TABLE 7.5

energy needs, transportation, Canada and USA	2027
passenger cars	333 TWh
motorcycles	1 TWh
buses	35 TWh
vans, pickups, SUVs	557 TWh
passenger trains[a]	204 TWh
freight trains[b]	121 TWh
trucks, class 4 and 5	86 TWh
Total	1,337 TWh

Notes: I assume that passenger cars require 0.19 kWh for 1 km of driving. Motorcycles require 0.04 kWh for 1 km of driving. Buses require 0.55 kWh for 1 km of driving. Vans and SUVs require 0.34 kWh for 1 km of driving. Trucks class 4 and 5 require 0.57 kWh for 1 km of driving. I assume and factor in a 12% loss of electricity en route to the battery. For passenger trains, I assume 0.08 kWh per passenger km. For freight trains, I assume 0.07 kWh per ton per kilometer. For power requirements of trains, the source is Matthew Wright and Patrick Hearps, *Australian Sustainable Energy: Zero Carbon Australia Stationary Energy Plan* (The University of Melbourne Energy Research Institute: Beyond Zero Emissions, 2010), 135.

[a]With the elimination of domestic flights and with the introduction of high-speed bullet trains I assume that in the United States, numbers will go from about 583 billion passenger miles to 1,492 billion passenger miles.

[b]Key assumptions: 38% of the existing freight traffic would be eliminated as it is currently used in the freight of coal.

For the above transportation table, there is no 2008 equivalent, as the motor vehicle fleet of 2008 is largely running on fuel combustion, not on electricity.

As a basis, I use existing numbers of total vehicle-miles[12] for both urban and rural roads adjusted to projected population growth. I make further adjustments to the various vehicle-miles categories based on the significant transportation changes described in the fifth chapter ("Transportation").

Table 7.5 can be split into two consumption categories: those vehicles that feed in real-time (rail-based, such as trains), and those that have batteries (e.g., cars and buses), which can be charged at off-peak times. This is a very important distinction when we come to ascertain the total power generation needs of the region.

TABLE 7.6

total hydrogen needs, USA + Canada	2027
boats	3,642,000 tons per year
trucks	14,204,000 tons per year
annual hydrogen / electricity to generate it	17,846,000.tons / 892,000 GWh
daily hydrogen / electricity to generate it	48,890 tons / 2,440 GWh

Regarding boats, I assume that marine fuel in the United States comes to about 7.4 billion gallons annually. Due to equipping some boats with a kite, I shaved 22.5% of the energy needs. I conservatively deducted 10% of the portion taken up by nuclear propulsion. Then I added the Canadian portion (6.6%) and computed the total needed fuel in tons of hydrogen. In the context of boats, the key assumption is that 1,000 gallons of diesel fuel is the energy equivalent of 609 kilograms of hydrogen. The total hydrogen requirement for all boats comes to about 3.64 million tons of hydrogen.

When it comes to heavy trucks that are to run on hydrogen, the pertinent assumptions are these: In 2009, heavy trucks in the United States required about 25.5 billion gallons of fuel. One gallon of diesel averages 4.8 highway miles for a class-8 truck, while one kilogram of hydrogen averages 6 highway miles. This is in effect the key for conversion of diesel fuel to hydrogen (1.25 gallons of gasoline is the energy equivalent, in this context, to 1 kilogram of hydrogen). I made a further adjustment to account for anticipated changes in transportation patterns and the anticipated introduction of fuel-saving measures. Finally, I added the Canadian portion and accounted for the energy required for transport of the hydrogen (1%) to the truck stops. It came to about 14.2 million tons of hydrogen. Assuming it takes 50 MWh to produce and duly compress one ton of hydrogen, it means that the total required electricity to produce all needed hydrogen is about 892 TWh.

However, all of it will come from excess electrical generation and won't need to be budgeted for in the generation capacity or grid demand. More on this later in the chapter.

TABLE 7.7

summary, projected power needs for the year 2027 for North America	
commercial, industrial, and residential sectors of USA	6,992 TWh
with the addition of the Canada portion[a]	7,910 TWh
with the addition of transportation sector of USA and Canada	9,247 TWh
minus Canada's and USA's outlying areas (i.e., Hawaii, Alaska, northern Canada)	9,173 TWh
added electricity to compensate for transmission losses[b]	13% of consumption
Grand Total for North America	**10,365 TWh**

[a]With the existing energy usage of Canada, I assume that the Canadian portion is 13.1% of the U.S. one.

[b]Largely due to local, existing AC transmission losses.

The projection is that we will need a total of 10,365 TWh in North America for the year 2027. This is well over twice our current consumption of electricity. This is to be expected, as we electrify transportation, heating, and practically everything else that is currently using combustion as a source of power. This also assumes that we will have a population larger than that of the present time.

In coming to determine our electric generation needs, the annual total (10,365 TWh) is of little use. What matters is how much we need at any given time. Had the required total (10,365 TWh) been spread evenly throughout all the 8,760 hours of the year, it would come to 1,162 GWh per hour. However, power consumption is never spread evenly. The hourly demand varies, depending on the time of the day, the day of the week, and the season.

While the volume may change, the pattern in western countries is fairly standard: evenings are busier than mid-days, and the hours after midnight are the least busy.

As there are no national hourly statistics for the United States, I have used as a basis the national hourly pattern of Ireland.

I took the existing consumption patterns and scaled them up so that in total they would come to 10,365 TWh. Once the total for the year matched that of our projection, I took note of the derived hourly usage values. As it turns out, demand on the North America grid would range from 600 GWh in low, off-peak hours to 2,000 GWh for the busiest hours of the year. (See figure 7.4.)

Therefore, whatever generation capacity we are to establish, it has to provide for this usage pattern, which ranges from 600 to 2,000 GWh. And on top of that, we need to factor in the unexpected.

As there is a considerable amount of flexibility around what battery is to be charged at what time, I scheduled the charging of batteries in the hours in which the grid has a lot of power with no takers. Basically, this lets us run the entire fleet of public and private battery-operated vehicles without the need for any additional generation capacity—about 1,012 TWh worth of energy. Thus, in truth, I scaled up the Ireland power consumption power to 9,353 TWh and than tacked the 1,012 TWh in key hours of each day.

FIGURE 7.4

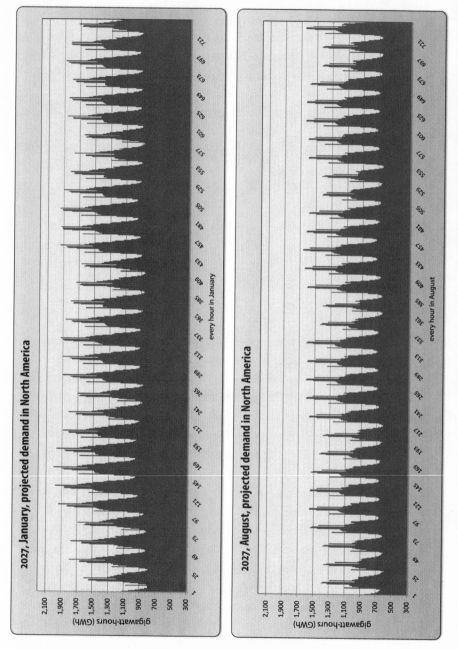

FIGURE 7.5

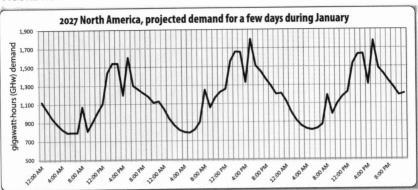

This very deliberate battery charging schedule accounts for the unusual spikes in demand in the daily-demands graph shown in figure 7.5. This is all behind-the-scenes, hands-off operation; the vehicle is plugged in for the night or otherwise for a certain amount of hours and the user does not need to get involved in the how and when. All they need to know or care about is that come morning or the end of the workday, the car is fully charged.

Now that we have arrived at an estimated power demand of North America, it is time to see how we are going to meet it. Let's start with the technologies that currently provide the vast majority of power in the region.

EXISTING POWER-GENERATION TECHNOLOGIES

At present, belching coal and natural-gas power plants generate the vast majority of electricity in North America. By 2027, all of these power plants would be shut down, get decommissioned, and perhaps be converted to memorials or penitentiaries. They already look the part.

Joining them would be the concrete walls bottling up the rivers; they are to be torn down.

The need to do away with coal and natural-gas power plants is clear as a bell in the context of climate change. The tearing down of dams requires some explanation.

Rivers are the veins and arteries of the planet. They carry off toxins and provide nourishment. Dams clog those arteries.

There are good reasons to be done with dams.[13] Their removal has shown to promote the rehabilitation of native species[14] with an overall upturn in species diversity.[15] Dam removal would restore low-flow periods of the river. This in turn would promote vegetation growth, which improves the spawning habitat for fish in the area.[16] A reservoir may separate into several layers of water with varying temperatures due to increase in water depth and decrease in flow velocity created by a dam. Dam removal can restore a river's natural water temperature range.[17] Sediment transport to the river system could also resume. The largest and coarser sediments facilitate population growth of some native fish species.[18] Gravel or stones that were previously covered under fine sediments would be re-exposed and provide new colonization habitats for aquatic insects and revitalized spawning habitats for fish. Lastly, reproduction success, which often depends on appropriate timing for reaching spawning or breeding habitats, would be improved by the removal of dams that prevent the migration of aquatic organisms.

In conclusion, natural gas would go; coal would go; dams would go. Nuclear would stay, though.

At the moment, there are 122 working nuclear plants in North America. With the additional 10.3 gigawatt capacity projected to come online by 2020, we can assume a total generation of 968,800 GWh from nuclear plants per year. More to the point, we can count on a fairly steady 110 GWh output each hour.

Providing but 8 percent of the projected needs, the existing nuclear plants won't make too much of a dent. But these workhorses are virtually emissions-free and have been reliably chugging along year-in, year-out; decade-in, decade-out. Waste not, want not. This excludes any nuclear plant that would not meet stringent safety standards—whether due to a potentially hazardous location, lack of a containment building, or the such.

We even have enough uranium to keep them running. The OECD Nuclear Energy Agency estimates total global uranium identified resources to be 6.3 million tons. The current annual worldwide output of all nuclear plants is about 2,731 TWh. Combined, the world's fleet of nuclear plants consumes 68,000 tons of uranium annually.

We would have finished the entire known resources of uranium in less than ten years had we run the world entirely with existing nuclear technology.[19] Nasty, but the plan does not call on nuclear plants beyond those in existence or under construction. Current uranium reserves will supply the existing nuclear plants for fifty years with a lot left over, which suits the plan. For the next fifty years, about half of the uranium will run the nuclear plants; a big chunk of the rest will fuel the larger marine vessels as discussed in the fifth chapter ("Transportation"); and the rest will be allocated to micro nuclear plants in the sparse regions outside the fourteen blocks where the sun and wind won't quite suffice—much as the pending proposal of a micro nuclear plant in Galina, Alaska. More on this later in the chapter.

After fifty years, it is highly likely we would find more uranium. But as will be explained toward the end of this chapter, whether we find more uranium or not may matter little.

WIND TURBINES

Almost always, electric power is generated by spinning a turbine. In the case of wind power technology, the wind does the spinning. The power-generation process of wind turbines is as straightforward as it gets.

A computer-controlled motor positions the three gigantic blades of a modern wind-turbine into the wind. The airfoil shape of the blades causes uneven air pressure, which in turn prompts the blades to rotate around the center of the turbine. The rotating blades spin an attached shaft. The shaft moves a series of gears at a greatly amplified speed, and these gears turn an electric generator. Voilà: electricity is produced. Alternatively, some manufacturers (e.g., Enercon) offer wind turbines with a gearless, direct-drive mechanism. Fewer rotating parts reduce mechanical stress and at the same time lengthen the service lifetime of the wind turbine.

Generally, a turbine will start producing power in 8-mph winds ($\approx$ 3.5 m/s) and reach maximum power output at about 27-mph winds ($\approx$ 12 m/s). Windier conditions won't increase electrical output. At around 56-mph winds ($\approx$ 25 m/s), the turbine calls it a day and shuts down until the strong gales subside. It can be described thus: the turbine starts generating electricity under gentle breezes; peak performance is reached and plateaus at strong-breeze conditions; and the wind turbine has to shut down under a full gale, that is, storm-force winds. A dozen sensors help the controlling software to regulate the power output and rotor speed in order to prevent overloading the structural components of the wind turbine. And in overly high winds, the controller software rotates the blades out of the wind's way.

A modern wind turbine is *large*: just the tower that the turbine is mounted on can reach 100 meters (≈ 330 feet). The height of the tower is not in order for the turbine to avail itself of a better view but to access the more steady and forceful winds of higher altitudes. The blades of a wind turbine are impressive in their own right; each may be 50 meters in length (≈ 165 feet). They are made of advanced fiberglass composites possibly reinforced with carbon. The tower itself is constructed of steel.

Fancy a 100-meter-tall steel pole with three massive blades rotating by wind power and getting a turbine to spin and generate electricity along the way. Fancy hundreds of thousands of such hulks next to each other. Had such an array been installed at sea along the shoreline, it would have formed what appears from shore to be an unbroken wall that irrevocably would alter our experience of the open sea. This cluster of wind turbines would be somewhat reminiscent of the towering fence structure in the movie *King Kong*.

Wind turbines widely interspersed throughout the vast farmland regions of the world would aesthetically be a different proposition. We can do farmlands. The only question is where ample winds are to be found.

There are good to excellent wind resources in some regions: in the Great Plains (North America), Patagonia (South America), the Horn of Africa, Scotland, southern Morocco, Gobi (northern China), Norway, and northern Chad.

Averaging data from dozens of modern wind farms, the permanent footprint of a wind turbine comes to 4,300 square meters (≈ one acre). About 80 percent of that is taken up by access roads, and most of the rest comes from the substation and the concrete foundation for the pole. In addition to the permanent footprint, temporary disturbances seem to average either around 29,000 square meters ($\approx$ 7 acres) or around 9,700 square meters ($\approx$ 2.3 acres) per wind turbine. This depends on the terrain and possibly on the design of the wind farm. The temporary footprint is associated with the wind-farm construction (e.g., temporary roads and staging).

Smaller turbines, such as 1 MW (megawatt) wind turbines, don't give one all the possible bang for the buck. Jumbo turbines like the 7.5 MW Enercon E-126 are too big for some sites. And within reason, the more turbines on the grid, the more their combined yield smoothes the erratic energy yields of the individual turbines. There are considerations either way for what particular size of turbine is to be used for a given location. For the purpose of arriving at projected estimates, I have relied upon a one-size-fits-all model: a mid-range turbine such as the ev100 model by Eviag with a 100-meter-high pole and 2.5 MW rated capacity.

After studying dozens of modern wind-farm configurations, I arrived at an average distribution of 1.85 wind turbines per square km for 2.5 MW wind turbines.[20] Based on the performance of the Eviag's ev100, it appears that a 2.5 MW turbine has a 42% capacity factor at 7.5 meters-per-second winds, and 50% capacity at 8.5 meters-per-second winds. For the North American region, the deployment of the wind turbines is to take place in the Great Plains, where wind speeds at 100 meter height typically range from 7.5 to 8.5 meters per second. Hence, I assume an average generation-output at 45% capacity. This means that a square kilometer, averaging 1.85 wind turbines, would yield 18.2 GWh annually (while each individual turbine yields around 9.9 GWh annually).

If the total annual electrical demands in North America would be 10,365 TWh, as projected, there would be a need for just over one million 2.5 GW capacity wind-turbines in the Great Plains, which, combined, would produce an average of 1,183 GWh each hour.

Unfortunately, an average is not cutting it.

Given the intermittent and highly variable nature of wind, the output pattern of a single wind turbine is just plain awful. Unfortunately, a field of wind turbines is better than a single wind turbine only in the sense that eight consecutive life sentences for a criminal are better than twenty consecutive life sentences. The best to be had is a transcontinental array of wind farms spanning thousands of miles, where the weather pattern of one area is independent and at times negates those of other areas. An array of hundreds of thousands of wind turbines scattered over vast distances and all feeding the power grid would smooth out some of the fluctuations.

To create a simulation of wind-turbine deployment in North America, I have used actual data of the entire array of wind farms in Denmark and that of Ireland and that of Australia for every hour of the year and then created a unified, harmonized composite of their total electrical output.

FIGURE 7.6

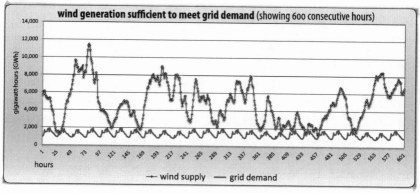

Studying the simulation in figure 7.6, the kindest thing that can be said is that the wind *always* blows somewhere. But otherwise, to provide sufficient energy at every hour of the year requires what for most of the time proves to be an incredible amount of excess generation-capacity. Many times the generated electricity could exceed the needs of the moment by a factor of three. However, a chain is as strong only as the weakest link. In this case, a grid comprised solely of wind turbines is only as viable as the lowest output it is likely to generate at any one point throughout the year.

In concrete terms, it means that to run North America solely on wind would require 3.6 million large wind turbines. This figure is based on the simulation above. At 1.85 turbines per square kilometer, this array will be deployed on over 1.9 million square kilometers—an area about the size of Mexico. Yes, at times the combined output of this array could reach staggering output levels of 7,000 to 8,000 GWh hourly, but far less than that can really be counted on.

Unfortunately, even if we were willing to commit to a 100% wind penetration scenario, it should not be done. And there are two very good reasons.

First, it is alleged that a few decades ago there was one week in which the wind did not blow in the United States at all, not enough to turn a turbine blade, at any rate. One does not design a grid with the prospect of shutting it down for a week—not even once every few decades. This entire setup is worse than useless to us if the wind does not blow for one straight week. We cannot shut down North America for seven days and make up for it the following week with some excess electricity.

Second, it seems that operating a lot of wind turbines is not all that good for the planet.

Wind turbine deployment at those levels would entail extraction of a significant portion of the energy embodied in winds. Millions of giant turbines collectively soaking up a big portion of the winds of our world, debilitating them, would have real repercussions on weather patterns, and none for the better.

A modeling study calculated the effect of covering the Midwest with an array of wind farms containing millions of wind turbines. On average, the study found that wind speeds were lowered by 5.5–6.7 miles per hour immediately downwind of the giant machines. More significantly, the wind turbines caused large-scale disruptions of air currents, which rippled out like waves leading to substantial changes in the strength, motion, and timing of storms over the entire North Atlantic. And indeed, scientists have noted considerable warming in west-central Texas—the home of massive wind farm installations.[21]

Much is unknown, but what is suggested in the study above urges prudence. The plan aims to stay far below the magnitude of the wind-turbine array modeled in the study. Making a judgment call, I put the cap on wind turbines installed in North America at the equivalent of 250,000 turbines, 2.5 MW capacity each. In other words, a combined total capacity is to be capped at 625,000 MW.

Another problem is that wind turbines have been killing bats in unprecedented numbers.[22] And as the bats die, the population of pest insects is likely to rise.[23] It is bad no matter how we care to look at it. Yet, I reckon there is more to gain than to lose if we keep the wind turbines. Incidentally, the weight of evidence suggests no association between noise from wind turbines and psychological distress or mental health problems in humans.[24]

As of 2011, the total installed capacity of wind turbines in the United States was 42,432 MW.

As it turns out, there is a creative way to put to good use the energy generated by the whimsical wind turbines, but we first have to introduce and set in place the other members of the energy-generation team for the year 2027.

SOLAR POWER TOWERS

Without a doubt, solar power towers are to be the backbone of the entire power-generation scheme. They have the incredibly useful capability to take in the sun's rays and provide energy at noon on an August summer day and also in the middle of a frigid night in January.

There are other concentrated solar power technologies; however, I deem the solar power tower to be the most suitable.[25] The solar power tower is a fairly recent technology, and it comes in a few flavors. At the moment, the only commercial plant that has the technology suitable for our needs is the Gemasolar plant in Spain, which became operational in 2011. However, the Gemasolar is a pint-sized installation. Furthermore, it is optimized and designed to have but a secondary role in the local grid. A few larger installations, more in line with what the plan calls for, such as the Rice Solar Energy Project and Ivanpah Solar Power Facility, have been announced or are under construction.

The mechanics of a power tower are straightforward enough.

Thousands of tracking, moveable mirrors—called heliostats—follow the sun's path throughout the day. In tandem, they reflect the sun's rays, directing them to a bank of tubes located on top of a central receiver tower at the heart of the installation. The tubes contain molten salt, which the converging rays of the sun heat up to a sizzling 565°C (≈1,050°F). Subsequently, the molten salt flows down into a storage tank. Later, the heat embodied in the salt is used to generate steam and electricity in the traditional fashion. However, in the interim, the tank stores the molten salt until it is time to generate electricity. This is a big deal. Essentially, this decouples power generation from the capture of solar energy. This makes it possible to have power on demand, both when the sun is shining and when it is not.

When electricity is to be generated by the solar power tower, the superhot salt is routed from the storage tank to heat exchangers. The resultant steam is then used to generate electricity in a conventional steam turbine cycle that is found in coal or natural gas power stations.[26] The heat energy extracted from the molten salt in the exchanger brings it down to 290°C (≈555°F), a temperature at which the salt still remains molten. After exiting the steam generation system, the cooler molten-salt is routed to a second insulated storage tank where it waits. When it is needed, the salt goes up the tower via pipes for reheating to blistering temperatures again.

The solar power tower is to have dual salt storage units that together provide up to 17 hours of storage, 17 hours of reserve power. The salt used in a solar power tower is a mix of 60% sodium nitrate with 40% potassium nitrate. These minerals are abundant. Nitrate salts are made by the oxidation of ammonia, while sodium and potassium are very common components of the Earth's crust. This is today. Halotechnics has discovered and is developing glass material that can operate at much higher temperatures (1,200°C), which would allow the turbines to operate at higher efficiencies and thus require fewer mirrors to produce the same yield. That may translate to 15%–20% less overall footprint for the solar power towers than reckoned for here.

The molten salt can be kept in reserve to be used as needed in a molten state for at least one week before it would inch down to dangerously low temperatures and turn solid within the pipes. In the liquid state, salt has a viscosity and appearance similar to that of water and has several highly beneficial properties in solar power applications. First, liquid salt has highly efficient heat transfer properties, and it retains heat for long periods with minimal losses. Second, the salt can be heated to high temperatures without any degradation, resulting in efficient energy storage and electricity production systems.

The following specifications scale up the tower technology as far as it can go without reaching diminishing returns.

Each heliostat is to be comprised of a multitude of mirror panels that are laid out together, forming one large 12.2 × 12.2 meter reflective expanse (148 m²).[27] This is about half the size of a standard movie theater screen. Each 6-ton heliostat unit would be mounted on a steel pedestal which in turn would be anchored down to a concrete foundation.

Each installation would have about 18,500 such heliostat units, arranged in 10 radial zones. The farthest zone is about 2 kilometers away from the central tower. The total power plant area is 1,340 hectare (≈ 3,300 acres), spanning about 4 kilometers from edge to edge.

The central receiver tower within each installation is to be a steel lattice tower,[28] much as the Eiffel Tower is. And as it happens, with a required 327 meters in height (≈ 1,070 ft.), the receiver tower would also be about the same height as the Eiffel. The height ensures that the reflected energy from the heliostats at the outermost edge of the solar array will have enough of an angle to reach the receiver on top of the tower.

The turbine is to have a gross capacity of 135 MW.

In order for the solar installation to provide power around the clock, we will need two fully independent facilities working in concert. So everything I have just described is to be multiplied by two. One facility would be tuned up and configured to generate energy during the night, the other would take care of the day—thus ensuring twenty-four hours of continuous power supply. To my knowledge, the idea of two solar towers working thus in tandem is novel.

Using the reserves of molten salt, the charge of the nighttime installation would be to provide juice during the night time, with a distant second goal of producing energy during the day. The charge of the daytime installation is the reverse. It kicks into a high gear during the hours that the night installation is at low ebb. Together, the two installations complement each other. Together, they provide continuous power.

During the summer months, the tower installations would generate enough energy to come out of our collective ears, day and night.

However, during the winter months, there is a need to control the amount of molten salt that is released in any given hour of the night, when the sun is not supplying the system with any additional source of heat.

It requires a creative and stringent regimen to coax electricity in the cold months from the relatively limited supply of pitch-hot molten salt that is in storage once the sun sets. Too much released at any one time will not leave molten salt in sufficient amounts for subsequent hours of the night. It is a balancing act.

Both the nighttime and the daytime installations aim to provide a steady generation of energy during the night hours and the day hours, respectively. In the night installation, you release but scant quantities of molten salt during the day, just enough to provide for internal operation requirements of the plant. You open up the machines in earnest in the late afternoon hours. Every month of the year calls for a somewhat different 24-hour schedule of molten-salt release, depending on the amount of sunlight available. In some winter months, a night installation may open the valves all the way around 4:00 p.m. and cut them to 25% output around midnight. Summer months have more play. It is possible to start with valves wide open around 5:00 p.m. and keep at it up until 1:00 a.m., dropping then just one notch, down to 90% output capacity. The amount of pitch-hot salt during the summer months is so considerable that even keeping the valves wide open will still see considerable levels of power output in the pre-dawn hours, when the reserves are at their lowest levels.

If every day, or every second day, was sunny, the above scenario would be sufficient. However, this is not the case. There are many consecutive days that are cloudy, at which time the reserves of salt would run down and with them the power output. To assure a continuous power supply throughout the year irrespective of local weather, there is a need for a network of such tower installations, scattered over many hundreds or thousands of kilometers. It may be rainy for a few days in one location or in two. But it is not cloudy *everywhere* for days on end, at least not in arid or semi-arid regions.

Thus, between many thousands of installations, each with a seventeen-hour reserve of molten salt, we can achieve year-round energy for the entire North American grid. Between weather variability spanning vast distances and the ability to store energy for many days at a time, this network of power tower plants would do the trick. I know, I modeled this.

I ran a simulation on every temperature, cloud coverage, solar radiation, humidity, and wind parameter of every hour of the year in a number of key locations around North America: in all, about sixty-five different meteorological parameters for each hour of the year for each location. I used a meteorological data set that had been carefully chosen to typify the weather in given locations sampled over decades.[29] I optimized the hourly power output for every hour of the day, for every month of the year, and for every one of the five different regions. Then I combined it all into one energy-generation composite and reviewed the resultant hourly energy outcome against anticipated demand for every hour of the year. The damn thing works.

FIGURE 7.7

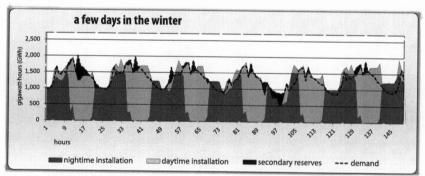

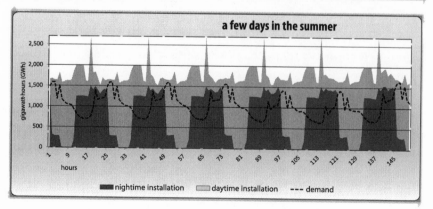

Note that in the winter-days graph (figure 7.7) the generation capacity doesn't quite meet the demand on a number of occasions without a secondary reserve kicking in.

While it is possible to meet demand by scaling up the installations, this is needlessly wasteful. We can pull off a better engineering job than this. The excess power on other days is channeled to a secondary reserve system, one in addition to the 17-hour tank noted earlier. The molten salt in the secondary tank is brought out when needed. The limits of the secondary tank, or rather, the combined limits of the thousands of secondary tanks across the entire grid are straightforward: given the number of required installations, combined, the secondary tanks can hold up to 42,816 GWh worth of molten salt, and their respective molten-salt reserves cannot be older than one week.

Installations in the vicinity of Imperial (CA), Blythe (CA), Daggett (CA), Phoenix (AZ), and El Paso (TX) in the deserts of the Southwest would provide all the power needed for North America during every hour of a given year.

That's the good news.

The bad news is that this would necessitate converting 281,000 square kilometers of pristine desert into a glorified parking lot for mirrors. The basic charter of this plan is that no natural habitat is to be degraded—be it 281,000 sq km or be it 281 sq km. So that's a no go. We don't have 281,000 sq km of already-degraded land in the desert to work with.

A fifteen-hundred page environmental-impact study of a power tower installation in an untouched desert habitat confirms the obvious. The tower and the heliostat field would eliminate or otherwise degrade native vegetation and wildlife habitat and would have both temporary and long-term effects on the land. Vegetation mowing, introduction of shade and added moisture from mirror washing, maintenance activity, and a possible invasion of weedy annuals would eliminate all but the most disturbance-tolerant native species. The installation would adversely affect wildlife and nesting birds due to ground disturbance, ongoing operation, and fencing around the perimeter of the solar generator site.[30]

Luckily the sun also shines outside the desert—albeit it would necessitate more mirrors and more land area to yield the same required power.

We shall overcome.

Numerous, sizeable regions bearing degraded land were tested in the model for suitability. In the end, the five regions that offer the most bang for the buck plus enough climatic variability to smooth out the weather fluctuations are regions in the vicinity of Oklahoma City (OK), Lubbock (TX), Bakersfield (CA), Limon (CO), and Dallas (TX).

This would require a total of 597,745 square kilometers, but it can be done. It is possible to power the entire future energy needs of North America with five mega installations of solar power towers. I do not mean that the combined array would provide enough generation capacity. I mean that it would provide sufficient amounts of electricity to satisfy the needs of the entire grid every hour of the year, come rain or come shine. It is a lot of land but we can accommodate it, as the ninth chapter ("Land Use") will illustrate.

Researchers at MIT, in collaboration with RWTH Aachen University in Germany, have devised a new configuration of the mirror array, along the lines of spirals on a sunflower, that cuts back on 20 percent of the land footprint.[31] This means a revised total area of 478,196 sq km. This is a step in the right direction.

The concentrated-solar power towers are, as stated, the backbone that makes the whole scheme viable; however, some sidekicks would go a long way toward trimming our footprint on land, degraded though the land may be.

Instead of having nighttime and daytime installations, we will take a different approach with the introduction of an additional player. Each power-tower installation would be optimized and geared to produce electricity for all hours of the day—except from around 10:00 a.m. to 5:00 p.m., when the sun is the strongest. That time slot would be taken care of by something that can handle that part of the day with a markedly decreased footprint: solar photovoltaic panels.

PHOTOVOLTAIC PANELS

Photovoltaic (PV) cells convert sunlight into electricity directly.

Unlike a concentrated solar power system, which requires direct sunlight to reflect and in turn generate heat, the PV panels can generate electricity from a diffused source of light. In other words, within reason, they can work under a cloudy sky. This is a big advantage, as the number of days in the year a PV installation will perform is significantly higher than that of a solar power tower installation.

Basically, when light strikes the cell, a certain portion of it is absorbed within the semiconductor material. This means that the energy of the absorbed light is transferred to the semiconductor, such as silicon.[32] The energy knocks electrons loose, enabling them to flow freely.

PV cells also have one or more electric fields that force the freed electrons to flow all in one direction. This flow of electrons is the electric current. Thus, a solar cell consists of two layers of materials, one that absorbs the light and the other that controls the direction of current flow.

By placing metal contacts on the top and bottom of the PV cell, we can draw that current off for external use. Photovoltaic panels are rugged and simple in design. The silicon used in PV cells is purified, melted, and cast into a brick. The bricks are then sliced into thin wafers to make individual cells. The silicon purification process releases potent GHG gases; however, the amounts are insignificant (less than 1 gram of CO_2e per 1 kWh).

Beyond the above, there are thin-film PV technologies, which don't scale up. It is possible to install PV panels on rooftops, which doesn't quite make sense, either. The reasons why are explained in Appendix A, for those who are curious.

A PV installation takes less footprint. In simulating the performance of PV in viable climatic regions, the PV averages an annual output of 68 GWh per square kilometer of installation—whereas the solar power tower averages 44 GWh per square kilometer of installation. This figure factors in the degradation of the PV and averages its performance over a 25 year period. Of course, it is comparing apples to oranges: PV cannot provide nighttime power, but when it comes to the daytime component of the plan, PV is the way to go.

TABLE 7.8

yield per year per one sq km of footprint	
wind turbine[a]	2,290 GWh
PV panels	68 GWh
solar power tower	44 GWh

[a]There is a sharp distinction between footprint and the total area over which wind turbines are being deployed. One sq km worth of net footprint—as concrete foundation and access roads—translates to many sq km over which a few wind turbines are deployed. Unlike solar-based technologies, wind turbines require a lot of space between the individual wind turbines.

In calculating needed area and output, I rely on the current specs of one of the foremost photovoltaic manufacturers, SunPower. Each SunPower single-crystal silicone solar PV panel is about 41″× 82″ (about 105 × 207 cm). A set of nine of those panels is mounted on a steel rack that tracks the sun throughout the day. A cluster of 48 of those racks turns to follow the sun by the action of a single ½ horsepower electric motor. Naturally, every row of racks is duly spaced so as not to cast shadows on neighboring rows.

This is the basic configuration. The technology is modular, and it is just a matter of how many such arrays are required.

A most thorough analysis was done on the environmental impacts of a large-scale photovoltaic installation in the Carrizo Plain, California. The study has concluded that such an installation will result in some permanent and some temporary habitat losses that support numerous animal species.[33] The analysis found other potential impacts from the project including the spread of noxious weeds and the loss of wetlands, foraging habitats, native vegetation, and bird nesting sites. So no natural environment is to be used.

For candidates, we have to look at some degraded open land. Given weather and current land usage, the most suitable regions to host PV farms are in the vicinity of Lubbock (TX), Bakersfield (CA), and Limon (CO). As it so happens, not all desert land is pristine. Notably, the Salton Sea in southern California and the stretch between Phoenix and Tucson are heavily degraded and therefore eligible for PV installations. All told, an estimated 12,000–12,500 sq km of desert land can be used. It is not much, but it adds some extra sunshine-boost on days we need it the most.

By using PV installations, we just shaved off an area the size of the state of Maine from the land we would have needed for power installations. This is 92,294 sq km less than would have been needed without the addition of PV installations. Combined nuclear, solar tower, and PV installations that would meet the grid requirements would take a total area of 385,902 sq km. This is today's technology. Tomorrow may offer us PV cells that are double the efficiency of existing panels—as it is possible to double the number of electrons harvested from one photon of sunlight. This was the groundbreaking discovery of Xiaoyang Zhu and his team at the university of Texas.[34] Another significant potential improvement may prove possible out of the work of Marco Bernardi and his team at MIT. They have modeled three dimensional photovoltaic structures that have both collected far more energy on the same land footprint and also provided much less variability in power generation throughout the day and throughout the various seasons.[35]

And more is possible yet.

WIND AND HYDROGEN

Now it is time to weave into the mix the wind energy discussed earlier. More specifically, 250,000 wind turbines at 2.5 MW capacity scattered about the Great Plains region. They are likely to generate a total of about 2,460 TWh for the year.

Adding the 250,000 wind turbines discussed earlier would reduce the number of needed PV and solar power tower installations—provided unruly wind generation is introduced in tandem with one more technology: hydrogen storage. When the grid could use the energy from the wind farms, it would. Otherwise, the energy would be routed to the generation of hydrogen.

North America's energy demand is projected to be 10,365 TWh a year. In this simulation, the total generated energy from nuclear, hydro, wind, and both forms of solar came to 15,321 TWh, out of which 9,615 TWh was feeding the grid directly. The rest, about 5,706 TWh, was excess—as the generation output in a given hour exceeded the needs of the moment. About 0.4 TWh of the excess was applied toward the secondary molten-salt tanks. However the vast majority is looking for a home, which we are to build.

FIGURE 7.8

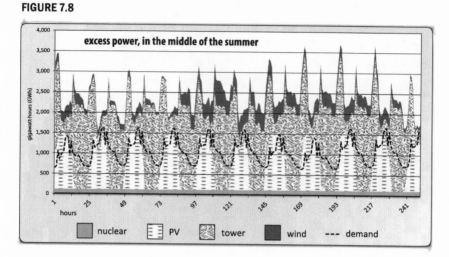

It is not just the energy from wind that is at times in excess. In order to accommodate all of North America's energy needs on wintry days, solar power plants have excess capacity for when the sun is out in force. This means the power installations would generate energy during summer months far in excess of need. The built-in excess cannot be helped. Once again, this is where an additional technology of hydrogen storage would come in handy.

We would put as much as we can of the extra and unruly streams of electric energy into vast storage facilities, after first converting it into hydrogen. Consequently, we would lower the total power installations needed as we can count on energy coming from storage in times when generation of electricity from wind and solar is low. Better yet, this would provide the projected 48,890 tons of hydrogen needed each day to run the entire fleet of hydrogen-powered heavy trucks and boats of North America.

Hydrogen is to be generated, compressed, and then stored in underground caverns: much as they do at the Clemens Terminal in Texas. Later, the hydrogen is to be brought up and made to perform. Here is the plan: A power-generation facility would be built next to each suitable underground cavern. Excess energy from the grid would come to such a facility and be used to generate hydrogen. Subsequently, the hydrogen would be compressed and piped underground. When the combined energy provided to the grid by the solar, wind, and nuclear facilities does not meet demand, hydrogen would be pumped back up and used to generate needed electricity.

There are many ways of generating hydrogen, but only one mature technology that does it without GHG emission: electrolysis. The only input would be electricity and deionized, filtered water. The electrolyzer breaks down water into its constituent elements: oxygen and hydrogen. This is done by passing an electric current between two electrodes submerged in the water. The largest commercially-available electrolyzer is a massive metal hulk produced by Hydrogen Technologies in Norway. It makes 1,000 kilograms of hydrogen in 22.5 hours.

An extensive array of fuel cell and electrolyzer units would make up such a facility. The electrolyzer generates hydrogen from water; and the fuel cell generates electricity from hydrogen and in the process, gets the hydrogen to combine with oxygen, forming water once again.

The price tag for the generation of one ton with due compression comes to 50 MWh worth of electricity. For the generation of electricity, we can use the largest available unit, which is made by Ballard. This massive, stationary unit can go through about 1.5 tons of hydrogen a day. Its yield is 15.9 MWh from each ton of hydrogen, or almost 24 MWh in one day of operation. This fuel cell can run continuously for over two years before its panels need to be sent back to the factory to be refurbished.

The production of hydrogen calls for water. Most of it ought to come from the recapturing of steam, generated when the stored hydrogen is put through a fuel cell to generate electricity.

In the end, this scheme is only as good as the amount of storage space we have to work with. The United States has 324 depleted natural gas fields, with a combined 3.58 trillion cubic feet of working gas storage capacity. Adding some salt caverns to this inventory brings it to a total of 3.81 trillion standard cubic feet.[36] The actual underground space in the caverns is about twice as large, but some gas has to stay in the reservoirs permanently as pressure has to be maintained. It is neither possible nor healthy to try to pump out most of the gas and create a vacuum in an underground cavern.

Canada produces a third of the natural gas that the United States does, and its territory is comparable in size to that of the United States. So I am going to assume it has the same volume for salt caverns and a third of the depleted natural-gas fields of the United States.

In accordance, the total estimated storage space for North America is 5.23 trillion standard cubic feet. It will do.

The ChevronPhillips Clemens Terminal, with many years of working experience with storing hydrogen for industrial use, provides reliable, pertinent data.

The actual working capacity of the Clemens Terminal cavern is 1,066 million standard cubic feet. Evidently, it is possible to store hydrogen at a pressure of 2,000 psi without causing metal embrittlement—something that does happen if the hydrogen is stored at a higher pressure. Thus, at full capacity, Clemens stores 2,540 tons of hydrogen in 1,066 million standard cubic feet. Given the total available storage in North America, this means that at any given moment about 12 million tons of hydrogen gas can be put under lid—spread over 500 underground caverns throughout the continent.[37] Each ton can yield 15.9 MWh. This comes to a total of 198,000 GWh worth of electricity that is to be stored in 500 caverns throughout the continent.

While this is so, we can pipe only so much hydrogen out of the ground at any given hour. Studying gas outflow rates from the caverns, it seems that the maximum possible hourly outflow is 661 GWh worth of hydrogen.[38] If we don't inject any hydrogen for a prolonged period and we just pump out the gas at full power, the hydrogen reserves will last for three hundred consecutive hours. Put another way, we can provide the grid 661 GWh hourly without missing a beat for about twelve and a half days—at which point we will run out of reserve gas. To put these output levels in perspective, the hourly demands of the grid would range from 600 to 2,000 GWh.

The hydrogen in reserve would also provide for all the heavy trucks and boats that would run on hydrogen. While we can suspend hydrogen production operation for a few hours or a day, in general, every hour of the year we need to provide 2,037 tons of hydrogen for our transportation needs.

The injection rate, the pumping of gas into the caverns, is about 3.5 times slower than the rate at which we can extract it. It takes 50 MWh to create, compress, and pump underground 1 ton of hydrogen. With the geo-physical constraints of the caverns being what they are, the maximum inflow rate is 11,960 tons of hydrogen per hour. Thus, even if we have more than 598 GWh excess in any given hour, this is all we can apply toward our hydrogen-producing operation—without exceeding the rate at which we can inject and store it underground.

In a simulation of one year, it was assumed that we start off with an almost-full tank, about 11 million tons of hydrogen underground out of the 12 million tons the reserves can hold. The results of the simulation were as follows:

FIGURE 7.9

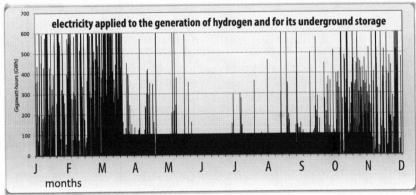

Note: Hourly generation of hydrogen can never exceed the rate at which hydrogen can be pumped underground. In other words, only up to 598 GWh can be used toward hydrogen generation in any given hour. In the summer (middle region of the graph), the volume of hydrogen pumped underground is very limited due to having the reservoir filled to capacity. The reason why there is a daily injection of hydrogen at all during the summer is to compensate for the daily output that is designated to delivery of hydrogen to the transportation sector.

Throughout the year, the grid summoned a total of about 5.9 million tons of hydrogen for electricity production, and hydrogen-powered vehicles required about 17.8 million tons of hydrogen. Hence, a total of about 23.8 million tons of gas were retrieved in total for the year. On the other side of the operation was the amount of hydrogen pumped into the storage facilities throughout the same period. At times, the storage facilities were full and could not accept any more gas even when we had energy to generate hydrogen and deliver it underground. At other times, we had space underground, but there was more hydrogen that could have been made than the rate of inflow would have supported. At times, we had no energy to spare for hydrogen generation at all. In total, the grid routed about 1,125 TWh into hydrogen generation. Thus, under all the restrictions listed above, we have infused 22.5 million tons of gas throughout the year on top of the 11 million already there; and we took out, as noted, 23.8 million tons of gas. At no time was the tank empty. In fact most of the time it was nearly full. The lowest amount, throughout the year, of total gas stored underground was 9 million tons of hydrogen out of 12 million tons possible total.[39]

There is no question that we lose a lot of energy in the process of creating hydrogen and then, later, generating electricity out of it. To end up with 1 GWh of electricity when we need it, we have to invest 3.14 GWh in times that we don't need it—in times when we can generate more electricity than is otherwise useful.[40]

Out of 8,760 hours in the year, the simulation indicates that the valves were open and hydrogen was being retrieved during 1,564 hours. (In truth, it was called upon every hour of the year for transportation needs; however, for electricity-generation needs the number was 1,564 hours.) In the vast majority of cases, the required hourly energy was a modest 20 or 30 GWh. Only during 35 hours of the whole year was hydrogen required to provide a vast amount of power—500 GWh or more out of the possible output of 661 GWh. Throughout the year, the grid requested a total of about 94,600 GWh worth of hydrogen from underground storage.

The reality is better than suggested here in one respect. Whenever there is an excess of electricity, some can be routed to truck stops and they can generate their own hydrogen and store it in giant cylinders. This cuts way down on numerous delivery trips. However, there would be times when there is no excess, during which the truck stops would rely on the central facilities as described above.

When it is all said and done, storing excess electricity in the form of hydrogen proves to be of significant benefit.

VEHICLE-TO-GRID (V2G)

The last source of power in the energy-generation package for 2027 is cars, that is, the entire battery-based fleet of electric vehicles in North America. There are going to be tens of millions. And each will contain a battery that stores…electricity.

We know with practical certainty that on average a car sits idle 22 hours each day. While the pattern of usage of an individual car is unpredictable, the pattern of the vehicle fleet as a whole would be highly predictable. At any time of the day, about 90 percent of them would be parked and, in rare cases of grid emergency, could provide energy to the grid from electricity stored in their batteries.[41] The intent is to draw upon only a portion of their stored energy.

TABLE 7.9

vehicle type	projected number of vehicles for the year 2027	average battery capacity	viable yield in a given day, per battery	total possible yield from entire fleet
pickups	45,820,372	40 kWh	2.83 kWh	129.76 GWh
vans	21,045,874	41 kWh	2.90 kWh	61.09 GWh
sport utilities	46,232,128	40 kWh	2.83 kWh	130.93 GWh
other lights	510,899	40 kWh	2.83 kWh	1.45 GWh
buses[a]	961,862	360 kWh	25.49 kWh	12.26 GWh
passenger vehicles	155,099,336	22 kWh	1.56 kWh	241.58 GWh
motorcycles	5,594,379	3.84 kWh	0.27 kWh	1.52 GWh
truck tractors (not including semi-trailers)[b]	2,202,558	100 kW	7.08 kW	7.49 GWhb

[a]Assume that buses are parked and plugged in to the grid only half the time of passenger vehicles.

[b]The connectivity of trucks to the grid is assumed to be less than that of passenger cars on two counts: trucks are on average on the road more than passenger cars, and some would be hydrogen powered.

The basis for energy availability from vehicles is an analysis done by Willet Kempton et al. A car battery capable of storing 27.4 kWh can kick to the grid 6.8 kWh. It can deliver this amount of energy over a period of one hour, or it can provide 1.7 kWh per hour over a period of 4 hours.[42]

The number of vehicles in table 7.9 is based on *Highway Statistics 2008*, which recorded the current number of registered vehicles in the United States. From the population projection for Canada and the United States for the year 2027 (minus the outlying areas), I extrapolated the number of vehicles on the road in 2027. Finally, I assumed 90 percent availability of vehicles at any given time, which is what studies have indicated to be the case.

This is but a rudimentary, crude projection. Yet, it does give us a sense of the amount of electric energy the vehicle fleet can inject into the grid.

Combined, the future fleet of electric vehicles packs a punch. It can provide the grid with a one hour massive infusion of 2,344 GWh, or for 4 consecutive hours it can provide 586 GWh for each hour.

This is truly the last line of defense. On those few occasions needed, massive, short-term amounts of energy can be brought into the mix to handle the unexpected—whether the needs measure in seconds and minutes or in hours.

The built-in electrical-generating capacity from wind, solar tower, PV, and stored hydrogen are designed to take care of the grid needs without resorting to power infusion from the vehicles. This is as it should be; extra generation capacity should be built into the grid not for the expected but for the need that is not anticipated to arise. It is a system with build-in redundancy. Indeed, out of 8,760 hours of the year, only 16 hours in the simulation drew on energy from the vehicle batteries. This is well below 1 percent of the time. The most extreme, dire hour of the year required the infusion of 294 GWh, which the fleet of vehicle batteries in the simulation accommodated without breaking a sweat.

PUTTING IT ALL TOGETHER

FIGURE 7.10

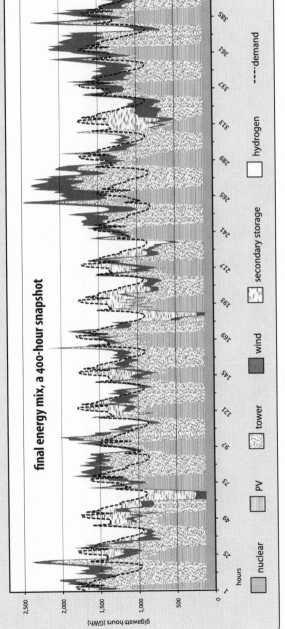

final energy mix, a 400-hour snapshot

Note: A chart showing the power supply and demand for every hour of the year is found at www.danielirdan.com

As usual, we are one step away from the next best technology. Be that as it may, the entire scheme of power generation is based on today's technology.

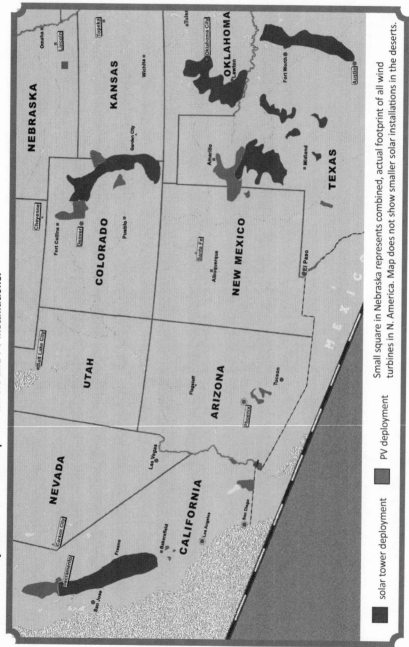

FIGURE 7.11. Projected areas for solar power towers and PV installations.

■ solar tower deployment   ▨ PV deployment

Small square in Nebraska represents combined, actual footprint of all wind turbines in N. America. Map does not show smaller solar installations in the deserts.

In broad brushstrokes, the envisioned power grid would function in this manner: The photovoltaic installations provide the bulk of the needed energy from the morning to afternoon hours and the solar towers provide for the rest of the hours of the day and night via molten salt which is utilized as needed. Wind blows when it's in the mood. And when it does, it would either be used directly and immediately or alternatively it would in effect be converted to hydrogen, to be stored underground—along with the excess generated by solar technologies. When the combined yields of solar tower, PV, nuclear, and wind do not suffice in a given period, the hydrogen gas is brought back up and converted to electricity that is pushed into the grid. In the very rare moments in which this still does not suffice, extremely potent, short bursts of energy can be brought online from the many millions of car batteries that are plugged into the grid at any given moment.

The final version of the plan to provide North America with all of its energy needs would require 224,284 square kilometers worth of installation. This is so for 2027 and somewhat less so in subsequent years—as the plan assumes and calls for a reduction in population numbers (see the twelfth chapter ["Consumption"]).

Had we put all installations together in one contiguous area, they would have taken up an area the size of the state of Minnesota. It is a good thing North America has so much land devoted at present to cultivating corn and soya for feeding cars and cattle. Those applications will no longer be required (more on this in the ninth chapter ["Land Use"]). Hence, finding 224,284 square kilometers is easy enough.

TABLE 7.10

area needed for power installations in N. America	224,284 sq km
maize under cultivation for ethanol consumption in USA	131,000 sq km
maize under cultivation for animal feed consumption in USA	122,000 sq km
soy under cultivation for animal feed consumption in USA	144,000 sq km
area made available	397,000 sq km

Sources: World Agricultural Supply and Demand Estimates (WASDE) for area planted. World Agricultural Outlook Board, "World Agricultural Supply and Demand

Estimates," WASDE 490, (January 12, 2011), http://usda01 .library.cornell.edu/usda/
waob/wasde//2010s/2011/wasde-01-12-2011.pdf; Breakdown of usage for corn and soy
was obtained in personal communication with USDA specialists. The total tally does
not account for wheat allocated for animal feed or other secondary sources of feed,
some of which may, unlike soy and corn, grow on the same land.

Naturally, a lot of shuffling will be required as, for instance, the
land in Texas required for power installations may not be the land
that is growing a superfluous crop.

Land can be found, but understand this: on average, the residential energy consumption per capita in North America is very high.[43]
Elsewhere, people have been getting along on much less energy—and
I don't refer to poverty-stricken regions. For instance, the average
residential consumption per capita in Spain is but 38 percent of that
in the United States, and the average in Venezuela is but 17 percent of
that in the United States.[44] Thus, altering the style of living of some
can cut down on energy consumption like nothing else.

It would be possible to severely cut back the energy consumption
in North America and proportionally cut the amount of land and
overall resources required. As it can be done on land already degraded
and with sun and wind that are abundant, we can be more indulgent
and shrink our footprint in baby steps through the decades.

Furthermore, the grid demands presented here for North America
also assume that the population will grow under a business-as-usual
trajectory, reaching a population of 391 million people in 2027.[45] As
will be discussed in the twelfth chapter ("Consumption"), this plan
in fact assumes otherwise: by the year 2027 the North American
population will increase only a little over its 2010 levels of 342 million
people. Therefore, the total electrical requirement would not be as
great as was assumed here in this chapter.

The analysis of the North American region is but a case in point
for other transcontinental regions. Some of those other regions will
have total energy needs comparable to those discussed here—and this
is assuming their existing, lower levels of energy consumption per
capita. It was important to ascertain that a region the size of North
America *can* yield in between 600 and 2,000 GWh each hour of the
year. If it can, maybe more populous regions of comparable size but
with lower per-capita consumption would be able to, as well.

Currently, twenty-four million people are served by a 6.4 GW line from Xiangjiaba to Shanghai. I will take it as a crude baseline and assume that this would serve only twelve million North American consumers, who gulp more power per capita.

The needed grid would be 31,000 kilometers long (see figure 7.2). More than one cable would be required to handle all the needed bandwidth. In reality it comes to 91,140 km worth of bipolar cables, each pair carrying 10 GW, combined. On most nodes, we would need a 25 GW bandwidth. This translates to a few pairs of bipolar cables. On the main loop we would need a 50 GW bandwidth, as it will draw in along its route most of the generated energy that is to serve the North America grid.

TABLE 7.11

superconducting DC grid for North America, raw materials needed (for 91,140 km worth of cable)	in tons
stainless steel	1,316,000
copper	999,000
liquid nitrogen	638,000
silver	875
YBCO[a]	964
kraft paper	556,000
extruded polypropylene	278,000

Sources: Computation is largely based on data provided by Nexans, a cable manufacturer, and American Superconductor, a superconducting-tape manufacturer.

[a]Yttrium barium copper oxide, the superconducting compound used in superconducting cables.

The plan calls for a total of 250,000 wind turbines, 2.5W capacity with 100-meter-high towers and 50-meter-long blades. The combined installation of all turbines would have a footprint of 1,075 sq km. In addition, the installations would have between 2,425 sq km and 7,250 sq km of temporarily disturbed area. Given proximity issues, the wind turbines within a given cluster will not be placed any closer than 1.85 wind turbines per sq km. The wind farms would be scattered throughout the Great Plains as widely as possible, to smooth out the fluctuation in winds, and with them the combined energy output of all wind turbines.

It seems that a turbine is commonly designed for 20 years; however, depending on maintenance and site conditions, it could be able to operate for even longer. After that period the turbine can be refurbished, or more likely its components recycled, and a next generation turbine can take its place out in the field.

TABLE 7.12

250,000 wind turbines, raw materials needed	in tons
adhesive	252,000
aluminum	223,000
balsa or low-density polymer foam	82,000
carbon fiber composite	148,000
concrete	279,742,000
fiberglass	1,245,000
permanent magnet	16,000
steel	21,685,000

Source: Data extrapolated from that presented in *20% Wind Energy by 2030: Increasing Wind Energy's Contribution to U.S. Electricity Supply* (US Department of Energy, 2008) and America's Future Panel on Energy from Renewable Resources, "Electricity from Renewable Resources: Status, Prospects, and Impediments," (Washington, D.C., The National Academies Press, 2010).

Concentrated solar tower installations would be established in five areas across the Southwest and South regions. They would be in the general vicinity of Oklahoma City (OK), Lubbock (TX), Bakersfield (CA), Limon (CO), and Dallas (TX). Each of these five regions would have 3,234 installations with a footprint of about 34,000 sq km per region. This comes to 16,170 installations throughout North America with a combined footprint of around 173,600 sq km—or a combined area the size of the state of Florida. (See figure 7.11.)

Each of the 16,170 installations would have around 18,500 heliostats: giant mirror arrays mounted on steel pedestals with concrete foundations. In total, there would be 44 billion square meters worth of mirrors in all installations.

The installations would use air-dry cooling versus the common water cooling for the steam turbine cycle. The combined total water consumption for all installations would be 550,000 acre feet of water annually.[46] As noted earlier, a city of half a million people withdraws 265,000 acre feet of water annually. A viable source of water will be discussed in the ninth chapter ("Land Use").

In the center of each installation would be three massive salt storage tanks, a steam turbine generator area, and an air-cooled condenser area. At the very center would be a 327-meter high ($\approx$ 1,070 ft.) solar receiver tower. The tower would be a steel-lattice tower. Its width, foundation depth, and composition are modeled after the specs of a 274-meter tall ($\approx$ 900 ft.) Self Supporting Tower built by Rohn Products International Holdings and installed in Jakarta, Indonesia.

Table 7.13 accounts for the raw materials needed for the towers and the heliostats. It does not account for the relatively small amount of raw materials needed for the wiring and machinery (e.g., the turbine and condenser), storage tanks, etc.

TABLE 7.13

solar power towers, raw materials needed	in tons
steel	1,425,025,215
glass	447,314,420
adhesive	21,687,970
silver	35,700
concrete	1,423,991,258
sodium nitrate	580,204,825
potassium nitrate	386,803,217

Sources: Extrapolated from *Assessment of Parabolic Trough and Power Tower Solar Technology Cost and Performance Forecasts* (Chicago: Sargent & Lundy LLC Consulting Group, 2003); Matthew Wright and Patrick Hearps, *Australian Sustainable Energy: Zero Carbon Australia Stationary Energy Plan* (The University of Melbourne Energy Research Institute: Beyond Zero Emissions, 2010); Gregory J. Kolb et al., "Heliostat Cost Reduction Study," (Albuquerque, NM: Sandia National Laboratories, 2007).

Photovoltaic power plants would be constructed in four areas: Limon (CO), Lubbock (TX), Bakersfield (CA), and in various degraded desert areas within the Southwest desert proper. (See figure 7.11.) Their combined area is to be 49,782 sq km. And they would be deployed in three clusters of roughly 12,500 sq km each, and an additional 12,500 sq km comprised of numerous smaller installation clusters. Within each of the four regions, the cluster of installations would require a combined 15,053 acre feet of water annually.[47]

We will need 4.6 billion solar panels, each containing 128 monocrystalline solar cells. In total, it comes to 108 billion square feet of solar cells.

Life expectancy of a PV installation is about twenty-five years. After this time, much of the PV panels can be recycled, and new silicon wafers can be produced from the old, as the German company SiC Processing GmbH does.

TABLE 7.14

photovoltaic panels, raw materials needed	in tons
glass	91,873,830
aluminum	30,624,610
purified, meteorological-grade silicon[a]	11,329,000
steel	33,923,760
concrete	217,775,000
silver[b]	103,900

Sources: Derived from specs of California Valley Solar Ranch, "Project Description," (January 2011), http://www.sloplanning.org/EIRs/CaliforniaValleySolarRanch/feir/b_proj_desc.pdf; Mariska J. de Wild-Scholten and Erik A. Alsema, "Environmental Life Cycle Inventory of Crystalline Silicon Photovoltaic Module Production" (paper presented at Materials Research Society Fall 2005 Meeting, Boston, MA, November 28–30, 2005); Niels Jungbluth, "Life Cycle Assessment of Crystalline Photovoltaics in the Swiss Ecoinvent Database," Progress in Photovoltaics 13, no. 5 (2005): 429–46;

[a]Assuming 7 grams of silicon per watt.

[b]Assuming 22.3 grams of silver for each 41.5″ × 81.5″ PV panel.

Hydrogen is next.

In all, five hundred underground storage caverns are scattered about, mostly in West Virginia, Texas, Pennsylvania, Ohio, New York, Michigan, and Illinois. Some allow higher pumping rates than others. Hence, the generation capacity of their respective aboveground facilities will vary from location to location. When all the valves are wide open, the five hundred facilities would have the capacity to generate 661 GWh hourly. On the other side, they would also have the ability to generate hydrogen corresponding to the maximum possible injection rate, which is 11,960 tons of hydrogen an hour. In short, it translates to a total array of 42,000 stationary fuel cell units and 269,000 large-scale electrolyzers. Thus, averaged, each facility would have 84 fuel cell units and 540 electrolyzers.

In the process of generating electricity with a fuel cell, about 10% of the water is used to keep the membrane moist. So this can be counted as lost. However, the water vapor coming out of a fuel cell can probably be captured to be re-used at the beginning of the loop, with the electrolyzer. The electrolyzers go through 9,990 liters ($\approx$ 2,639 gallons) of water to produce a ton of hydrogen. At maximum production rates, it would come to about 2.8 million cubic meters of water ($\approx$ 2,300 acre feet) over a peak 24 hour period for the North American grid. As stated, most of the needed water ought to come from the steam coming from the back end, a byproduct during the generation of electricity from hydrogen.

TABLE 7.15

needed hydrogen-reserve setup	No. of units
electrolyzers	269,000
stationary fuel cells	42,000
facilities	500

Table 7.16 is a tally of the needed total raw materials for the North American grid.

TABLE 7.16

North America, raw material requirement for power-generation array, combined	total, in tons	annually averaged over the construction lifespan (15 years), in tons
nitrogen	637,980	42,532
copper	998,665	66,578
concrete	1,921,509,208	128,100,614
permanent magnet	15,961	1,064
aluminum	30,847,966	2,056,531
steel	1,481,950,901	98,796,727
glass	539,188,250	35,945,883
silver	140,475	9,365
adhesive	21,939,522	1,462,635
balsa or low-density polymer foam	82,405	5,494
carbon fiber composite	147,895	9,860
fiberglass	1,244,748	82,983
sodium nitrate	580,204,825	38,680,322
potassium nitrate	386,803,217	25,786,881
purified, meteorological-grade silicon	11,329,000	755,267
YBCO	964	64
kraft paper	555,954	37,064
polyethylene	277,977	18,532
water (routine usage during operation)	...	565,053 acre feet[a]

[a]The equivalent of 697 million cubic meters or 184 billion gallons. This is about 0.5% of the current net consumption of freshwater in North America. This does not include water usage for hydrogen production, but only water required for the ongoing operation of solar towers and PV installations. Water requirements are based on Rice Solar Energy Project (see endnote 46) and on California Valley Solar Ranch (see endnote 47).

Figure 7.12 compares the annual production needs versus the current annual production figures for a given commodity. The concrete requirements for the power grid are but a drop in the bucket. On the other hand, when we spread our raw-material needs over a fifteen year period, the annual requirements for glass and silver exceed the entire current production rates.

Silver is the one raw material that gives me pause. As of 2011, world reserves of silver are estimated at 510,000 tons.

FIGURE 7.12

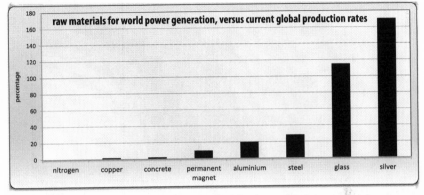

Note: I arrived at an educated guess of the worldwide needs of raw materials per year for all of the fourteen transcontinental grids by multiplying the North American grid's requirements by a factor of four—reflecting assumed energy consumption of the rest of the world in relation to that of North America.

The plan requires 140,475 tons just for North America and conceivably about 562,000 tons for the entire world—which is a bit more than the entire known silver reserves. If carried through, it would leave us with zero silver and with one big question: what to do next, beyond the first generation of PV and solar power tower installations. It would be the height of folly to commit the entire known reserves of silver on the planet to provide for the electrical needs of one generation of people.

There is a very real possibility that far more silver resources may be identified, or the extraction process would improve, or we would manage to make do with less. Yet, with relatively little silver to go around, it is short-sighted to use it for anything other than what is absolutely essential and cannot be accomplished by any other substance.

Surgical pins and plates could be made with tantalum and titanium instead of silver. Stainless steel utensils can take the place of silver ones. Aluminum and rhodium may replace silver in mirrors and other reflecting surfaces. In addition to using viable substitutes whenever possible, we ought to recycle silver as much as possible. Insofar as trinkets are concerned, there are enough other pretty metals to do the job. As it stands, we should not use silver frivolously.

Ultimately, the PV panels are not essential to the plan. They don't have the essential storage capacity nor do they offer the small footprint of wind turbines. In accordance, PV will be a part of the plan only if we can get the silver consumption to go down from the thick 50 microns that are the PV industry standard—and assumed in table 7.16—to 2 or 3 microns, which the micro-electronics industry uses. Using more silver than that for PV manufacturing is utterly irresponsible. If PV would consume twenty-five times less silver, the silver required for all the various power generation installations in North America would go down to 40,731 tons or, by inference, for the entire world it would come to about 163,000 tons.

———— •————

The bright side of having 7,000 million people on board the Earth ship is the fact that, well, there are 7,000 million people. That's a lot of people. A measly 8.3 million of these are employed in the construction of 60 million vehicles annually. Both facts should give us some indication of the number of people needed for energy-grid construction and for the numerous people that are available in general.

I do not have an estimate for all the workers required. However, I can assess the number of people required for the solar tower power installations. This would likely make up a substantial portion of the power grid construction. Extrapolating from the detailed construction plans of the Rice Solar Project and Ivanpah Solar Power Facility, it works out that if we assume a few years of production ramp-up and that the construction of the 16,170 installations is spread over a twelve year period—from 2015 thru 2027—it would require about 4,000 construction teams working simultaneously on 4,000 installations. The estimated construction time is 36 months per installation. Each team needs to have about 440 people, for a total workforce of 1.76 million people.

As of 2011, over thirty-seven million people in the United States are either registered as unemployed or engage in vocational pursuits designed to extract money from consumers—and little else. It is time for them to get *real* jobs and have something genuinely meaningful to do. More on this in the last chapter ("Economic and Political

Paradigm"); however, suffice it to say that we have more than enough man power.

This energy plan is not all that clean, green, or renewable. Nothing that keeps seven billion people afloat can possibly be. For instance, by one set of calculations, the related emissions of solar tower construction (i.e., concrete and steel work) is the equivalent to anything from six months to six years of emissions from the construction and operation of a coal plant of a comparable generation capacity.[48]

However, the proposed scheme has an overall lower impact than that of the present setup by a wide margin. For the next few decades, this would suffice.

What has been described at some length up to this point makes up the first-generation solution. It will serve us for a few decades and then, in the natural course of events, the installations will reach the end of their lifecycles.

I wish to introduce what potentially can come online in the second generation, say around 2070.

In total, three technologies. Each of them is overall superior to any solutions of the first generation. However, none of them is quite ready for prime time and cannot be relied upon to come through at the moment. In the next few decades we will find out if they will deliver.

ENHANCED GEOTHERMAL SYSTEMS (EGS)

With a measly projected hourly output of 6.76 GWh,[49] geothermal power is not something that needs to be bothered with in North America. The technology of Enhanced Geothermal Systems (EGS), however, is another matter.

The heat contained deep underground, within the Earth's crust, can be put on active duty. We can heat up water, pressurize the resultant steam, and then generate electricity in a conventional fashion. However, the hot rocks deep under are rarely porous. And so a bore hole is drilled and highly pressurized water is pumped underground. As more water is pumped, pressure mounts, and the rock fractures. The water can then travel through the rock, get heated in the process, and be brought back up to drive an electricity-generating steam turbine.

The US Geological Survey estimates the EGS potential to be a 518 GW capacity for the contiguous United States.[50] However, a more recent and comprehensive study reckons the capacity to be 2,980 GW.[51] I will go with the more recent assessment. Assuming a 90% yield, it would mean the potential generation of 23,494 TWh annually. This is over twice the entire future electricity needs of North America. Yet, if we were to run water through all known EGS resources, we would have a substantial influx of energy and then nothing for many decades—as the cooled rocks will need a lot of time to regain their natural heat.

To harvest energy from deep underground year after year, we need to do the equivalent of crop rotation. The total resource is to be divided to four groups, each of which has a thirty-year shift. After thirty years, the first batch can regain its heat for a century or so, while a second group is engaged, then the third, and then the fourth. With a crop-rotation style of energy harvesting, we would have an annual output of 5,873 TWh in North America that would last for many thousands of years, after which the surrounding rocks will cool and there will be need for a longer rotation cycle.

In North America, we would be able to count on, well, a rock-steady output of 670 GWh each hour of the year.[52] This is not an average 670 GWh in the sense that a wind power is: 900 GWh one hour and 3 GWh the next. This is a constant, hour-in, hour-out power output. And oftentimes this would account for the demand of the entire grid in a given hour.

Beyond the steady nature of the power generation, the footprint of EGS is one of its greatest advantages. An EGS plant with a 100 MW generation capacity is estimated to take 2.1 sq km.[53] Ergo, the annual generation of 5,873 TWh will require a series of power plants with a combined capacity of 744.6 GW and a total footprint of 15,634 sq km. It comes to 375 GWh in annual output per square kilometer of facilities. To put it in context, PV averages an annual output of 68 GWh per square kilometer of footprint and a solar tower averages 44 GWh per square kilometer.

Experimental work on EGS is being conducted in various locations around the world, with the facility at Cooper Basin, Australia, arguably leading the way. However, we are not quite there. We may get there in a few years, in a few decades, or never.

HIGH-ALTITUDE WIND

Winds at higher altitudes are steadier, more persistent, and of higher velocity—notably around 7,000 through 10,000 meters above ground (≈ 23,000–33,000 ft.). It is expected that high-altitude kites will have a generated power density dramatically higher than that of ground-based wind farms and at much lower energy production costs.[54]

While the air density drastically decreases with rising elevation, usually the strength of the wind high up more than makes up for it. However, the more significant difference between a ground-level wind and a high-altitude wind is the portion of time in which the wind is blowing. It turns out that miles above most land areas, the wind practically always blows, and we have at least 0.1 kW to 0.2 kW per square meter worth of wind power.

High up, virtually all the time (>95%), the wind power density is 0.5 kW/m² to 1.5 kW/m² above the southern part of South America; the broader region around the Midwest through Northeast in North America; northeastern China through Japan; southern Australia; and North Egypt through Syria. Better yet, most of the time (>68%), these same regions experience power density upward of 3 kW/m². Half of the time they experience power density over 5 Kw/m². It is most notable at the Northeast seashore, North America; Japan; Northeast China; and some parts of Australia—which experience over 10 kW/m² half of the time.[55]

To put this in context, I will describe in a comparable manner the situation over land at 80 meters above ground—the height most contemporary ground-based wind turbines tap into. Virtually all the time (>95%), the wind power density is a measly 0.01 kW/m² (around 1% available at high altitude). Most of the time (>68%), the better regions enjoy between 0.05 and 0.1 kW/m² (around 1.5%–2% of that available at high altitude). Half of the time, the better regions enjoy around 0.1 kW/m² (around 1.3% of high altitude wind potential).[56]

Is that very significant? Perhaps, yes. We have yet to find out the extent to which future air-borne wind turbines would be able to harvest wind energy. What we do know is that the wind high above is far stronger and—just as significant—blows far more often. The basic idea is to capture wind energy using tethered kites. Electricity is generated aloft and transmitted to the surface via the tether. In a design proposed by Sky Windpower Corporation, four rotors are mounted on an airframe, tethered to the ground via insulated aluminum conductors wound with Kevlar-type cords. The rotors provide both lift and power generation. The aircraft can be lifted with supplied electricity to reach the desired altitude at which point it starts to generate power.[57]

Due to the meandering and unsteady nature of the jet streams, the high-altitude wind resource is less than rock steady. Although in general wind speed increases with height, altitudes at which winds are strongest can vary depending on the weather conditions. Obvious benefits would arise if a high-altitude technology were able to dynamically change the altitude of the kite to adjust to wind conditions.

Needless to say, the footprint on the ground is nil. Nothing else can aspire to make such a claim.

INTEGRAL FAST REACTORS

If performed as advertised, a particular nuclear power technology might be the silver bullet.

But first things first. Let's first see if we can learn to stop worrying and love nuclear power. Let's look at what has been the worst about nuclear energy and review the bottom of the barrel: the Chernobyl Disaster.

The accident occurred on April 26, 1986 during an ill-executed engineering test. Improper, unstable operation of the reactor caused an uncontrollable power surge to occur, resulting in successive steam explosions, which severely damaged the reactor building and completely destroyed the reactor. As the nuclear reactor had no containment structure, the power plant released the largest radioactive emission ever recorded for any civilian operation. The radioactive cloud dispersed over the entire northern hemisphere and deposited substantial amounts of radioactive material over large areas of the former Soviet Union and Western Europe.

The United Nations Scientific Committee on the Effects of Atomic Radiation (UNSCEAR) has reviewed myriad scientific papers related to the aftermath of the accident and has summarized its findings. The findings are based on more than two decades of experimental and analytical studies of the radiation consequences of the incident.[58]

Two workers died in the immediate aftermath. High doses of radiation to 134 staff members and emergency personnel resulted in acute radiation syndrome, which proved fatal for 28 of them and resulted in cataracts and skin injuries to some of the remaining members of this group. The consumption of milk contaminated with iodine-131 has adversely affected the general public. Out of 6,848 thyroid cancers observed to date among people who were children or adolescents at the time of the accident, a substantial portion is assumed to be due to intake of that contaminated milk. Out of these, 15 cases have proven fatal. However, most of the workers and members of the public were exposed to low level radiation comparable to or, at most, a few times higher than the annual natural background levels. In total, forty-five people died as a result of the accident, some immediately, some years later.

Forty-five fatalities is the number of coal mine related fatalities that take place every week. Had thousands of people died from the nuclear-meltdown, and had such a meltdown occurred every year, it would have been comparable to the present-day, annual fatalities related to coal mining—which we readily accept as part of doing business.

Two years before the Chernobyl accident, a liquid petroleum tank in Mexico exploded. About 6,000 people suffered severe burns and about 550 died. During that same year, a chemical plant in India leaked methyl isocyanate gas. It resulted in 3,900 severely and permanently disabling injuries. In that accident, 3,787 people died.

In a way, Chernobyl is not much of an example, as no one in his right mind would build nuclear plants in the future without a containment shell or with ill-conceived control rods.[59] In a way, old Fukushima, which does have a containment and an overall safer design, is a better example of what can happen when you put an aging nuclear plant right on a geological fault, subject it to one of the most severe earthquakes possible, and then get a colossal ocean wave from multiple merging tsunamis to flood generators and cut the plant off from any source of power. That is to say, when it comes to injuries and fatalities, not *that* much would happen.[60] The main toll is on the land of the surrounding region, which got extensively contaminated.[61]

In April 26, 1986, the coolant was cut off in Chernobyl, and the subsequent explosion tore the roof off. About three weeks earlier, a similar experiment was initiated in Argonne National Laboratory West in Idaho with the IBR–II reactor. The reactor was brought to 100% power and the cooling pumps were turned off. Subsequently, nothing undue happened.

The power came down rapidly while temperature shot up from 477°C to 704°C; however, that was well below the coolant's—liquid sodium—boiling point. After a while, the core temperature came down until it reached normal operating temperatures. There was no damage to any of the systems or the fuel. The reactor smoothly returned to safe conditions without bringing into play the control rods and without any human intervention. When the coolant flow stopped, the reaction stopped and the reactor just shut itself down. This was a matter of physics, not operator's alertness. For all that it mattered, the operator could have gone out fishing in the back pond while all of this was happening. For all that we know, the operator could have tried to fry the nuclear reactor every day of the year and the reactor had remained, well, cool.

Water-based reactors have lots of valves and pumps and mechanical things that can go wrong. The reactor in the Argonne Lab had a simpler design and a far safer one. The composition of its fuel pins was such that if they began to overheat, the resulting expansion lowered their density to the point where the fission reaction simply shut down. In the same absolute certainty that sufficiently cold water turns to ice on a pavement, overheated fuel rods of that kind expand and the fission reaction comes to an end.

The two 1986 experiments had very different outcomes, both physically inevitable for each of the two profoundly different reactor technologies. The Chernobyl reactor was a garden-variety light water reactor. The IBR–II was an entirely different animal.

The story of the IBR–II started years earlier as an impressive ensemble of scientists came together at Argonne National Lab and devised a new kind of reactor, the Integral Fast Reactor (IFR).[62] After ten years of development, Congress shut down the project in 1994 due to its concerns pertaining to nuclear proliferation and, in general, having the opinion that there is no longer any need for advanced nuclear reactor technology.

However, we have gained enough knowledge to recognize that IFR might very well be our ticket. Passive safety, as described above, is one of its main safety advantages. Then there is another. Conventional reactors operate with pressurized water, which magnifies the chances of vessel and coolant leaks. Fast reactors, on the other hand, usually use liquid sodium metal as the coolant, and it does not have to be pressurized. Having said that, liquid sodium and air do not get along all that well. Fires have been known to occur. Hence the IFR design calls for a vessel that is not only sealed but also contains argon instead of air.

The IFR is a type of a fast reactor. Neutrons slow down when they go through the water medium of conventional reactors. With sodium molecules, they just bounce off and do not lose energy.

Existing, conventional reactors use less than 1 percent of the energy in uranium. The rest is discarded. In contrast, IFR utilizes 99 percent of the energy in uranium. It is really quite incredible to contemplate how much energy an IFR can yield out of uranium. Imagine fifty households around you, which may make up your immediate neighborhood. Within an IFR, a lump of fuel the size of a large marble (sixty-two grams of fuel) will suffice to run all the power in those houses for one year.[63]

Better yet, an IFR can employ as fuel all the so-called depleted uranium and nuclear waste. In other words, with IFRs providing our energy, we don't need to mine uranium for a while. We have about 1.8 million tons of nuclear waste and depleted uranium. Once we put them through IFR plants, they will yield about 15.7 million TWh.[64] If we assume that the thoroughly electrified world of 2027 will consume 40,000 TWh a year, just the stockpiles of nuclear waste we now have sitting around, coupled with the depleted uranium, will run the world for about 400 years. In the process, we will get rid of the vast majority of the nuclear waste that is just lying around.

Every eighteen to twenty-four months, one third of the fuel would be removed from the reactor and replaced by new fuel. The used fuel would be recycled. In an IFR, the nuclear material that comes out of the reactor is processed on site, made into new fuel rods, and off it goes, running through the reactor again and again, till nothing but fission products remain. For every one hundred tons that go into an IFR, only five tons would come out, and those five will have a high level of radioactivity for three hundred years—not for tens of thousands of years.[65] The discards from an IFR plant would be a mélange of lighter elements such as krypton, zirconium, and barium. These can be stored for the needed duration in an underground storage facility described at length in the fifth chapter ("Transportation").

At least that is the idea. IFR development was terminated before the principal element in the fuel processing could be proven successful; before they did a full-scale separation and collection of the new fuel product mixture from the spent fuel. This process was demonstrated to work only on a small, laboratory scale when the projected was aborted.

As a matter of course, there is the political element, the terrorists and the unstable regimes of the world. This has been addressed in the fifth chapter ("Transportation") and will be fully addressed, albeit indirectly, in the thirteenth chapter ("Economic and Poilitical Paradigm"). However, one thing will be stated here.

Nothing in the IFR cycle allows for the technical possibility of separating pure plutonium or other weapon-grade material from the rest of the fuel. In this respect, it is not any better or worse than the existing, light-water reactors. The only way to distill weapon-grade fuel is to build another, dedicated facility with heavy shielding and robotic arms and have a go at it. But then again, it may be easier just to mine uranium and enrich it.

If proven viable on a commercial scale across decades of service, I find IFR to be the best technology we have along with the harvesting of high-altitude wind. I find these two technologies best because they appear to have the least footprint. In other words, they are the most environmentally friendly of the bunch. In the case of IFR, it is also the single best way to take care of the high-level nuclear waste we have already accrued.

The power plants coming online in 2027 will have a lifetime of about forty years. Come 2070, we can set about re-establishing the same wind mill and mirror gig. Alternatively, with high competence and a bit of technological luck, we can eliminate hundreds of thousands of square kilometers of footprint and replace them with thousands of small IFR plants or with high-altitude wind collectors.

If we assume that a full-scale deployment will take 15-20 years, we have between now and 2050 to set up a dozen or two IFR reactors, put them through their paces, test their mettle under commercial-scale installations and draw conclusions. Come 2050, if all checks out, we can replicate thousands of them, which will reduce both costs and the unexpected. Earlier, I summarized the findings of the United Nations Scientific Committee on the Effects of Atomic Radiation on the Chernobyl Accident. I assume that they got the gist of the story right. However, I cannot rule out in good conscience the possibility that they are wrong, dead wrong, and in fact many thousands died as a result of the Chernobyl Accident, as some claim. This is an additional issue to be looked at in the coming few decades.

In 2004, a few years before the Fukushima Accident, *Time* magazine wrote: "Of all the places in all the world where no one in their right mind would build scores of nuclear power plants, Japan would be pretty near the top of the list."[66] Well, with a superconducting DC grid, we don't need to put IFR reactors on geological fault lines nor anywhere remotely near a population center. For good measure, we can also explore the possibility of constructing nuclear plants underground. Not much can break through fifty meters of solid earth.

OFF-GRID SOLUTIONS

Over 95 percent of the world population falls within one or another of the proposed fourteen energy grids. As it stands, there are practical limits to the deployment of the principal power lines. It makes little sense to lay down a dedicated underground superconducting DC cable over many hundreds of miles just to serve a community of 50,000 people in a remote region.

Population centers in remote regions will largely be in the same position they are in now: off-grid. In fact, as of 2008, 20 percent of the world's population had no access to electricity.[67] With about 95 percent of the people within one of the fourteen grids, this stands to be a vast improvement over the situation at the present.

There is an array of technological solutions that fits individual homesteads, entire communities, and everything in between. These solutions provide everything from erratic afternoon power all the way to around-the-clock, on-demand power. In short, these solutions are comparable to those of the present age—minus the accompanying pollution from the burning of kerosene or the cutting of tropical forests for wood fuel. A few lower-impact options are available. Their exact combination is a matter of the local wind and sun conditions and budgetary allowances.

Photovoltaic panels and small wind turbines are obvious choices for standalone, off-the-grid communities. However, wind turbines will not function well in urban settings, due to air turbulence, and solar panels will not operate in areas that are cloudy for many days of the year. A big help would come in the form of NAS batteries, which can assure round-the-clock supply. The NAS battery unit is made up of twenty 50-kilowatt modules. The existing unit is roughly the size of two semi-trailers and weighs approximately 80 tons. The battery is able to store about 7.2 MWh worth of electricity. When the wind blows or the sun shines, the NAS battery is charged. At other times, fully charged, the battery could power about 300 homes for 12 hours. One can include as many NAS battery rigs, solar panels, or wind turbines as is practical.

Whether wind turbines or PV panels are used, if a community wishes to supplement with batteries, there is a need to install extra generation capacity. Hence, when the wind or PV panels generate electricity, a part of the power can be used directly, on the fly, and a part of the power can be used to charge the NAS batteries.

For larger towns that are off the grid, a nuclear micro-reactor may be a more suitable solution.

Toshiba has developed a new class of micro-size 10 MW nuclear reactors that are designed for small towns, in the order of a few thousand households. The micro-reactor is only 6 × 2 meters (about 20 × 6 ft.) designed to be buried deep underground. The micro-reactor is engineered to be fail-safe; it is wholly automatic; and it cannot overheat. Unlike traditional nuclear reactors, the new micro-reactor uses no control rods to initiate the reaction. The whole process is self-sustaining and can last for up to forty years. There are a number of designs for nuclear batteries such as this, and although they have been shown around, there have been no takers so far.

Instead of the ubiquitous, GHG-emitting kerosene used for illumination and the environmentally degrading wood fuel used for heating and cooking, lower-impact alternatives would be mandated for the 35%–40% of the world population who currently burns trees, dung, and kerosene for illumination and for cooking.

An off-grid cooking technology that is both sustainable and viable seems to be available in only two flavors: solar and solar with heat storage. In Zambia, SunFire Solutions offers parabolic solar concentrator fixtures for boiling water, solar box cookers for actual cooking, and pads to retain heat or cook in semi-cloudy conditions.

A more viable design, offering 24-hour use, is the solar kitchen designed by the engineers of the intentional community Gaviotas, Colombia. The sun heats low-viscosity cottonseed oil to 180°C (about 355°F), and then the super-heated oil is siphoned into a holding tank. At the flip of a switch, a forty-watt micro-pump runs on batteries charged by photovoltaic cells and pushes the hot cottonseed oil through top cooktop coils, then back up to the roof to re-heat. The closed system is hot enough to operate twenty-four hours a day. In fact, not only off-grid places would benefit from a solar cooktop of this kind; all houses in sunny locations could enjoy considerable energy savings.

Lights can also be powered directly by the energy from the sun. Predominantly in Sub-Saharan Africa and the Indian subcontinent, Barefoot Power and Greenlight Planet offer a range of ultra-bright LED-based light fixtures powered by diminutive PV panels that are charged during the day. The Canadian Carmanah offers a self-contained solar LED street light for off-grid situations.

8

NUTRIENT AND WATER RECYCLING

≫≪

focusing on living spaces

SOLIDS

WE PRODUCE ABOUT 1.4 MILLION tons of shit each day.[1] The nutrient-filled, homemade turds could have been one of humanity's most important—if not single—contributions to the planetary ecosystem. Alas, we squander this possibility.

We defecate into drinking water, pipe it away along with toxic chemicals, add more chemicals, dry the resulting sludge and burn this mélange or bury it deep. Alternatively, we dump raw sewage into the waterways or make deposits into a hole, as in the case of using a pit latrine. The foods we harvest contain nutrients from the soil. Instead of returning these nutrients to the fields, we bury them, burn them, or dump them into the waterways—ecological abominations, each of them.

These practices need to be replaced with closed-loop nutrient recycling.

Two things ought to change: one, a halt to polluting the waterways with chlorine and with open sewage, as the case may be; two, returning the nutrients contained in human excreta to the soil and resuming the nutrient cycle.

Not surprisingly, the excreta contain essential macro-nutrients and micro-nutrients in the same proportions as is found in food. Excreta make great fertilizer. And it is best served as compost.

Compost is the end product of controlled biological decomposition of organic material. It is sanitized through the generation of heat, and it is stable.

Co-composting human excreta and organic waste is advantageous because the two things complement each other. Human waste is relatively high in nitrogen content and moisture, and organic waste is relatively high in organic carbon content and possesses good bulking quality. High temperatures attained in the composting process are effective in eradicating pathogens contained in the fecal matter and will convert both forms of waste into a hygienically safe fertilizer. A good example of human-manure composting on a large scale is done at the Edmonton Composting Facility in Canada.

Compostable items are sorted and separated from the rest of the incoming garbage. At the same time, sewage sludge is piped to the facility and then put into centrifuges where it is drained of liquid, much like clothing in the spin cycle of a washing machine. The now relatively dry excreta are mixed with the organic discards in giant cylinders and tumble there for about two days, giving the composting process a good head start. From there, the compostable material is conveyed to the aeration building, where it remains for 14 to 21 days. Mobile augers, resembling giant drills, are used to turn the mixture, fluffing it, and thus admitting the needed oxygen. The system takes in water as required. Compost reaches temperatures of over 55°C (about 130°F) for at least 3 days, killing potentially harmful bacteria. From there the compost is moved to another area where it is cured for between 4 and 6 months on outdoor curing pads. This lengthy period also serves to completely destroy ascaris eggs. And presto, we have first-class, hygienic fertilizer and soil additive.

The Edmonton facility can process annually 200,000 tons of residential food waste and 25,000 dry tons of sewage sludge. Combined, those yield 60,000 tons worth of compost annually.

In a series of experiments in Zimbabwe, spinach, lettuce, green peppers, tomatoes, and onions were grown in buckets with poor local topsoil. Their growth was compared to that of plants grown in similar containers filled with a 50:50 mix of the same poor local topsoil mixed with an equal volume of humus derived from co-composted human feces and urine. Results indicated about a four-fold increase in biomass in the plants grown with the aid of the compost.[2]

Compost can greatly enhance the physical structure of the soil.[3] In fine-textured (clay, clay loam) soils, compost will improve workability and porosity of the soil. When used in sufficient quantities, the compost has both an immediate and long-term positive impact on the soil structure. It resists compaction in fine textured soils and increases water holding capacity and improves soil aggregation in coarse-textured (sandy) soils. The soil-binding properties of compost are due to its humus content. Humus is a stable residue resulting from a high degree of organic matter decomposition. The constituents of the humus act as soil 'glue,' holding soil particles together, making them more resistant to erosion, and improving the soil's ability to retain moisture.

The incorporation of compost also stabilizes soil pH,[4] so that soil will more effectively resist pH changes. This allows crops to use nutrients more effectively, while at the same time reducing nutrient loss by leaching.

Compost products contain a considerable variety of macro- and micro-nutrients. Although often seen as a good source of nitrogen, phosphorous, and potassium, compost also contains micro-nutrients essential for plant growth.[5] Since compost contains relatively stable sources of organic matter, these nutrients are supplied in a slow-release form.

Pound for pound, large quantities of nutrients are not typically present in compost when compared to what is found in most commercial fertilizers. However, compost is usually applied at much greater rates; therefore, it can have a significant cumulative effect on nutrient availability.

Compost has the ability to bind heavy metals and other contaminants, reducing both their leaching properties and absorption by plants.[6] Therefore, sites contaminated with various pollutants may often benefit from the application of compost. The same binding effect enables us to use compost as a filter medium for stormwater treatment and has been shown to minimize leaching of pesticides in soil systems.

The nutrient composition of excreta is primarily a function of one's diet. Averaged across a few representing countries, our feces contain 383 grams of nitrogen and 164 grams of phosphorous per person per year. Combined, assuming a 75 percent utilization rate, human excreta may yield 2.68 million tons of nitrogen and 1.1 million tons of phosphorous per year.[7] The Edmonton facility produces 60,000 tons of compost out of organic waste with the aid of 25,000 tons of dry solids. I assume that 100 kg of wet fecal mass makes 22 kg of dry mass.[8] In accordance, with a 75 percent utilization rate of all the human manure in the world, we can manufacture about 200 million tons of compost a year.

LIQUIDS

In Burkina Faso, human urine is sold in jerry-cans to be used as a liquid fertilizer.

Scientists fertilized cabbage with urine collected months earlier from urinals. The conclusion is that human urine could be used as a fertilizer, and it does not pose any hygienic threats or leave any distinctive flavor in food products. Growth, biomass, and levels of chloride were slightly higher in urine-fertilized cabbage than in industrial-fertilized cabbage.[9] The taste mildly differed in both batches; however, different testers varied in their flavor preferences.

One study showed that urine increased the yield of red beets.[10] Yet, another found that banana trees fertilized with urine had growth superior to those trees that received industrial fertilizers.[11] The same thing has been noted with barley.[12]

Urine has been used to fertilize cucumbers, maize, wheat, pumpkins, and tomatoes to great effect[13] and without any microbiological risks.[14] While people with certain severe infections may release microbes into their urine, those are destroyed with the natural pH increase during six months of storage. And at any rate, these microbes cannot survive in soil for a long time.[15]

Pharmaceutical residues are a problem not unique to urine; they are also found in present-day wastewater that is being fed to rivers and used in irrigation. It does seem that many medical compounds are degraded in the soil, but there are uncertainties and potential concerns in utilizing the urine of people on medication.[16]

Urine contains mostly nitrogen (N), phosphorus (P), and potassium (K). It has an NPK mineral ratio of 18:2:5, that is to say, comparable to the ratio present in commercial mineral fertilizers. Urine can be used safely if it is applied onto the soil around the root area. Experience shows there is little smell if the urine is watered down and is spread close to the plant and directly onto the soil.

The fertilization value of urine is higher than that of feces.[17] Indeed, the majority of beneficial nutrients leave the body through urine rather than through feces. In addition, the majority of the highly available nutrients in urine exist in a form that plants can easily use.[18] If ventilation is kept to a minimum during the transport and storage of urine, nitrogen loss is often less than 5 percent.

About 90 percent of the plant-available nitrogen in urine is lost when it is mixed with feces and composted together. The answer is urine-diverting toilets, such as the NoMix toilet. A survey in seven European countries found that the technology of urine-diverting toilets could be well received.[19] For men, there is also the option of waterless residential urinals, which incidentally are as odor-free as their flush-based counterparts.

Applying urine via drip irrigation in fields of rice, vegetables, and yams has been tested successfully in the Ivory Coast. It was carried out during a rainfall to facilitate urine introduction into the soil.

Instead of applying the urine in liquid form, it is possible to add struvite to the urine and sun-dry the resultant compound into a white odorless powder. The powder is mostly comprised of phosphorus found in the urine. With the utilization of powder, weight and volume are reduced to a strict minimum. This makes it a lot easier to transport and otherwise handle. Furthermore, it has a longer shelf life than urine and no unpleasant odors. Due to the slow nutrient release, the powdered fertilizer can be applied to plants at superior rates without risk of burning the roots or leaves. The leftovers contain most of the urine's nitrogen and potassium, which can be applied to crops through drip irrigation. Finally, struvite precipitation produces a fertilizer largely free of any pharmaceutical residues.[20]

Averaged across a few representing countries, one year's worth of urine from one person contains 2,820 grams of nitrogen, 310 grams of phosphorus, and 1,100 grams of potassium. At a 75 percent utilization rate, we make available 20 million tons of nitrogen, 2.7 million tons of phosphorus, and 7.7 million tons of potassium annually.[21]

The annual urine of one person has enough usable nitrogen to fertilizer 300–400 square meters of crops per year and enough phosphorus to fertilize up to 600 square meters of crops per year.[22] Combined, we can fertilize in this manner upward of 184 million hectares if we assume application of 75 percent of the urine we generate collectively. To put it in context, the total land under cultivation is estimated at 1,500 million hectares.[23]

TABLE 8.1

pertinent worldwide figures (annually, whenever applicable)	
land under cultivation	1,500,000,000 hectares
area that human urine can fertilize	184,000,000 hectares
amount of compost that can be produced by human manure	200,000,000 tons
amount of nitrogen produced in 2010	131,000,000 tons
amount of nitrogen in human excreta	22,680,000 tons
amount of phosphorus used in 2010	39,900,000 tons
amount of phosphorus in human excreta	3,800,000 tons

Note: Assumes 75% utilization rate of human excreta. The 184 million ha cultivated by urine and is based on its nitrogen content. It is almost twice the volume of its phosphorous content.

The amount of synthetically-produced nitrogen applied in agriculture is substantially higher than that available in our combined poop. However, the amount of nitrogen currently applied is excessive, and much of it ends up leaching away. Better practices, discussed in the ninth chapter ("Land Use"), would require substantially less nitrogen.

OFF-GRID SOLUTIONS

At present, 17 percent of people have no access to a toilet of any kind and practice open defecation—next to houses, along river banks, in the fields, and by roadsides. An additional 11 percent or more of the world's population is not connected to any sewage pipe system, doing their business in buckets, pit latrines, or otherwise unceremoniously dumping their feces outside.[24] Ecologically, this is a crappy state of affairs. We need to render our excreta into composts and dispense it where it would do good.

A composting toilet is the way to go for those in rural areas not linked to any sewage pipe system. The toilet must be thought of not so much as a disposal mechanism, but as a processing unit.

Urine in such a composting toilet is diverted and then funneled to a dedicated container. Feces fall into a bucket held in a brick vault. Soil and ash are added to the bucket after every deposit is made. The contents of the bucket are removed regularly and placed in another site to stew between six and twelve months until the composting process is complete. Two shallow pits, about 1.5 meters deep (about 5 feet) and dug close to each other, are used alternately. For a medium-sized family, a pit takes about twelve months to fill up and this same period allows sufficient time for the mix of excreta, soil, ash and leaves to form compost. Every year, one pit is excavated as the other becomes full. Regular addition of soil, wood ash, and leaves to the pit helps to reduce odor. An investment in a screened ventilation-pipe made of PVC can reduce odors and flies. The vent pipe draws out air from the pit or vault, mostly by the action of air passing across the top of the pipe. The air that flows out of the pipe is replaced by air passing down the squat hole or pedestal.[25]

Composting fecal matter can make sanitary and economic sense as well.

The resultant fertilizer can be sold for a profit or used directly by people to cultivate food. For instance, in rural Niger two bags of chemical fertilizer cost roughly USD 80, or about 10 percent of the average annual income in that country. A family can produce a comparable amount of toilet-made fertilizer in a year.[26] The price of fertilizer can easily defray the costs both of constructing the composting toilet and the labor of emptying and otherwise making the circuit—cleaning the toilet system routinely and providing a fresh supply of organic matter or ash as needed. Hence, organizations can install the composting toilet free of charge and recoup the cost through the sale of fertilizers. Basically, users would be contractually committed to provide the raw material for fertilizers.

For those living in abject poverty in urban environments, where there is no possibility for a compost toilet, or for any kind of toilet, the best solution is something along the lines of the biodegradable Peepoo toilet bag. It is used by placing the bag inside a rigid container, such as a bucket.

The Peepoo bag is made by the Swedish company Peepoople. Its two-layer design ensures that the bacteria in human excreta do not come into contact with the skin. Closure after use isolates the excreta. Each bag contains four grams of urea, which works to rapidly destroy pathogens in the urine and feces (including the hard-to-destroy helminth eggs). The Peepoo bag makes the excreta safe to use as a fertilizer within a relatively short period of time: two to four weeks if the bag is maintained at 20°C (about 68°F) or higher.

In a pilot program in Bangladesh, the Peepoo bag was well-received. After using it a few times, most people were comfortable with the bag and reported that they liked it mostly because it was easy to use and it was not necessary to leave the house. In fact, the Peepoo bag could be used anywhere. The privacy and immediacy of using the bag is a definite step up from the existing practice of journeying to the foul-smelling communal latrine or from the practice of some women of going out in the wee hours of the night, out of eyesight of leering men.[27]

In destitute urban environments, such biodegradable bags are to be tossed after usage into specially-designated dumpsters. One organization or another may empty the dumpsters and deliver them for co-composting. If not paying for the provided manure, the organization can dispense bags for free upon receiving a bag full of nutrients. It's a win-win situation both for the organization, having the raw material for compost; and for the bag user, the alternatives being what they are.

BIOGAS

We can make biogas out of our excreta and out of organic waste in general.

Biogas is comprised mostly of methane with some carbon dioxide thrown in along with traces of water and hydrogen sulfide. It is produced when organic matter decomposes in the absence of oxygen. Biogas can be recovered and then used either directly for cooking and lighting, or it can be upgraded to natural gas and used for a wider range of applications. And the best thing about it: it is carbon neutral with zero land footprint.

All types of organic matter that contain carbohydrates, proteins, and fats as main components can be used as substrate for the production of biogas. Possible feedstocks are cattle dung and manure, goat dung, chicken droppings, meat leftovers, kitchen waste, factory wastes, and human excreta.

Throughout the world, myriad designs of biogas plants have been developed under specific climatic and socio-economic conditions. One of the most common types is a dome plant. It has a round tank, built usually from bricks and mortar in an underground pit, which conserves aboveground space and shelters the unit.

A pipe leads from a nearby waste latrine to the underground unit. The feed material is mixed with water in the collecting tank. The slurry of manure and water ferments in the tank with the aid of bacteria. It may spend three weeks in the bowels of the digester. The resultant gas is piped from the top of the tank to a biogas cooking stove and biogas lights. The slurry comes at the other end and can be retrieved and composted.

There are tens of millions of biogas units in rural China, a few million throughout the Indian subcontinent, a few tens of thousands in Southeast Asia, and a few thousand units in Sub-Saharan Africa. A number of industrial, centralized digester facilities exist in developed countries. An example of a large-scale biogas plant is in Lille, France. It is designed to treat about 108,000 tons of organic waste per year and subsequently compost the solid outcome. Another example is the Henriksdal plant in Sweden. With seven giant digesters, it treats wastewater from about 800,000 people.

While at present the predominant feedstock is dung, livestock manure would have another fate. This is described in the ninth and tenth chapters ("Land Use" and "Nature Restoration").

In the context of biogas production, the plan calls upon human manure instead. We can make electricity out of biogas and poop our way to 46 TWh worth of electricity annually.[28] This would meet about 0.1% of the anticipated electrical needs, which matters very little. For a typical household in an impoverished rural area, where a biogas stove can play a big role, 1,500–2,000 liters of biogas per day may meet its cooking needs. A household of five generates daily human waste for only 60 liters worth of biogas, which matters not at all.

Although household biosolid waste can provide only an insignificant amount of cooking fuel or electricity, billions of us produce a respectable combined amount of manure that can be put to use in other ways that do matter. Let's first see what is the biogas volume we have to work with.

One kilogram of human feces may yield 60 liters of biogas. This means that if we assume a 75% utilization rate, we can generate 23 billion cubic meters of biogas annually. The actual biogas volume will be higher though. Organic waste, such as food scraps, will add to this figure substantially. We generate about 0.5 kg of organic waste for every 0.2 kg of feces.[29] Thus, the volume of feedstock is about 3.5 times larger once we account for the organic waste. Better yet, there is a synergetic effect here. A biogas facility in Luzern yields about 2.5 times less methane than a biogas facility in Bern per kilogram of feedstock.[30] The difference is that the Luzern facility digests solely sewage sludge, while the other facility co-digests organic waste alongside sludge. Hence, I am going to assume we can get a total of 4 times the amount available strictly due to sludge: 92 billion cubic meters of biogas in total, annually.

In order to make it useful, the first step is to clean up the gas, extracting the methane and relieving it of the other trace gases. The extraction process is thus: pressurized biogas flows into a water column where carbon dioxide and other trace elements are absorbed by running water. As a second step, the resultant methane needs to be dried, as it is now saturated with water. Given an average concentration of

60% methane in biogas plus slight methane loss in the process of upgrading,[31] 1,000 cubic meters of biogas will yield, on average, 594 cubic meters of methane. So our collective manure combined with organic scraps will amount to 54.6 billion cubic meters of pure methane, annually. Now that we have all of that gas, the question is what under this plan is the best use for it. The foremost answer is jet fuel.

Once we get pure methane, we can put some of it through the Fischer-Tropsch process, much as is currently done in a gas-to-liquid plant—such as Shell's plants in Malaysia or Qatar. It is not possible to convert all of the methane to jet fuel; naphtha is coming along for the ride. The nature of the process is such that 9 Btu worth of methane would yield approximately 3.9 Btu worth of jet fuel and 1.9 Btu of naphtha.[32]

As laid out in the fifth chapter ("Transportation"), it would take 38 million barrels of jet fuel a year to provide for all our future aviation needs. Out of the 54.6 billion cubic meters of pure methane that our biogas would yield, 13.4 billion cubic meters would suffice to generate all the needed fuel for aviation—along with generating, in the process, the equivalent of 22.7 million barrels of naphtha.[33] Thus, without the need to set a dedicated area for biomass cultivation, we can have carbon-neutral fuel that will take care of all our aviation needs.

In addition, biogas would provide a small fraction of our naphtha needs. Naphtha is used as an industrial solvent for cleaning, as a base for oil based paints, lighter fluid, shoe polish, and olefin. And once we have olefin we are talking about polypropylene. Once we talk about polypropylene we are talking about packaging and textiles.

And this is not all; we still have a bit more than 41 billion cubic meters annually of carbon-neutral methane to work with. This could be used as a feedstock for some of the things natural gas is now being used for. In the Unites States, 0.6 billion cubic meters of natural gas is used as a feedstock for the production of organic and inorganic chemicals.[34] The list of chemicals is long and includes ammonium compounds, aluminum compounds, chromium salts, hydrogen sulfide, nickel carbonate, sodium compounds, acetone, calcium nitrate, ethyl, lactic acid, methyl, sodium compounds, and methanol. Albeit the total amount of natural gas used globally towards that end is unknown, it seems that biogas-derived methane may sufficiently meet these needs, worldwide. With some possible leftovers, we can also use methane for the production of the carbon black pigment used for tires and as a feedstock in the production of steel and aluminum.

While all the biogas we can possibly aspire to will not be nearly enough to take the place of the natural gas, fuel oil, and coal used to fulfill those functions, our produced biogas could still play a respectable role. And each cubic meter of biogas being used for that purpose is a square meter less of biomass we have to grow on earth.

RECYCLING THE REST

The Edmonton Composting Facility, mentioned earlier, has a few screening stations that remove non-biodegradable materials. This is a necessary evil if we dump all our waste products into a single bin, carrot peelings and plastic toys alike. Separation of waste at the source will go a long way.

A good example of a sensible waste-collection system is found at the Hammarby Sjöstad development in Stockholm. Every apartment building has three chutes: for paper waste, for food waste, and for what is mostly plastic waste. All chutes are linked by underground pipes to a central collection-station, where the refuse is siphoned via vacuum suction. The central collection-station houses an advanced control system that sends the garbage to the right container. There is a large container for each refuse type.

Every year, mountains of refuse are unloaded in numerous gargantuan diapers we call landfills. More mountains of stuff are delivered to centralized crematoriums we call incinerators. These batty practices will cease. The goal is for every component in every product and industrial process to either return to nature through biodegradation or be recycled indefinitely in the manmade world.

Everything that can be recycled will be. For everything else, there's plasma arc.[35]

Plasma arc is not fooling around. Essentially, we are talking about a torch generating 5,000°C–11,000°C (about 9,000°F–19,800°F). These are higher temperatures than those on the surface of the sun. The heat from the plasma breaks the molecular bond asunder. The waste is broken down to its basic elemental components. This is a clean process. No burning, no oxidation. You put cyanide in and out comes innocent carbon and nitrogen.

Everything we cannot or should not recycle would pay a visit to the local plasma arc gasification plant. Candidates for a one-way trip include pharmaceutical waste, chemical waste, heavy metals, and tires.

Basically, we drop the toxic waste into a giant hopper, where a shredder breaks the stuff to small chunks and feeds the resultant mélange to the plasma combustion chamber. The organic waste would turn into carbon-neutral syngas. Properly cleaned, the resultant gas can be used as a source of fuel—and in fact may more than suffice as the sole power source for running the plasma plants. Indeed, it seems we won't need to provide the plasma plants with an external source of power. Having said that, as the organic waste has a higher calling and ought to be recycled back into the natural system, most of the elements going into the plasma converter plant would be inorganic. Thus, what comes out would mainly be slag.

The slag can take different forms depending on how it is cooled. If slag is air-cooled, it forms black, glassy rocks that look and feel like obsidian, which can be used in the making of modernistic tables and jewelry. Molten slag can be funneled into molds for casting bricks and paving stones and then air cooled. If compressed air blows through a stream of the molten slag, we end up with rock wool. Much akin to fiberglass insulation, rock wool can be used in a similar fashion.[36]

Going through the plasma process, metals melt and clump together. Afterward, they either can be extracted from the slag or be vitrified, entombed with no real danger of leaching.

Every hospital and all applicable industrial facilities will need to be equipped with a plasma unit or have a contract with a local plasma facility. The same goes for municipal garbage-collection services, for their non-recyclable stuff. Thus, anything from oily waste to pharmaceutical medicine is to be converted to inert bricks, construction aggregates, or insulation material. As a road material, antibiotics and cough medicines can actually do some *real* good. So basically we permanently rid ourselves of high-toxic waste and get a plentiful supply of bricks, insulation material, and paving stones.

One commercial plasma facility of that kind is the Trail Road Facility in Ottawa, constructed by Plasco Energy.

RECYCLING AND HARVESTING WATER IN BUILDINGS

Much of the water for residential and commercial use can be collected on site.

At present, most of the rain that falls on the approximate one million square kilometers of paved area of the world goes down the drain, quite literally.[37] Much of this rainwater merges with sewage and industrial waste and then flushes into the oceans. This cavalier practice ought to come to an end. Rainwater may be collected much as it is collected now from city roads, sidewalks, and roofs of buildings. However, it should not be thrown together with sewage and pollutants.

The collected rainwater may be used as a source of potable water, channeled to a nearby lake, applied in agriculture, or be used to recharge groundwater.

Whether rainwater ends up in a city-wide drainage system or is captured in a tank, every building ought to have roof setup designed to capture rainwater.

The first flush of rainwater after a dry season contains most of the contaminants that build up on a roof, such as dust and bird droppings. A first-flush diverter prevents this polluted water from entering the tank. This aside, the roof area should be kept clean. This involves cutting back vegetation and the periodic washing of the roof to reduce debris such as leaves, pollen, and animal droppings. Rust or other metal contaminants should not leach into the harvested water. Therefore, roofs should be clad in a nonreactive material such as enameled metal. Once brought down from the roof, water should pass through a trap to filter out both large and minute debris.

Rainwater storage tanks, whenever these are called for, can be made from wood, fiberglass, or galvanized steel. These can be above ground or underground. They can be small, serving individual households, or large, serving an entire community. Overflow can be routed to vegetable gardens or to a water facility that would direct the water wherever it may be needed. Collected rainwater is often pressurized for use in plumbing or irrigation. When gravity-fed systems are not possible, mechanical pumps are used. Rainwater destined for potable use needs to be disinfected to remove possible pathogens as well as organic and chemical compounds. Ultraviolet (UV) light removes microorganisms from rainwater, while carbon filtration absorbs organic pollutants and chemicals.

The second major source of water to be used in buildings is reclaimed graywater.

Graywater is water that may have been used for washing dishes, laundering clothes, or bathing. Other than toilet wastes, any water draining from a residential or commercial building is considered graywater. It contains traces of skin, hair, oils, salts, fibers, soap residue, cooking oil, and fecal matter.

To start the process of graywater reclamation, we would need separate piping for graywater. It is technically possible to process the sewage alongside the graywater, but with this we would lose the excreta—for which have an important use. It is a pain to install an additional, dedicated plumbing system to existing construction, but it needs to be done.

The graywater would be channeled to a water reclamation system. As good an example as any of such a system is the one built at the Omega Center for Sustainable Living in Rhinebeck, New York, which can process up to 197,000 liters (about 52,000 gallons) of water every day. This system has been designed by John Todd Ecological Design and is termed the Eco Machine. It is in fact designed to handle all wastewater of a residential or commercial building, sewage included. This is more than is called for under this plan. Be that as it may, the process would be largely the same if the Eco Machine processed only graywater.

The graywater goes through a number of steps.

First, the wastewater flows out to tanks where naturally-occurring microbes feed on the nutrients in the water. They digest ammonia, phosphorus, nitrogen, potassium, and many other substances in the water. This process happens with very little oxygen and produces a modest amount of methane gas.

From there, the derivative liquid flows to a series of wetlands. Each wetland is three feet deep, lined with rubber, and filled with gravel to the top. The wetlands use microorganisms and plants, such as cattails and bulrushes, to eliminate smells, to reduce nitrates, and to remove phosphorus. As the wastewater flows through the wetlands, the microorganisms and plants are fed.

There is a 75 percent increase in the water's clarity by the time it is ready to leave the wetlands and move on to the next phase. En route, some of the water evaporates and some is absorbed by the plants.

From the constructed wetlands, the water is pumped into two highly oxygenated aerated lagoons, each ten feet deep. At this stage, the water looks and smells clean, but it still contains toxins. The plants, fungi, algae, snails, and microorganisms within the aerated lagoons are busily converting ammonia into nitrates and converting toxins into harmless base elements. Beautiful tropical plants thrive in the lagoons. The roots of the plants act as a habitat for the other organisms in the lagoons and are sustained by them. The flowers of these tropical plants illustrate the beauty that naturally treated wastewater can yield.

From there, the water is sent to a recirculating sand filter. Sand and microorganisms absorb and digest any remaining particulates and small amounts of nitrates that may still linger. After the water has moved through the recirculating sand filter, it meets advanced wastewater standards and is as clean as water from one's kitchen faucet at home. The system not only purifies the waste water but also sports minnows and goldfish.[38]

Such a system works well year-round. Even during the coldest winter months, the microorganisms within produce enough heat to keep the water from freezing, as can be seen from another engineered ecological system in Vermont.

A comparable technology is the NextGen Living Machine, such as the installed wastewater facility at Furman University in Greenville, South Carolina, or that at the Esalen Institute in Big Sur, California. The Living Machine is a product of Worrell Water Technologies.

The folks at Worrell estimate that the effective capacity of such a reclamation waste system is in the service of between 20 and 200 households. They estimate a footprint of about 12.5 square feet per household. This means that 200 households may have a treatment area that is about 2,500 square feet (about 232 m²). This can be set in narrow, 4-foot-wide corridors along sidewalks.

Such biological wastewater treatment systems can be set up world-wide, each serving up to two hundred households. This may mean each wastewater treatment unit serving a housing subdivision in the suburbs or a large apartment building in the city.

The purified, reclaimed water will be recycled back into the building for toilet flushing. The excess may be used either to water neighboring vegetable gardens or to recharge groundwater. Alternatively, with an addition of UV treatment, it may be used as a source of potable water. With rainwater samples found to contain the hormone-disruptor atrazine,[39] I am unsure how much rainwater we ought to drink—or groundwater, for that matter—if we have alternatives. As it happens, we do. More on this in the next chapter ("Land Use").

Finally, there is the issue of heavy metals in water. Engineers at Brown University have developed a novel electrowinning system that removes heavy metals, such as copper and cadmium from water—effectively and markedly. Such systems should be commercialized and then installed in places such as textile factories and metalworking plants.[40]

The entire schematic outlined here is shown in figure 8.1 at www.danielrirdan.com

9

⚍

LAND USE

reinventing the food system
bringing a new water source online

TIME TO SWITCH GEARS.

No more harvesting or otherwise tampering with the seas; somehow we have to cultivate on land the equivalent of the 118 million tons of fish harvested annually from the ocean.[1] In addition, under no circumstances will new land go under cultivation. The degradation of natural habitats is to come to a stop and reverse its course. That also means no more hacking down forests. As these are to be off limits, we will have to provide for our lumber needs from elsewhere. Furthermore, in the case of drained wetlands converted to agricultural land, the crops have to go and the wetlands must be restored—enabling them to slowly recapture their vast stores of carbon which were released when they were drained. We would need to relinquish some land back to nature, such as the palm plantations in Indonesia that are to revert to a tropical rainforest and be made into a sanctuary for orangutans. Last, we are to leave the rivers and aquifers alone. This means that we let the rivers be rivers once again, and not engineered, vast water and sewage canals.

While we are backing off and reducing our ecological footprint across the land, we also need to do the seemingly contradictory: find space for the expansive power-generation infrastructure described in the seventh chapter ("Energy").

Pulling all of this off while feeding everyone on board *would* be a neat trick. Well no, not really; there is no trick, no magic wand. We will have to do it the hard way: by tightening the belt and acting wise.

Before we get down to specifics, I trust you have a sense of what's coming. Some things *will* have to give; some things *will* have to change. No sacred cows. In the ensuing analysis, nothing bearing a significant footprint will be spared consideration. For the first time in our history, we will have to set caps on our collective consumption and footprint. Seven billion people are assumed. This is what we have now, and this is what we've got to work with. However, it is also assumed we would slowly bring our numbers down. The subject of overpopulation is treated in the twelfth chapter ("Consumption").

Let's scrap the existing system. Let's dismantle the centralized feeding lots, the fecal concentration-camps for livestock. Let's dispense with the injection of hormones to animals, the outpouring of ammonia, the herbicides, and the insecticides.

CUTTING DOWN OUR CALORIES

Taking all the food available for human consumption and dividing it by the number of people comes to 2,770 calories available per day for every woman, man, and child on the planet. Peru has calorie availability that averages 2,430 per person. About 16% of its population is undernourished and about 16% are overweight. Let's call it a wash between those in Peru who are overstuffing themselves and those who don't have enough to meet basic dietary requirements. Let's assume that 2,430 calories availability per person will keep us within operating specs. This analysis is simplistic almost to the point of being useless. However, it does suggest that if some of the calories in the world are to be pulled away from the donuts and soft drinks of the estimated 12% that are overweight, we will be able to more than adequately feed the estimated 13% that are undernourished. With 2,770 calories available for each person and 2,430 calories needed, we have, literally, more than enough food to feed everyone.[2]

FOOD WASTES

Calorie availability is just that; it is not the amount that is actually being eaten. A sizeable portion of the food never makes it to our mouths; it is being thrown away.

A USDA study assessed the avoidable food loss by food retailers, consumers, and the food service industry. In 1995, about 96 billion pounds of food were wasted in the United States. This was 27 percent of the edible food available for human consumption during that year in the United States.[3]

When in doubt, we take the food we stored in the fridge and discard it. Often, we opt not to keep leftovers, and when we do save them, they eventually still end up in the trash can. A case in point is the situation in the United Kingdom. Every year, items that are thrown away whole and unused include 4.8 billion grapes, 1.9 billion potatoes, and 1 billion tomatoes. To those, one ought to add 161,000 tons of fish and meat and 45,000 tons of rice. Over one quarter of the food tossed out is still in its packaging. Some of the discarded food is left on the plate after a meal, some has passed its expiration date, and some has gone moldy. The study that aggregated these data surmised that close to a fifth of all the food purchased by a household is wasted and would have been eaten if it had been managed better.[4]

All food ought to have expiration date labels that would reflect genuine food-safety risks. The same label would have clear and graphically inviting instructions on the optimal storage of the given food commodity. A label that carries both a realistic expiration date and storage instructions would help to minimize the amount of food wasted and ultimately purchased in the first place. Meanwhile, the folks at the University of Strathclyde in Glasgow are working on food packaging that has an intelligent plastic indicator, letting the consumer know when the food is actually going bad—versus a speculated, generic date.

We can do more. As a society, we can both promote and subsidize the use of vacuum sealers. Vacuum packaging of food does prolong the shelf life of a wide range of food products, from dry cereals and nuts to vegetables and meats.

Through the decades, portion sizes have increased across the board, whether in the pages of cookbooks or in fast food places. Hamburgers have gone supersize; portions of popcorn and soda have acquired epic proportions, and even apples are bred large, too large to make a good portion for many. All of this drives either overeating or waste, which comes down to the same thing. At many restaurants, jumbo portions are often left on the plate unfinished. Furthermore, food stores refuse to run out of anything during business hours. This builds excess into the system. It ensures surpluses by closing time.

Much of the perfectly edible produce targeted for retail is never picked as it does not conform to one shape or another. Be it cucumbers or bananas, carrots or eggplants—they may not be straight enough, not curvy enough, not long enough, not short enough, not skinny enough. There is not a problem with less-than-perfect produce when it can be used in canning or juicing. As for the rest, we should revert back to embracing fruits and vegetables that are eccentrically shaped and also those that have bad spots, which can very easily be removed with a paring knife. Those with blemishes can be discounted at the store.[5]

PETS

At any given time, we entertain in our homes and backyards hundreds of millions of dogs. Five to ten percent of these are served as food in restaurants and homes, mostly in East Asia. Other dogs herd cattle, work as guards, or assist the blind. However, the vast majority are neither raised for food nor used for work. The vast majority of dogs have the role of pets, of companions, and many are provided with food better than some children have access to.

We don't do those dogs any favors when we own them as pets. The majority of the dogs are confined to small quarters, whether indoors or in a backyard, with a daily entitlement of a twenty-minute walk around the block. There is no doubt; we *own* the dogs. We cut off their balls—figuratively and in many cases literally. They will never be allowed to become elders or even know adulthood. They will never lead a pack nor experience true parenthood. Their growth is arrested. They are a lowly member of the household pack, and thus they remain until their last breath. A pampered girl in a gilded harem is still a slave. It was never really about the dogs; it was always about us. In the role of pets, dogs (*Canis lupus familiaris*) are prevented from realizing, from actualizing themselves as such. If one mulls over this long enough, one reaches the inescapable conclusion that this *is* animal cruelty. We can find our needs met, such as the need for companionship, from within our own ranks.

In the end, in the context of this blueprint, my final argument lies elsewhere. In the end, it comes down to the environmental equivalent of dollars and cents. We have overdrawn and need to cut back our food consumption. That's the issue at hand. As far as pets go, if we stop breeding them, we have hundreds of millions fewer mouths to feed. This addresses dogs used strictly as playthings, not working dogs such as those trained to work with the severely depressed, to render assistance to the blind, or to work at the police department or with graziers.

The same goes for cats, to the extent they do not take a part in holding in check the population of mice and other pests in the urban environment.

Next are horses. Worldwide, we maintain control over about sixty million horses. Some of these are working horses; some are used as a source of meat. The majority of these noble beasts, though, serve humans as a source of entertainment—in races, leisure riding, and rodeos. All considered, we should cut out these extra-curricular activities. It is not that the footprint of horses in the wild, where they truly belong, would be vastly different from their existing footprint. However, it is one thing to mechanically mow the grasslands down for the feed of horses in our collective backyards. It is something else to populate the barren grasslands of the world with horses, which would dung and aerate the land and overall take an active part in the broader food-chain.

CROP RAISING

Considering we have been at it for about ten millennia, those few who have developed regenerative food-growing practices are not so much a testimony to the innovativeness and ingenuity of humankind as a testimony to the questionable state of mind of the rest.

Our prevalent monoculture method of crop raising is removed from the rhythms of nature. Hence, it can be maintained only with colossal life-support systems of manufactured fertilizers and hormones and held together by the flooding of land with chemicals and irrigated water. This food-production system is the brainchild of buffoons and monkeys with wrenches. With its pruning and gassing and weeding, it is ludicrous. With its land dedicated to the growth of grain expressly to feed animals, it is deranged. With its crowding of animals into giant closed buildings, it is monstrous.

This environmental freak show is to be scraped away along with its required vast energy inputs. The blueprint calls a halt to the usage of chemical fertilizers, herbicides, and pesticides. There won't be a need for their services any longer anyway. And unlike the application of compost, no matter how much synthetic chemical fertilizer is used, it does not improve nitrogen levels in the soil.[6]

What follows are the building blocks of *another* food system. These are to be just the opening moves. When the considerable resources and ingenuity of millions of people shift along these new pathways, what is about to be discussed will evolve in myriad ways across the various economic and physical landscapes of the world.

First and foremost, the principle underlying all that is to follow is the return of organic matter into the soil. No more landfilling it, incinerating it, or flushing it to the waterways. The nutrients within excreta, both that of humans and animals, will find their way to the soil. To the extent that is logistically possible, all food waste and crop residue, all timber residue and scraps, and all slaughterhouse waste will be duly processed and then routed to the earth.

The obvious, immediate derivative of this precept would be the hasty demise of biofuel. We are not going to burn up organic matter:[7] not for warmth, not for cars, and not for power generation. The energy-generation plan presented in the seventh chapter ("Energy") has deliberately rendered biofuel unnecessary. And it is not even carbon neutral; at least insofar as wood fuel is concerned, it results in a net increase of GHG emissions.[8] Among other things, this refers to the tens of thousands of square kilometers in the United States of grasslands now dedicated to the growth of maize for the making of ethanol. This refers to the tens of thousands of square kilometers of rainforest in Indonesia converted to palm oil plantations to be used in part as a feedstock for biofuel, which incidentally causes GHG emissions no better than that of petroleum.[9]

This all ought to be over and done with.

———————

Organic agriculture should be the basis for any crop-raising method.

Since 1981, the Rodale Institute, Pennsylvania, has conducted the longest-running scientifically controlled comparison of organic and conventional crop production practices in the United States. Had the study stopped after two or three years, its results would have affirmed the conventional wisdom that organic agriculture cannot compete with the yield of high-input conventional agriculture. However, five years into the study, the narrative of this study changed significantly. Past the first five years of building up the soil in the organic patch, yields were comparable to crops grown conventionally in the next patch. In fact, organic yields exceeded conventional yields during severe droughts.[10] The high levels of organic matter helped conserve soil and water resources and proved beneficial during drought years.

Then there is the matter of soil fertilization. The main limiting macro-nutrient for agricultural production is biologically available nitrogen. To maintain sufficient levels of nitrogen in the soil, we either synthesize ammonia and administer it to the land in the form of a fertilizer, or we make the nitrogen bioavailable the organic way.

Industrial synthesis of ammonia takes hydrogen and gaseous nonreactive nitrogen and combines them into ammonia under high temperatures, high pressure, and with the aid of an iron catalyst. The ammonia, in turn, is an important precursor for nitrogen-based fertilizers. We produce about 130 million tons of ammonia per year. Alas, overloading the planetary ecosystem with biological nitrogen comes at a price.[11] It leads to loss of diversity, compromises air and water, and poses threats to human health across large areas of the planet.[12] It is also the cause for the dramatic rise in atmospheric nitrous oxide, a greenhouse gas.[13]

The second way to sustain bioavailable-nitrogen in the soil is the old-fashioned way. Nitrogen amendments in organic farming derive from crop residues, animal manure, compost, and leguminous plants.

Leguminous plants can be grown during the winter fallow period, between food crops, or concurrently with the food crops during the growing season. At a later date, they are plowed back into the soil for the benefit of subsequent crops. The plot at the Rodale Institute has achieved its yield without outside fertilizer input and while producing a cash crop every year. That is to say, all of its nitrogen and nutrient needs were obtained from red clover under sown with the main crop or through the cultivation of hairy vetch as a winter crop.[14] The rotations of cash crop and cover crop,[15] typical to organic agriculture, have been shown to reduce soil erosion, pest problems, and pesticide use.[16] At the Rodale Institute, continuous soil improvements after two decades have resulted in marked environmental benefits and production resiliency during weather extremes.

In recent years, hundreds of thousands of smallholder farmers in southern Africa have been incorporating nitrogen-fixing trees into their lands. These unique trees, such as the long-lived acacia tree *Faidherbia albida* replenish soil fertility by transforming atmospheric nitrogen and making it available in the soil through their roots, leaf litter, and eventually the entire biomass of the tree—after it is cut and left to decompose in the soil. Farmers practicing this fertilizer-tree system have reported a doubling of their maize yields.[17]

The principal objections to making a wholesale switch to organic agriculture are claims for lower yields and insufficient quantities of organically acceptable fertilizers. A major study evaluated the universality of both claims.[18]

The data for this study came from dozens of different sources originating in the United States and in prosperous countries in Europe. That is to say, these are case studies of organic agriculture pitted against the high-input, high-yield crops found in their conventional agriculture counterparts.

From a dataset of 160 cases, the average yield for organic farming came to 92% of that grown by their conventional counterparts. It ranged from 87% yield for vegetables to 106% in the case of eggs grown organically. Yet, organic farming yields may be higher than those figures suggest. Observational data in some of those organic farms goes back twenty consecutive years. However, many of the case studies that made up the average were comprised of only one season. Production initially declines on the heels of a conversion from conventional to organic farming. Had all case studies been of established farms, the yield is likely to have been higher across the board.

Some papers dispute the ability of organic agriculture to feed the world. For instance, one study surmises that only 62 percent of the current population could be fed without the use of manufactured nitrogen fertilizer.[19] It is not the only study to infer that organic farming cannot feed the world.[20]

A more recent study re-evaluated the yields of organic and conventional agriculture.[21] The study found that under best management practices the organic yield in established organic farms averaged 87% of their conventional counterpart.

Data from temperate and tropical organic farms suggest that crops such as peas, beans, or clover could make enough nitrogen bioavailable to replace the amount of synthetic fertilizer currently in use—with no additional land area needed to grow the nitrogen-generating crop. This can be done during winter fallow periods or can be grown between plantings of other crops. In arid regions, drought-resistant crops such as pigeon peas can be used to provide the needed nitrogen. None of this accounts for the additional sources of nitrogen available from compost discussed at length in the previous chapter ("Nutrient and Water Recycling") or from the rotation of livestock with annual crops, to be discussed later in this chapter. A quick, back-of-the-envelope calculation shows that the combined annual manure produced by our stock of cattle, hogs, and broilers contains 149 million tons of nitrogen.[22] This is more nitrogen than our current, extravagant practices utilize in producing food for the human family, which this plan posits is at peak, and its numbers would decline in subsequent decades.

But ultimately it is of no concern whether existing organic practices can provide the 131 million tons of nitrogen in use today. Current non-organic practices flood the fields with nitrogen. Assuring that the roots have access to nutrients in this manner is akin to washing our feet by flooding our apartment. The word "excess" does not begin to describe it. The studies that find synthetic fertilizers to be essential for feeding the world population are of limited relevancy to this plan. Large-scale organic agriculture as is practiced today is still but a simplistic monoculture with plants that are annuals. Many of the nutrients never make it back to the soil. It is not much of an ecological system.

Much better agricultural systems than these are possible and indeed are called for.

PERENNIAL GRAINS

Annual cereal, legume, and oilseed crops are the foundations of the global food supply. As the annuals are short-lived and have less extensive root systems than do perennial species, they cannot retain as many nutrients or filter as many pollutants. Annuals are also rough on soil life, as they have to be planted every year anew. This causes more soil erosion and droughts than would happen otherwise. This explains much about the state of the soils of our planet—and none for the better.

With their roots commonly exceeding depths of two meters (over six feet), perennial plants have essential roles to play, as they manage water and help cycle nitrogen and carbon. A widespread perennial polyculture would be an effective tool to deal with persistent droughts and prevent erosion of land.[23] They would hold the potential to rectify hydrological imbalances, nutrient issues, and carbon losses.[24]

In a study based on data encompassing one century, annual crops were found to be 50 times more susceptible to soil erosion than were perennial pasture crops.[25] Scientists have also documented a 5-fold increase in water loss and a 35-fold increase in nitrate loss from soil planted with corn and soybeans as compared to that with alfalfa and other perennial grasses.[26] What's more, as rainwater trickles below the relatively shallow root zone, it results in soil salinity in some places. One study projected that the potential salinity could be sharply reduced by replacing annual wheat with a perennial counterpart, if such was to be developed.[27]

Plants may face physiological tradeoffs between seed productivity and longevity. Resources that otherwise would have been diverted to seeds may instead go below ground to maintain the long roots of a perennial. It is a balancing act. However, this would not necessarily prevent perennial grain crops from achieving yields comparable to those of annual counterparts. Compared with annuals, perennial crops tend to have longer growing seasons and deeper root systems. They intercept, retain, and utilize more precipitation. Thus, all considered, perennials may very well end up producing yields comparable to those of their annual counterparts.[28]

Right now, we don't have perennial vegetables and grains. We do have a few perennial vegetables on the sidelines, such as Udo (*Aralia cordata*), and watercress (*Nasturtium officinale*), but that's about it.

A number of research centers are working on developing grain species that will function much more like the natural ecosystems that were originally displaced by agriculture. The researchers aim to replace the major annual grains with perennial counterparts, which they are in the process of breeding. At the forefront of this revolution have been two research centers: The Land Institute in the United States and Future Farm Industries Cooperative Research Centre in Australia. So far, they have produced hybrids of wild and cultivated species of wheat, wheatgrass, sorghum, and sunflowers. To get us all the way there will take a few more decades. This would be a first step in turning the world away from ten-thousand-year-old practices and making it possible for perennials to assume, once more, their dominant role in the grassland biome.

PASTURE CROPPING

We don't yet have perennial grains for consumption. However, we do have something else, something that cashes in on the benefits of perennials: pasture cropping.

Pasture cropping was pioneered in Australia. It is a technique of sowing crops into living perennial pastures and having these crops grow symbiotically with the pastures. Usually, the perennial plants of the pasture are summer-growing plants, and after the first frost of autumn they become dormant. Hence, there is little overlap between the growth periods of the perennials and the annuals. In pasture cropping, the landscape has permanent plant cover all year;[29] there is another layer of energy capture. Pasture cropping relies on perennial grasses to maintain organic carbon in the soil. With year-round cover, waterlogging is reduced.[30] The livestock can graze right up until sowing, accommodating an additional six months of grazing. This system constantly improves pasture performance and biodiversity.

Soil benefits in many ways from the year-round presence of living plants: reduced erosion, buffered temperatures, enhanced infiltration, and a markedly improved habitat for soil biota.[31] It has been found that one of the greatest benefits of pasture cropping is the stimulation of the number and diversity of native perennial pastures. Crops and pastures benefit each other.

Having said that, at least one study found the perennial grasses to somewhat inhibit the annuals' crop performance. Growing through the summer, perennial grass prevents, to some extent, the accumulation of nutrients and moisture made available for the annuals.[32] However, in total, there is a greater useful biomass generated throughout the year than for either cropping or pasture treatments alone.

Livestock are a critical part of the pasture cropping system. Large mobs of sheep and cattle are rotated to manage weeds, create litter and mulch, and prepare areas for cropping. Sheep are put into a pasture at a density of 70–80 head per hectare for up to 6 days to trim down the grass. This is repeated 30 days later. The sheep control weeds, open the grass canopy, mulch the grass, and help feed soil microbes. One week after the sheep are removed, the thick litter holds the weeds in check.

Pasture cropping holds a remarkable promise: with some possible compromises in yield, we may be able to grow crops and have pasture to feed our livestock—all on the same piece of real estate.

MASANOBU FUKUOKA

In nature, plants live and thrive together. It is not a cutthroat competition for nutrients but a dynamic based on mutual benefits. In the context of crop growing, weeds may have their role. Their stronger and deeper roots break the hard subsoil and bring nutrients to the topsoil. As weeds open up the soil and build up its fiber content, crop roots can more easily gain additional access to nutrients. Many of the weeds are a source of food for birds, which in turn enrich the soil with their droppings. Finally, when the weeds die, their deep roots decompose and enrich the soil. If some plants are truly overbearing, it is wiser to control their population with other plants rather than seek to uproot them. If the farmer were to grow cover plants beneficial to him, then he would no longer have to weed; the cover plants would assume dominance over potential weeds.

These have been some of the insights gained by Masanobu Fukuoka in his many decades of working with the land in Japan. Fukuoka practiced a form of rice-barley cropping, in which he seeded barley together with clover over the standing heads of rice. He scattered rice seeds and crop residue while the barley was up. Through the decades, he came to see that grains can be grown without fertilizer or weeding and can attain yields comparable to those achieved by conventional high-input methods.

What follows is a simple, yet carefully ordered cycle that Fukuoka refined and articulated through the decades. This system offers us the means to dramatically reduce our footprint as we cultivate the two cardinal crops of the world: rice and wheat.

In the autumn, Fukuoka broadcast white clover over the mature rice field.

About a month later, he scattered the winter grain, either wheat or barley, over the same field in a similar fashion. When seeded over the standing heads of rice, both the clover and barley seeds readily germinated because of the high moisture in the ground.

A couple of weeks after the winter grain was sown, Fukuoka harvested the rice. At that stage, the young winter grain shoots emerged from the soil.

After leaving the rice to dry for three days, he threshed it and then scattered the residue straw, uncut, over the entire field. Fukuoka followed this with a layer of chicken manure. He found no need to prepare compost. Left to its own, the straw with the chicken manure would completely decompose within six months.

One month later, Fukuoka used clay pellets containing rice seeds and scattered the pellets over the straw. Once scattered, the seeds within the hard clay pellets did not sprout until rain had fallen and conditions were just right for germination. Nor were they eaten by mice and birds. These clay pellets were like protective shells or time-release capsules. The scattering of these clay pellets containing rice took place toward the end of November. Nothing much was required of him until May. In the spring, a thick layer of clover grew at the foot of the maturing barley, and beneath the clover, rice seedlings began to emerge. Fukuoka's system did away with the intermediate, labor-intensive step of seedling beds.

In late May, he harvested the winter grain. Fukuoka left the barley on the ground to dry for a few days and then gathered it into bundles, threshed it, and cleaned it. Subsequently, he scattered the residue straw uncut over the field. At this stage, he also spread over it a layer of dried chicken manure.

Shortly thereafter, he released forty ducklings per acre onto the field. Not only did the ducks weed and pick off insects, they turned the soil and fertilized it. Fukuoka then flooded the field for four to seven days to weaken the clover and give the rice shoots a chance to break through the cover of clover. The fields were then surface-drained in order to grow as hardy rice plants as possible.

In the months that followed, irrigation was not really needed. However, depending on how the plants were doing, he might water the field once every week to ten days.[33] In this manner, he continued to irrigate intermittently but to an extent where the land did not retain water for more than five days at a stretch. A soil moisture level of about 80 percent would suffice.

This completed the annual cycle of sowing and harvesting.

TABLE 9.1

Fukuoka's sowing and harvesting schedule, in brief	
autumn	broadcast white clover
1 month later	broadcast winter grain (wheat or barley)
2 weeks later	harvest rice, leaving it on ground to dry
3 days later	thresh rice and scatter residue straw + chicken manure
1 month later (end of November)	scatter rice seeds
late May	harvest winter grain, leaving barley on ground to dry
a few days later	gather to bundles, scatter residue straw + chicken manure
a bit later	release ducklings to pick off insects, turn soil, fertilize it
a bit later	flood field for seven to ten days

After decades, Fukuoka managed to reduce his labor to the broadcasting of seeds, the spreading of straw, and limited water-management. The required labor and energy input is but a fraction of that in conventional farming methods.

As the earth cultivates itself naturally via roots of plants, earthworms, and such, there is no need for cultivation. If left to itself, the soil maintains its fertility naturally. Weeds should be controlled, not eliminated. They play their part in building soil fertility. Very few agricultural practices are actually necessary. Fukuoka maintained that cultivating the land is in fact harmful. It removes the green cover and exposes the bacteria to sunshine. Just as we need clothes to protect our skin from sunshine, the soil needs a green cover. He had worked his land for over fifty consecutive years. There had not been a need for the land to rest. The soil stayed fertile perpetually.

Plowing or weeding was eliminated. Herbicide and pesticides were avoided. Transplanting rice seedlings was found to be unnecessary, and so was the preparation of compost. Last but not least, Fukuoka did away with the permanently submerged rice cultivation of the paddy—and with it the harmful greenhouse gas emissions in the form of methane.

Fukuoka held that true high-yields come about through the spirited activity of nature, never apart from nature. The bio system that is nature cannot be dissected or broken into its parts. What he achieved is a balanced rice field ecosystem, with insects, plants, small animals, and earthworms.

Without any chemical boosts, Fukuoka achieved 88 bushels per acre with 60 pounds a bushel for a total of about 5,200 pounds of rice per acre (about 5.8 tons per hectare) and a comparable yield of barley. This compared favorably with his neighbors who practiced conventional rice-growing methods under comparable soil and climate conditions. In Japan, across various soil conditions, the average yield is about 5,900 pounds per acre for rice and about 3,000 pounds for barley. However, the common practice in the region is to grow barley on one piece of land and rice on another. The norm is to dedicate a field just for the growth of rice.

Year in, year out, Fukuoka went out to the field armed with nothing more than uncut straw and dried chicken manure. And year in, year out, he reaped the equivalent of 5,200 pounds of rice plus 5,200 pounds of barley per acre.

SEPP HOLZER

In Kenya, researchers and farmers developed the push-pull strategy to control parasitic weeds and insects that damage the crops. The strategy consists of pushing away pests from corn by inter-planting corn with insect-repellent crops like *Desmodium*, while at the same time pulling the pests toward small plots of napier grass, a plant that excretes a sticky gum which both attracts and traps pests. The *Desmodium* can also be used as fodder for livestock.

The net outcome of the push-pull strategy doubles maize yields and milk production while, at the same time, improving the soil.

In a polyculture system, many elements supply the energy input for others, and the system can be mostly self-regulating. The design of such a system relies on species interaction, maximizing the benefits of each species to the others. The system inherently requires less input in energy and labor as it routes the forces of nature rather than pushing against them.

Sepp Holzer tends 45 hectares of land (about 111 acres) amid the forested mountains of Austria. His land includes 60 ponds and thousands of fruit trees, shrubs, vines and highly productive vegetables and herbs at an altitude of 1,500 meters (about 4,900 feet). He has brought into existence a self-sustaining farm in which he tends to many varieties of fish, fruits, nuts, vegetables, mushrooms, pork, and poultry. All of this without irrigation, fertilizers, pesticides, or weeding.

His work is a good case in point of what is possible when one hits on all notes. Holzer employs and at times orchestrates the flow of water, the heat released from boulders, the induced fog of dawn, the slope of the land, and the characteristics of the soil—weaving all of these into a regenerative living tapestry.

For decades, Holzer has kept brown trout, arctic char, carp, pike, and minnows. The ponds provide habitats for innumerable useful creatures such as snakes and amphibians. These animals play an important part in the regulation of so-called pests. An example of this is the toad whose favored prey is the Spanish slug (*Arion vulgaris*).

The pond provides additional services to the broader ecosystem Holzer has fashioned. The bodies of water throughout the land help moderate temperature fluctuations on the neighboring hills. This effect can be augmented with boulders, which retain heat and then radiate it slowly to the surrounding area when they cool. The ponds magnify the moisture levels in the soil and create useful microclimates through evaporation. This generates proper moisture, light, and temperature conditions for differing kinds of plants in a relatively small area. Every such microclimate harbors a particular community of plants, animals, and insects. The diversity of plants and animals helps to create a system in which every species finds its natural place. This is the only way to prevent the population of any one species from becoming dominant and undermining the integrity of the broader ecosystem. Microclimates let plants grow in regions that otherwise would not support them. In the case of Holzer's land in the Alps, examples include the prickly-pear cactus and kiwi fruit.

Holzer's raised beds are made of organic material. As the inner parts of the beds decompose, they release heat, which in turn improves the conditions for germination and plant growth. The decomposition also releases nutrients, making it possible to cultivate more demanding varieties of vegetables without using fertilizers. Holzer incorporates entire tree trunks and shrubs into the raised beds. Those rot slowly. Thus, over a long period of time they provide sustained heat, nutrients, and a balanced level of moisture within the system. Raised beds such as these last for ten years.

In areas that need a build-up of humus in the soil, Holzer grows plants expressly selected for this, such as sunflowers or peas. After they die, the dead organic matter slowly decomposes and builds up a productive layer of soil. In autumn, Holzer usually just lets these plants be. Nature takes over and does all the work. The first heavy snow pushes down the plants. Subsequently, they die and shortly thereafter begin to decompose.

Holzer found that hybrid seeds are unsuited for a polyculture system. These overly bred seeds are more susceptible to disease because they are not suited to any given local conditions. They are one-size-fits-all.

Livestock play an integral part in the ecosystem Holzer has established. By allowing livestock to be free-ranging and using a system of paddocks, Holzer reduces to a minimum his workload and the amount of feed required. His pigs loosen compact soil and till his terraces. Large amounts of fallen fruit in orchards can lead to the spread of mold and fungus. If pigs are brought over to pasture at the right time, the issue is generally averted. Pigs also help to hold in check the number of snails on the land. It is easy enough to direct the pigs by scattering maize in the desired areas. However, if needed, moveable pens can be used to set the pigs exactly where they are required. In the case of pesky weeds, Holzer sends out his pig crew to the area. Ahead of time, he broadcasts beans or corn amid the unwanted plants. The pigs will either uproot the weeds or eat some of them.

After he removes his pig crew from that area, Holzer plants Jerusalem artichokes, sunflowers, or hemp. These beneficial plants absorb all of the excess nutrients and make the conditions hard on the weeds. They thrive on the nutrient-rich soils, so they overshadow and kill off any remaining problem-plants.

Many of the modern breeds of pigs no longer have the natural instincts they need to be good workers. Consequently, Holzer keeps old pig breeds on his farm, such as the curly-hair mangalitza. Holzer keeps livestock outside and in family groups. He built them simple earth cellars and open shelters for the cold winter nights. As it turns out, when pigs have the room to move, they keep their pigsty clean. The pigs are largely self-sufficient. Even in winter they find enough food beneath the snow. And in the end, Holzer also gets to have fine bacon.

This manmade ecosystem is also a bird haven. Insectivorous birds like robins are more than happy to oblige in keeping in check the population of beetles, butterflies, greenflies, and whiteflies. Good forage plants for local bird species are the elderberry and wild cherry, among others. The birds do not need to be fed; they find enough food amongst the diverse plant life on the farm even during the winter. All it takes is a measure of graciousness, leaving some of the garden's offerings for the birds to pick on.

Holzer has built into his system a deliberate excess of food, accommodating the needs of all denizens of his land. When he leaves a modest portion of the harvest for animals, those can live within his world and play an essential role in maintaining its stability. Beyond the promise for a regenerative, self-sustaining food growing practice, such a polyculture offers the promise to integrate the manmade world and blend it into wild nature. Diversity is at the heart of this polyculture farm. This spells stability and the confidence that no parasites or pests could gain dominance.

This polyculture microcosm is a basis for reinventing farming as a complex system that is, above all, anchored in the rhythms of nature. Such an ecosystem holds the potential to convert large swaths of Earth from vast desolate monocultures with dead soil into places that require minimal input and reconnect us with the dynamics of our planet.

FOOD CULTIVATION IN POPULATION CENTERS

If out there, in the vast open areas, the aim is the rewilding of nature, the closer we get to suburban and from there to urban centers, the more cultivated the land will get—while still harboring rich diversity of life.

The Grow Biointensive method is taught and propagated by Ecology Action based out of California. It is self-sustained, self-contained organic farming in hyperdrive, expressly designed for the dense, urban cityscape.

A five thousand square foot area ($\approx$ 465 m²), which includes access pathways, could provide one person with a complete vegan diet.

The total *unpaved* urban area is estimated at about 815,000 sq km.[34] If we assume that we would grow vegetables of one kind or another on 70 percent of that area, it would encompass a total area of about 570,000 sq km. With this area in mind, the potential is to grow within city limits enough to provide all the vegetable, grain, and fruit needs of about 1.2 billion adults—right there within city limits. This would provide for about 20 percent of the world's adults.[35]

This is significant. It directly cuts down on our cropland area beyond the bounds of the city, making that much more room for the natural environment. However, the principle of growing food locally as an end in itself may or may not make ecological sense—depending on the biome in question. Either way, the long-distance transportation footprint is not as significant as may be imagined.

At the heart of the Grow Biointensive method are double dug beds, with soil loosened to a depth of 24 inches ($\approx$ 60 cm). Other key principles include the use of compost and plants grown in close proximity to each other. Plants are grown in combinations that allow them to enhance each other, for instance, beets and onions or peas and carrots. A careful ratio of various types of plants is maintained to assure both a wholesome diet and enough carbonaceous material for the needed compost.

Grow Biointensive includes in its garden scheme a large variety of herbs to keep some of the pesky insects at bay. Marigolds may deter whiteflies, pennyroyals keep ants at bay, and nettles repel aphids.

The system is self-contained and self-sustaining. To make it work, it mandates certain planting schedules that vary from year to year. One template developed under the Grow Biointensive method calls for the planting in the first year of heavy feeders that draw upon large amounts of nutrients, such as tomatoes or corn. To return nitrogen to the soil, the second year calls for nitrogen-fixing plants such as peas and beans. In the third year, under this schedule, light feeders (root crops such as turnips and green peppers) would be planted to give the soil relief.

One template of a food garden 1,300 square feet in size ($\approx$ 120 m^2) contains dwarf plum, pear, peach, apple, and cherry trees. In addition, it has beds of potatoes, tomatoes, sunflowers, pumpkins, zucchini, green peppers, carrots, eggplants, spinach, peas, beets, corn, lettuce, cucumbers, and various herbs. Ecology Action offers viable templates for areas as small as a patio.

This is exactly what Grow Biointensive is: a template. One kind of a template. In fact, there are endless routes to the creation of ecological fruit and vegetable gardens. All ecological gardens, however, strive to be mature ecological systems. This mean systems that have a high diversity of plants and animals that interconnect, are largely self-sustaining, and have multiple layers of plants—from the low herb-layer, through shrubs, and all the way to the large trees that make up the overstory. All ecological gardens have rich soil biota that can support the wide array of plants and insects. All have minimal irrigation requirements as their soils act as sponges.

Vertical farming offers another possibility.

The vision of vertical farming is that of growing food in dedicated skyscrapers, in essence a series of greenhouses stacked sky high, where food is grown in a tightly controlled environment. With this technology, it would be possible to grow any crop anywhere at any time of the year: mangos in Moscow in the dead of the winter and watermelons in the deserts of Arizona. So far it is but a vision, and it is hard to assess its merit without a detailed lifecycle analysis. Yet, it is worthy of mention.

Dickson Despommier has provided a hypothetical example to suggest the land savings of such vertical farms. A 30-story vertical farm straddling 5 acres of land ($\approx$ 2 ha), may furnish the equivalent of 150 acres ($\approx$ 61 ha) of farming area. In other words, an insignificant amount. Yet, what we may be able to pack in such small quarters changes the equation. First, this is a year-round operation; a multitude of crops can be grown within a 12-month period. Second, we can grow dwarf varieties that are high in nutrition. Third, we can grow the plants in far higher proximity than they are grown on open land. When we combine all of those factors, Despommier reckons that we can produce the equivalent of 2,400 acres (≈ 971 ha) of food within a single building. The number may be higher yet, once we factor in the drought, floods, and infestation—all of which are potentially eliminated under a vertical farming technology.[36] The total combined savings in land area may prove significant. If we assume the equivalent of 2,400 acres of cropland per building, it means that we can fit the entire crop area of humanity (≈ 3.7 billion acres, which is an area almost the size of Russia) into a vertical-farming city that is comprised of about 1.5 million such buildings, taking a total area of about 7.36 million acres (≈ 2.98 million ha) or an area the size of Maryland or that of Belgium. Add to this a disturbed area for access roads, for power generation, and for construction materials, and we may still come up with very significant land savings—which in turn could allow us to restore vast tracts of natural habitat. This can be a big deal.

I will briefly address the other 30 percent of the unpaved urban area, which is not going to have vegetable gardens: the parks, recreational areas, and decorative landscapes per se.

The objective is to set up landscapes that are able to thrive in a given biome without the constant input of pesticides, irrigation, and fertilizers. This may mean drought-tolerant species in some regions, such as cacti. It may mean plants native to a given area, which have adapted and evolved to flourish under local climate conditions and cope with the local pests and diseases.

Much as a bonsai tree can distill the essence of a landscape and miniaturize it, natural gardens can distill and manifest the essence of one biome or another—be it a wildflower garden, a grassy savannah, or a dunescape.

HOLISTIC MANAGEMENT

To attain the same amount of biomass, insects need over four times less feed than cows do.[37] Insects are high in protein and calcium. Just one hundred grams of caterpillars can provide all of an adult's recommended daily protein, along with iron, B vitamins, and other essential nutrients.[38]

With over 1,000 edible species, insects offer plenty of delicious variety. For approximately 2.5 billion of us—mainly in Africa, Asia, and Latin America—eating insects is a part of our diet, as much as eating meat or fish. A number of insect species are considered to be an exquisite meal, such as roasted termites and barbecued palm weevil larvae.

The potential may be there for incorporating insects into our diet, but as it stands that's all it may ever be. No one has come up with any design for an insect farm, and harvesting billions of insects daily from the wild is ecologically unsound. Moreover, pound for pound, a cow may consume a lot more feed than insects, but a grass-finished cow is fully integrated into its grassland environment in a benign cycle. It is yet to be seen what the net footprint of an insect farm is.

When it comes to livestock husbandry there is no such question. If done right, it really works.

Chad Peterson, a ranger in Newport, Nebraska, has transformed four thousand acres of sandy sparse grassland into a lush pasture. His cows have accounted for this transformation. A mob of cows.

Holistic Management International has conducted a survey of forty-three ranchers who have practiced high-density rotational grazing for two or more years. Nearly three-quarters of the graziers have reported an escalation in the amount of wildlife on their land. The vast majority reported increases in new seedling success, plant diversity, water infiltration into the soil, and litter cover. In terms of livestock performance, there was a marked increase in production per acre.

What underlies the method of these graziers and accounts for its dramatic impact on the land are a few biological principles.

For the various vegetative regimes of the world, rapid biological decay is an essential part of their nutrient cycle.

In tropical wet forests or other humid environments, plant matter falls off and is decomposed by the vast army of insects and microorganisms. In the savannahs and semi-arid grasslands of the world, the nutrient cycle is somewhat different. A very high percentage of the plant matter dies off at the end of the wet season. This is when the soil dries out and the air loses its humidity. This is when most insects and microorganisms die or become dormant. The perennial plant transfers the nutrient to its base. The part of the plant that is above ground dies and just remains standing, waiting. This is where the herds of large herbivores play their role.

In these seasonally arid environments, it is essential that a high proportion of the leafy matter, once dead, be consumed by herbivores. The grazing animals metabolize the dead organic matter and deposit its nutrients on the ground in the form of dung and urine. At the return of the wet season, the microorganisms decompose the excreta and thus return the nutrients to the soil. Without the herbivores, the dead, dried vegetation either washes away along with the nutrients it holds, or the dead stems remain standing for decades like an army of fossil plants. What's more, in the absence of grazing animals, the dead matter aboveground may eventually kill the perennial plant, as new growth cannot break through the mass of dead, ungrazed matter.

The wellbeing of the semi-arid grasslands depends on the periodic grazing, trampling, dunging, and urinating of herbivore herds.

That which is eaten decomposes in the gut of the animals. That which is trampled down provides soil cover and has an opportunity to be biologically broken down rapidly. As they dung and trample and urinate, a giant mass of large herbivores returns standing grass plant material to the soil, starting its process of decomposition. In addition, the stamping of the ground by a mob of large animals breaks up the hardened surface, providing better moisture-penetration and aeration.

A mob reduces individuality and foments less selective grazing by the individual members. The individual animals are forced to be less discriminate with what is in front of them. If they pass up on some plant, it will be gone by the time they come back to it later. The result is an even, thorough mowing across the entire grazed area. Had only the choicest species been grazed, they would have been weakened and disadvantaged. This would eventually create grasslands with fewer and fewer species, with less succulent plants and more woody ones.

The role of the predator is essential in this system. The fear of predation has kept the large grazers bunched. This bunching is what provides the beneficial high-impact trampling, fertilizing, and even grazing of the land. As no animals like to feed on their own dung and urine, the herd keeps moving. Movement keeps the plants from being nibbled to death. This is why the millions of bison and horses that roamed the Great Plains did not overgraze the land. They did not stay in one area long enough to do so.

These have been the findings of and principles set by Andre Voisin in the fifties, later expounded, systematized, and popularized by Allan Savory under the framework he developed and termed Holistic Management.

At its essence, the rotational grazing fashioned by Holistic Management seeks to mimic the rhythm and dynamics of herds of grazing animals in the wild accompanied by packs of predators. In the end, the grazing rotation, the confines, and the number of animals espoused by Holistic Management all strive to arrive at the dynamics of a herd of large herbivores in the wild.

This is no mere theory; graziers have been practicing Holistic Management principles and have attested to their benign effects. One of the graziers practicing high density rotational grazing is Chad Peterson, briefly mentioned earlier.

Peterson keeps his cattle bunched, aiming for one thousand cows per acre per grazing period. A grazing period is short; Peterson does not want the cattle to be in the paddock long enough to get a second bite. He does not want them to start grazing the regrowth. Whenever an animal bites a plant, it shocks the system of that plant, and in turn it starts to slough off some roots. It takes energy from the plant's carbohydrate reserves to get a new leaf out and start photosynthesis going again. If the cattle were to repeatedly graze new shoots, the plant would run out of carbohydrate reserves and start to weaken and likely die.

Peterson moves his herd to a fresh patch of grass several times each day. One of the side benefits of moving the cattle so often is that his herd is moving ahead of the fly and parasite cycle. By the time these hatch in the manure, the cows are long gone. One of Peterson's goals is to impact every plant in the paddock. Every plant needs to be eaten or trampled. When the mob moves on, the grazed ground is entirely covered in an evenly spread layer of fertilizing manure. Once a paddock is grazed off, Peterson does not graze it again that year. Given the rainfall pattern in his part of the world, he gauges that this is the rate of grazing his meadow can sustain.

Greg Judy also practices his own brand of high-density grazing management. He honed it to the point that he has quit buying seeds and fertilizer. Judy runs his cattle in a ranch near Clark, Missouri. Once the animals learn to expect fresh pasture when the four-wheeler arrives, moving the herd to a new paddock takes less than thirty minutes. Over the weekend, Judy lays out seven paddocks, five to ten acres each, for the week's grazing pattern.

The benefits of mob grazing are realized through the short episodes of high animal impact followed by long rest periods. During these grazing episodes, Judy's goal is to graze 60% of the forage, trample 30%, and leave 10% of the forage standing. Grazing only 60% ensures that animals can select the highest quality forage. Trampling 30% provides a large amount of litter to soil-building earthworms and microbes. And, leaving 10% of the forage standing provides protection from wind and water evaporation while also providing wildlife habitats.

In addition to deer on his land, Judy sees increased populations of quail, wild turkey, and songbirds that flock to the mob-grazing area to eat insects stirred up by the cattle.

Neil Dennis is a cattle farmer who is always experimenting and innovating along those lines. He is the owner of Sunnybrae Farm in Saskatoon, Canada. Neil grazes as many as 800 head of cattle in half-acre paddocks, moving them every few hours. He has learned that overgrazing is a function of time spent in one area and not the number of cattle per se. During those few intense hours of grazing, his cows either eat all the plant's top-growth or trample it into the ground—either way, building up organic matter on the soil.

The soil buildup due to mob grazing is remarkable at Sunnybrae Farm. Ample amounts of litter keep soil microorganisms active at a later period in the autumn, and let grasses green up earlier in the spring. This yields a soil with a better water-holding capacity and more microorganisms, earthworms, and dung beetles. This healthier soil eventually also translates to higher plant diversity. Dennis has more than forty species of plants in his pastures, providing higher quality and more consistent forage throughout the grazing season. Most new grasses come in naturally, although he occasionally spreads seed on the surface and lets cattle push it in with their hooves. All of this without the use of any outside fertilizer.

As the animals are moved on so often, there are neither ammonia patches on the grass nor swarms of flies. By the time the eggs hatch, the herd has already moved on.

Dennis varies the grazing and recovery periods of a given paddock. To plan his grazing strategy, he closely monitors his land and cattle. If a paddock needs attention, he gets his herd to graze it heavily in the spring when the soil is soft, removing all the top growth. Then he rolls out bales of hay for the cattle to eat or trample into the ground. He does that twice, staggering the hay placement so the entire paddock is covered. This turns out to be a highly effective way to add new seeds, boost organic matter, and provide food for the soil microbes. Healthier soil means better water infiltration, and that means when it rains or snows, more of that water is available for growth. In fact, Dennis maintains that if he gets rain before mid-July, he has enough soil moisture to carry him through the rest of the grazing period.

The implications of practicing Holistic Management worldwide are far reaching.

Almost one and a half billion tons of food, notably corn, are produced every year expressly to feed cattle. This comprises about 15 percent of all global food production. That is one and a half billion tons of feed that is mostly unnecessary. Cows, buffalo, sheep, and goats can do what they always did: graze, move on to a new area, and graze again.

The cattle love it; the grassland biome needs it. There are billions of hectares of grassland out there that cry out to be rejuvenated and once again know the hoofs of millions of large animals. In one swoop, the portion of our 1.5 billion cows and buffalo that is grain-fed turns from an ecological liability to an asset, and in addition, fully 15 percent of our food production can be shed. The livestock would henceforth do all the work. They would move themselves, feed themselves, and revitalize the land in the process.

FIGURE 9.1

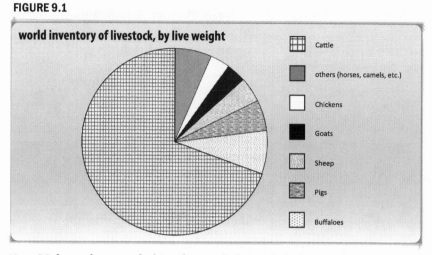

world inventory of livestock, by live weight

Cattle

others (horses, camels, etc.)

Chickens

Goats

Sheep

Pigs

Buffaloes

Note: We have a lot more chickens than cattle, but each chicken weighs far less than a cow. This chart puts things in perspective. In the end, what counts is the amount of meat—both as a source of calculating amount of food and grazing needs. A more nuanced view is available in color at www.danielrirdan.com

Source of data: FAOstat database, under Production, Live Animals, categories "World" and "Live Animals," http://faostat.fao.org

Our entire inventory of grazing animals—from goats, to horses, to camels, to cattle—is about two and a half billion standard animal units. That is to say, their living mass is equivalent to that of many animals each weighing 1,000 pounds.

With assistance from Ann Adams, Director of Education at Holistic Management International, I have estimated what area of grassland would be required to sustain the lot of them. It is but a tentative estimate, as the number of animals land can support varies tremendously from one type of land to another. Worse yet, the problem is that we don't know how much rangeland we have, let alone its breakdown into distinct types.[39]

In an attempt to arrive at some semblance of an estimate, I extrapolated from the sketchy estimates present in "Livestock Grazing Systems & the Environment."[40] I put the data together, and the results are shown in table 9.2. The figures are not start-up values but peak producitity values the land would reach after a decade of employing such rotational practices. I assumed the doubling of land productivity in the case of rangeland and the tripling of cropland producivity.

TABLE 9.2

biome type	hectares needed to support one cow for a year	total area in existence, (in hectares)	in total, the number of cattle that a given biome can sustain
Rangeland	15	5 billion	.33 billion
Cropland/pasture	0.15	1.5 billion	10 billion
Grand Total	...	...	10.33 billion

In the case of rangeland, I assume the cattle can graze the same area only once a year. In the case of pasture cropland, I assume cattle can graze the same area of pasture twice a year.

Hence, all the cropland and rangelands of the world would eventually be able to support the equivalent of over 10 billion head of cattle, while at first, the land would be able to sustain just over 3 billion head of cattle. The level of confidence in those numbers is so low that no conclusion can be reached but the one that matters most: we have significantly more than enough land to sustain our entire livestock inventory for an indefinite period (In total, once we account for camels, mules, donkeys, horses, goats, sheep, buffalo, and cattle, our current inventory is the equivalent of 2.24 billion head of cattle.)

Yet, the notion that livestock will predominate the open rangelands of the planet is repugnant. What is of far more interest is the possibility of pasture cropping as described earlier in this chapter. It seems that half of our cropland area can sustain the entire livestock inventory—while at the same time supporting crop growth.

With large herds and big expanses of land to work with, it is possible to have one human herder per 1,000 head of cattle. This translates to about 2.5 million herders worldwide. This is not really a big number of people; it is far less than the number of people currently unemployed. The herding itself is straightforward enough. Beyond the planning, it mostly entails relocating light electric fence daily. Such mobile fences have short posts and light aluminum wires. Cattle farmers that engage in rotational grazing report that it takes between fifteen and thirty minutes to relocate the herd to a new paddock, as the herd is more than happy to move on to a lush, ungrazed area.

Thus, the gates of the fecal feedlot, the gates of the livestock concentration camps, will be flung open, and hundreds of millions of animals may move out to the open pasturelands of the world. Insofar as their death is concerned, it will come as it comes to all. It will be cut short, and by our hands, but they would actually get to experience life. They would live for what they are. They would know parenthood, bonding, and siblings.

POLYFACE FARM

Polyface Farm in Shenandoah Valley, Virginia, has taken rotational grazing and layered it further: incorporating turkeys, chickens, and hogs into a larger, richer living community. The diversity of animal farms utilizes the nutrients of the land to a greater extent, breaks pathogen cycles, and by and large has the animals on the farm act in their best self-interest while in the process aiding the farm operation as a whole.

Under the rotational grazing regimen, cattle are moved every day to a fresh spot. The hens follow suit a few days later, eating bugs and scratching through cattle droppings, and in this manner sanitizing the pasture. While scratching for parasites and fly larvae, the chickens also distribute the cow manure—a fertilizer—across the field. The hens eat the grasshoppers, crickets, and worms of the field, converting them in effect to the eggs that they lay.[41] Thus, the chickens not only lay valuable eggs but also play the role of a sanitizing crew.

During the day, the hens pick and move about within the confines of an electrified poultry netting. The electrified poultry netting keeps the chickens in and predators and cows out. At the center is a portable structure built on skids. It has nest boxes where the hens can lay their eggs. The structure is equipped with bulk feeders that can hold enough feed for three weeks. It also comes with water supply units connected to the water system of the farm.

Every few days the structure is hitched to a tractor or pickup and is moved along with the chickens to a new pasture area. If chickens are confined to one area, they can strip down a lush pasture in a matter of weeks. Moving the flock of chickens frequently has another great benefit. These frequent moves also ward off disease outbreaks among the chickens. One of the benefits of doing away with routine dispensation of antibiotics to chickens is the significantly lower levels of drug-resistant bacteria around.[42]

In winter, the cows spend most of their time lounging in a shed filled with dry wood chips, sawdust, and old hay to absorb the excrement. The source of the bedding can be whatever biomass happens to be available. It can be sawdust, or wood shaving residue, or something out of a municipal leaf dump. It can be comprised of peanut hulls, crop residue, or food scraps. When the nitrogen in the manure and urine comes in contact with carbon, they form a chemical reaction that locks down nutrients. The ratio Polyface maintains is 25 or 30 parts carbon to 1 part nitrogen. Additional, dry bedding is laid down once every few days.

The bedding stays warm as it ferments, and this provides the cows with a heated bed. This translates to more comfort and less feed as the cows burn off fewer calories to stay warm. This bedding ferments in the anaerobic conditions created by the heavy cows walking on it.

In preparing the bedding, oats and corn are added to the mix. When the cows come out in the spring, Polyface purchases hogs and turns them loose in the compact mix of bedding and manure. The pigs root for the fermented corn and aerate the bedding in the process. Oxygenation turns all the deep bedding into a compost pile, which is the source of the farm's fertilizer. When the pigs have turned the entire anaerobic stew into compost, their work is complete. And the farm has pork to sell.

By itself, mechanically harvesting grassland is a necessary evil—as it is a poor substitute for the grazing of live animals, which graze but also fertilize the field in the process. However, Polyface Farm has demonstrated that it is possible to retain the winter nutrients generated by the animals and return them to the land at spring time. Furthermore, it is possible to use haymaking as a bank, storing excess grass in times there is more of it than there are cows to graze and utilizing it during the winter when the natural supply dwindles.

Polyface constructed a few ponds that feed the farm. If a drought hits, the farm has millions of gallons of water in reserve that are utterly dependable. They have water year-round. The ponds only catch surface runoff, and even though it may take all winter for the ponds to fill up, the reserves last throughout the summer.

The aquatic environment of the ponds offers biodiversity, which spells higher ecological stability. For example, the ponds promote growth in the number of frogs and toads, which are insect predators. Birds that eat bugs are drawn to ponds. Cattails and other hydrological plants filter out toxins from surface runoff. In addition, the ponds help control flooding and erosion. By providing a place for surface runoff to stop in its downward plunge to creeks, the pond lets vital sediment and nutrients settle out and remain on the farm. Over time, algae, salamanders, cattails, and other plants and animals live and die, falling to the bottom of the pond. The resulting muck is a rich fertilizing material, which Polyface Farm puts to great use.

AQUACULTURE

Combined, ducks, geese, turkeys, chickens, and pigs make up about 10 percent of the total live weight of the global livestock inventory. This is the 10 percent of our inventory that requires higher-octane food supplements. They cannot live off grass alone.

With a bit of creative accounting we won't need to dedicate land for the growth of their feed.

To start off with, some of their feed comes out of byproducts of our food and textile system: feathers, bones, molasses, and the seeds of the cotton plant. Some additional sources of feed are made available in the field in the form of insects, larvae and such. The last source of feed is in the form of grain.

In most years we grow more grain than we need. A surplus is built into our food system as a simple precaution. It seems we would always have some surplus of grain to divert to pigs and chicken. And when, for instance, the total amount of barley hitting the world's markets is low, we would just use the surplus of another grain as a source of feed. It will work out. In fact, it will more than work out; it's perfect.

In good years when we have bumper crops, we feed the excess to animals to convert them to valuable sources of protein and fat. Better yet, we store this protein in their living person until we need to convert it to food. In really bad harvest years across the world, we just don't breed as many pigs or chickens, or alternatively we slaughter more of them. If push comes to shove, we can cut back on some of the chicken meat and ham and possibly opt for more beef. The only thing that has no real substitute is the six to seven billion egg-laying birds.

In the case of ruminant cattle, the accounting looks just as good in its own way. No grain is ever needed; the cows just live in symbiotic relation with the grassland.

Fish are another story, though.

———

In 2009, we harvested 55 million tons of fish from aquaculture, and we harvested 90 million tons of wild fish, that is, 145 million tons in total. Out of this total, about 118 million tons were destined for human consumption.[43]

As oceans are off-limits, we are saddled with the task of setting up aquaculture operations to provide us with 118 million tons of fish annually. But it cannot be aqua farming as we know it. As it is currently practiced, aquaculture has its own set of environmental troubles.

At present, many of the fish are farmed in nets or cages in the marine environment, which makes the fish especially susceptible to a host of pathogen-borne diseases. Both for treatment and prevention, antibiotics are either included in the feed or administered in the form of a bath treatment. Given the open structure of most fish farms, a significant portion of the applied antibiotics end up in the open sea, and at times may drift afar.[44] These antibiotics are potentially toxic for some marine species such as algae,[45] and they contribute to the growth of antibiotic-resistant bacteria.[46] In all, the aquaculture industry uses an estimated 5.5 million kilograms of antibiotics annually.

Coating is applied to marine net pens to prevent some parasitic organisms from colonizing and clogging up the nets. Copper is the primary active ingredient in the vast majority of these coating applications, and over time it leaches into the surrounding sea. This is in effect the equivalent of administering a slow-release substance that is highly toxic to algae,[47] tiny crustacean species, and microbes.[48] Above and beyond this, over 16 million kilograms of parasiticides are deployed annually, the majority of which are toxic to invertebrates of the ocean.[49]

By the same token, uneaten feed and fish waste also end up in the open sea. This results in hazardous levels of nitrogen, carbon, and phosphorous in the water neighboring the aquaculture farms.[50] Most of the impact is at the bottom of the sea, where oxygen is reduced.[51]

A new kind of aquaculture technology is called for.

The Israeli company Grow Fish Anywhere and the Center for Marine Biotechnology at the University of Maryland have both developed sophisticated zero-discharge tank systems to grow both marine and freshwater fish. The fish tanks themselves are made from plastic, and the water used is ordinary tap water with some salt added. However, there is nothing simplistic about the zero-discharge systems.

The feces of fish are metabolized and rendered harmless with bio-filters. Those filters contain specialized bacteria that break down the nitrogen and carbon compounds in fish waste and convert it into carbon dioxide and gaseous nitrogen, which are discharged harmlessly into the atmosphere. This eliminates the need to replace the water in the fish tanks; water is only added to compensate for evaporation losses. The zero-discharge system can raise fish more quickly because both the minerals and the temperature of the water are fully controlled and hence optimized.

Most fish do not reproduce in captivity due to the absence of environmental triggers. To induce reproduction, the team at the Center for Marine Biotechnology invented a pellet that mimics the hormone necessary to spur the fish's natural reproduction process. The researchers were able to develop a highly potent, synthetic version of the hormone. They incorporated it into polymers that, once administered, led the fish to release the hormone continuously, inducing spawning at any time throughout the year. Once fish are induced to spawn, the exercise is of little value if too few of the larvae survive. Through a careful study of nutritional requirements for larvae, the team was able to develop a diet that provided the essential components required for their sustenance.

Around 5 to 7 cubic meters of water are needed for the production of 1 ton of fish annually. This is about four times less than the amount of water required in fish pens. For the needed 118 million tons, we ought to budget close to 9 billion cubic meters of water annually. This would provide water for all the needed zero-discharge systems in the world. As it is now, the United States alone uses ten times this amount for irrigation of its fields. In fact, the 9 billion cubic meters of water that would be required worldwide is the amount of water the United States currently uses for its existing aquaculture.

Once we factor in the tank dimensions, needed pathways, and space allocated for pumps, 100 tons of fish require 800 square meters. To support our 118-million-tons-of-fish diet, we will need aquaculture that takes a combined area of 944 square kilometers (about 365 sq. mi.). This is an insignificant amount of land; this is as much area as we allocate at present for the growth of our kiwifruit. Better yet, we can incorporate the innovative aquaculture farms into our existing footprint of warehouses and fields. With a bit of planning and luck, we won't incur any extra, dedicated footprint for their production.

For the first time, we can have fresh, locally-grown fish in the desert of Arizona, in the Alps, and in Moscow. Aquaculture facilities could be established near transportation hubs, within urban areas, and next to sources of fish feed, such as barley fields. These innovative fish farms could be set anywhere that people need them. They can be an integral part of fish markets. They can be scaled to giant facilities, or they can be a part and parcel of neighborhoods. Basements of large apartment buildings can host tanks that would provide all the fish needs of the hundreds of residents of the building.

About 1.3 kg of feed is commonly needed for each 1 kg of fish being harvested. Hence, for 118 million tons of fish, we will need about 153 million tons worth of feed annually. To put it in context, this amount of food is equivalent to 2 percent of the current total food production.[52] The really nice thing about it is that the vast majority of the feed can come from existing waste in the food system without the need to set aside land dedicated for the growth of fish feed.

Grain, animal byproducts, microalgae, yeast, and bacteria—these are the ingredients of the total fish diet. All of them, with the exception of the microalgae, can be byproducts that do not require special, dedicated space to cultivate.

The technology of Oberon FMR is a case in point. The byproduct, the organic waste from a food manufacturing plant or even a brewery, goes to a wastewater treatment plant. With the introduction of a certain bacteria, which provides essential fatty acids, the sludge is transmuted to a premium protein feed. Another source of fatty acids and omega 3 can come from microalgae. In that case, this is the only crop that would need to be cultivated expressly for fish meal. However, it can be grown in desert-like, bleak, and salty conditions—the truly marginal land.

A large portion of feed can be comprised of poultry meal. Notably, the portions of the poultry unfit for human consumption can be ground up and otherwise converted to food for fish. Up to 60 to 70 percent of the feed can be made of poultry bound with wheat. While the bone and meat of livestock are not very suitable, their blood can make 5 to 10 percent of fish food. Fat is currently extracted from soya to make oil. However the unused protein portion can go to the fish.[53]

It is possible, and at this stage necessary, to do away with fish oil and fish meal. By and large, the sea is to be off-limits. Having said that, the unchecked growth of certain life forms in the ocean is in effect infestation, and in our effort to help existing, higher level ecosystems recover, harvesting some of those infestations will have a net beneficial effect. Whether it is jelly fish in the Sea of Japan or carp in Lake Michigan, those can be captured and used as fish oil and fish meal—as long as our harvesting is highly selective, having negligible collateral damage.

The real footprint of the zero-discharge aquaculture is not going to be its water, its feed, or its tanks. It is in its electricity requirements, which are considerable.

This is the one advantage cows have. They may consume more, but it is all tax-free. In contrast, fish grown in a closed-tank system require prodigious amounts of electricity, and there are no two ways about it.

Most of the power is applied to recirculating water in the tanks. Smaller amounts of power are required to generate purified oxygen gas to further the carrying capacity of the tanks. Additional power yet is called for to operate blowers and fans to aerate water. More power goes to gear motors, to turn drum filters, and for high-pressure pumps to backwash the drum filters. Finally, a substantial amount of power is drawn by UV irradiation that is occasionally used for disinfecting the water.

Six megawatt-hours are needed for every ton of fish harvested. It means that to grow the entire fish harvest of 118 million tons would require a bit more than 700 TWh annually. In the envisioned thoroughly-electrified world of the future, this would comprise 1.7 percent of its electrical generation. With the expansive, lower-impact technologies outlined in the seventh chapter ("Energy"), to provide for the world's aquaculture needs would require power generation facilities whose combined area is about 18,000 square kilometers ($\approx$ 6,950 sq. mi.) This is the area currently used to grow almonds or somewhat over 1 percent of the area under cultivation for rice. From this, we need to deduct the savings in energy, hence footprint, due to reduced energy needs for transportation and related refrigeration.

In conclusion, most of the zero-discharge tank systems could be integrated into existing urban spaces and woven into agricultural spaces and industrial parks and facilities. The feed source would largely if not entirely be comprised of infestations at sea and derivatives of current waste products of the food industry. Water is a non-issue, to be discussed later in this chapter. Therefore, the predominant footprint of aquaculture would be the estimated 18,000 square kilometers of power plants added to accommodate them. With this system, we can have fresh, antibiotic-free, local fish anywhere and, poetic justice aside, without ingesting along the way the plastic debris and mercury the fish of the sea carry in their bodies.

All in all, considering that this setup would provide for all our dietary fish needs, I regard the zero-discharge tank arrangement as somewhat of a bargain.

One more subject needs to be addressed and resolved. Practically all the fish adorning our saltwater aquariums in homes are wild caught. Eleven million fish each year are caught, largely in the coral reef, and end up in aquariums in the United States.[54] We capture the fish by squirting cyanide into the crevices of a reef. Anesthetized fish float on top, and we grab those desired. The cyanide inflicts collateral damage: the coral is bleached and many fish in the coral habitat die or otherwise get in harm's way. All of this ought to come to a stop. For years now, marine biologists at The University of Texas at Austin Marine Science Institute have been developing means to breed saltwater aquarium fish. It's landmark work. If it proves successful, saltwater ornamental fish would stay open for business.[55]

WOOD USAGE

Each year, we harvest and use 3,300 million cubic meters worth of trees: About 1,900 million cubic meters are used for fuel, 930 million are used as lumber, and the remainder, 490 million, are pulped for paper.[56]

So fuel, lumber, and paper.

Wood fuel is out, to be replaced by technologies with lower ecological footprints, as outlined in the seventh chapter ("Energy").

Lumber remains, as it can be a low-impact practice. As a tree grows, it draws in carbon dioxide from the air. Once we harvest it, we lock in the carbon dioxide after a fashion.[57] As long as the wood is not decomposing, we may have a bona-fide carbon sequestration method. Under sustainable tree harvesting practices, we are indeed better off substituting steel and concrete with wood whenever it makes sense. A great case in point is the office building of The Finnish Forest Research Institute. Wood is the main material used throughout the building, from the post beam slab system to the exterior cladding.

Paper usage can be cut back considerably, though.

I am not going to make the case that cutting back on printed matter is going to advance our standard of living. I *am* making the case that we have exceeded sustainable consumption rates, and we need to cut back. When it comes to paper, we can cut back on many fronts and get the services we need in other ways.

Newspapers can exist entirely online. The vast majority of those who read the newspaper on a regular basis can find Internet access on personal computers and hand-held devices.

Undoubtedly, many have been looking for a good excuse to do away with junk mail, to do away with the never-ending stream of catalogs and unsolicited mail—and in the process save trees and ultimately reclaim natural habitats. So out they go.

Many books can also have only virtual existence on the screens of e-readers such as the Kindle. These devices mimic the look of ordinary ink on paper. Reflecting ambient light rather than emitting their own light, the e-readers are more comfortable to read than the backlit displays of conventional monitors. The screen of an e-reader can be read even under direct sunlight. Their images are rock solid and offer a wide viewing angle. Books that do not lend themselves to this format (coffee-table books, pop-up books, and children's books) can be printed on something other than fiber-based paper. Yupo paper comes to mind. It is primarily made from polypropylene. It does not use any wood pulp or cotton fiber. It prints beautifully and is utterly waterproof and tear and stain resistant.[58] This synthetic paper can be melted down and reused over and over again, indefinitely.

Along the same lines, an incredible amount of paperwork is generated at physicians' offices. We can get rid of most of the paperwork with the introduction of a universal database with the patient's full medical information, available to one's physicians. With such a database, there would no longer be a need to fill out forms describing medical and personal history over and over and over again. For that matter, all paperwork of that kind—insurance, government related, monthly bills—can be sharply curtailed with such a master database. Obviously, data access will be context sensitive: a bank would not be able to access medical records; a hospital would not be able to access banking information. Beyond that, offices can cut down to bare minimum the use of paper, striving to achieve a paperless office.

In all, I assume that fully half of all the biomass-based paper we use today would be replaced with one alternative or another. With fuel wood gone and half of the paper requirement eliminated, we are down to a total annual consumption of about 1,200 million cubic meters of wood—down from 3,300.

As forests are out of bounds, the obvious source for lumber is forest plantations of one variety or another. At the moment, we have about 225 million hectares worth of wood plantations.[59] While a natural forest has a yield of 1 to 3 cubic meters per hectare per year, the yield potential of a plantation is far higher. This is our lucky break. The annual harvest potential of all plantations is assessed at 1,300 million cubic meters,[60] which is a bit more than we actually would need under this plan. Better yet, with improved site preparation and harvesting practices, experts believe that productivity will grow markedly. Therefore, we have the potential for 1,300 million cubic meters from wood plantations today. Tomorrow, the number would be higher under the same cultivated wood plantation area.[61]

In conclusion, by taking out wood fuel and using paper prudently, we can get all our lumber needs strictly from existing forest plantations, and in subsequent decades reduce the footprint of the plantations as those become more productive, letting nature reclaim some of the land they are on.

PROVIDING US WITH WATER

It is mighty nice to have waterless urinals and to recycle the graywater of a building, as discussed in the eighth chapter ("Nutrient and Water Recycling"). It is also just plain dandy to have city parks with

FIGURE 9.2

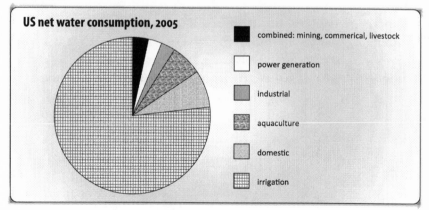

native vegetation, which take little or no irrigation water. Pity that all of these measures are but a drop in our water-consumption bucket.

Irrigation is where the action is; it accounts for 92 percent of the water consumption in the world.[62] And if we can but cut a smidgen of it, this may save more water than all the waterless urinals and low-flow showerheads combined. As it turns out, we *can* cut current irrigation water needs by a prodigious amount.

The vast majority of irrigation is still done by flooding or through open channels, that is to say, an indiscriminate and wasteful water use in the extreme.

Subsurface drip irrigation uses drip lines buried underground that release small amounts of water into the plant root zone. Hence, in addition to very substantial water savings, this method has a number of added benefits. It can work with uneven topography and minimize the effect of salinization and top soil erosion due to water runoff. It also eliminates crop water-stress with the ability to apply water and nutrients to the most active parts of the root zone. This translates to higher yields,[63] better crop health, and harvest flexibility for many specialty crops.[64] Such a system can also be automated and monitored by a computer, incorporating weather forecasts and existing water conditions to avoid overwatering.

Some studies indicate that with a carefully controlled, economical irrigation regimen, it is possible to reach about 80% water savings.[65] This one straightforward measure can cut our water consumption like nothing else. With the use of subsurface drip irrigation we can save perhaps 1,000 billion cubic meters of water every year if we postulate 35%–40% water savings.[66] If you take this 1,000 billion cubic meters and use it to fill five glasses of water daily for every one of the billion inhabitants of Africa and also for every one of the billion inhabitants of India, you could keep doing this, year in, year out, for about eight hundred years.

It does take some doing to bring about a water crisis aboard a water planet and have the leading cause be the indiscriminate flooding of a field instead of the utilization of a piece of plastic with little watering holes.

We can probe further the soundness of current agriculture practices such as the prudence of growing water-hungry crops (e.g., cotton) in arid, water-strapped areas of Pakistan. Yet, there is only a point in investigating the matter so far. We can provide *all* our water needs from the oceans of the world—and at no energy cost. In other words, seawater desalination is to assume a paramount position, displacing by and large waterways and groundwater as sources of our irrigation and potable water. It would take care of our needs and get us out of the rivers.

Healthy ecosystems require natural patterns of water flow. Species migrate, spawn, nest, and feed when nature cues them to do so. By disrupting natural flow patterns through the construction of dams, reservoirs, and diversion projects, we are throwing it all out of whack.[67] What's more, the waterways are the arteries of the world. They keep land sound by taking in toxins and carrying them into the oceans, and the water they contain nourishes the life on land. We need to let rivers be rivers once again—without pumping water from them and without polluting their waters.

Most of the existing desalination plants are located in the Middle East, North Africa, and southern Europe. By the end of 2006, the worldwide capacity for seawater desalination was 10 billion cubic meters a year.

There are a number of desalination technologies. The one that seems most promising all-around is based on reverse osmosis. In essence, pressurized water is pushed through a membrane whose holes are so small that even salt particles are kept back. This process was even found to remove organic matter, disinfectants, and endocrine-disrupting compounds.[68]

As may be guessed, existing seawater desalination practices contribute their share to marine environmental degradation.

Aquatic organisms collide with the intake screens or otherwise are drawn into the facility along with the sea water. The remedy for this is straightforward. The intake rate would be slow enough to permit organisms to escape, and the incoming filters would be fine enough to assure that marine animals won't get through.

There are other ecological problems afflicting reverse-osmosis technology of the present. Once the fresh water is extracted, what is discharged to the ocean is brine. The high concentration of salts hurts marine populations near the outlet. Once again, the remedy is straightforward. The brine would be pre-diluted as much as needed with a stream of ocean water that is drawn in, and its composition would not be modified.

A third form of ecological problem with the existing reverse-osmosis technology is the pollutants contained within the discharge. The desalination plant uses chemicals to prevent corrosion and the build-up of organisms on the equipment. The residue of those chemicals washes out to sea. For this I found no satisfactory solution. However, I am confident that it is nothing that a few thousand engineers, chemists, biologists, and marine ecologists working in tandem cannot handle in a few years of focused research and development.

Using energy recovery systems, the power consumption of reverse osmosis is about 3 kWh per cubic meter of processed water.[69] On top of this, I will throw in 7 kWh per cubic meter of water for a grand total of 10 kWh per cubic meter. This considerable amount of added power is apportioned to all the processes that would make the desalination operation ecologically sound. This may include power for pumping the water through very long discharge pipes away from sensitive coastal areas. This may include power for running a secondary flow of seawater, mixing it with the brine and diluting it to safe levels.

As a case in point, I will take the area that makes up the contiguous United States and address the power required to provide for all its water needs based solely on the desalinated seawater throughout the land. But first, let's assess possible water requirements of the future.

TABLE 9.3

annual US water consumption	present usage, in cubic meters	projected needs, in cubic meters
mining	865,478,714	865,478,714
power generation	3,899,628,230	2,136,982,622
industrial	4,052,724,451	4,052,724,451
livestock	2,243,833,704	2,243,833,704
aquaculture	8,842,694,400	396,000,000
domestic	10,633,340,016	5,316,670,008
irrigation	106,026,669,198	31,808,000,759
commercial	1,379,818,456	1,379,818,456

annual US water consumption	present usage, in cubic meters	projected needs, in cubic meters
total	137,944,187,170	48,199,508,715
compensating for 5% evaporation loss en route	...	50,609,484,150

Note: The volume of annual withdrawal of water is considerably higher than that of net consumption. This aside, the exact figures of projected water use do not mean to suggest an abnormally high degree of confidence in these figures.

Under this blueprint, water usage patterns are expected to change:[70] Power generation requirements will reflect the needs of the new power grid.[71] Livestock requirements will remain the same.[72] When it comes to aquaculture, it is assumed the region gets all of its related dietary needs from zero-discharge systems. Given graywater recycling, rainwater harvesting, and other measures introduced in the household, it would be assumed that the domestic sector will draw from outside

FIGURE 9.3. Water canals providing water to the contiguous United States, using a series of desalination plants.

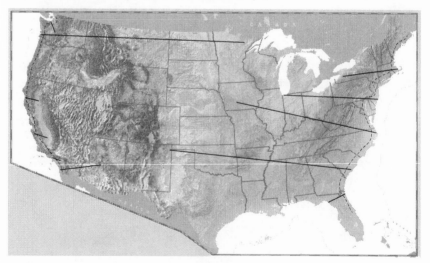

sources only half of the current amount.

Currently 1% of irrigation in the United States is subsurface drip irrigation. And a total of 7% is some sort of drip, trickle, or low-flow micro sprinkler system.[73] I am going to assume a total of 30% in water savings can be had with the full utilization of subsurface drip irrigation.

I will assume that the entire water requirement would come from desalination, leaving the entire restored waterway network untouched. Furthermore, I will add 5% to the total required amount, accounting for evaporation en route from the coastal desalination plants to the interior of the country.[74] It comes to projected use of a bit under 51 billion cubic meters annually.

To desalinate this amount of water at 10 kWh per cubic meter would require a total of 506 TWh each year. Before we see how the electric power is to be had, let's ascertain the rest of the power requirements. After all, getting the water desalinated is only the first step. We also need to create a network of water canals to convey it to the interior.

The suggested scheme shown above relies on the extensive network of rivers and creeks and local water pipes in existence. The water canals proposed here do not deliver water directly to the myriad end consumers. This would take a canal network that rivals in extent that of our roads. Instead, the water canals would deliver water into all the major, southbound rivers. Naturally, the piping of water into the rivers will be carefully balanced against the amount of water withdrawn from the rivers. That is to say, the canals will not charge the rivers to the point of overflow and will not draw water beyond what was brought into the rivers in the first place by the canals.

As a basis for calculating the power needs for routing the water I used the Arizona Central Project. It is a concrete-lined, steel-reinforced 336 mile-long canal (≈ 540 km). The aqueduct is over 16 feet deep and is 80 feet wide at the top. It has the capacity to move up to 2.7 billion cubic meters a year. In a series of steps, the canal works its way up the mountainous terrain, making a 3,000-foot climb (≈ 915 m) in total.

It takes the Arizona canal 2.8 TWh to move 1.9 billion cubic meters of water. In other words, it takes 1.48 kWh to move one cubic meter of water 336 miles and lift it 3,000 feet high.

I will assume that in most cases there will be a climb, but not so steep. Consequently, I reckon power requirements of 1.2 kWh per cubic meter of water per 300 miles of travel.

The combined length of the proposed eleven inland water-canals comes to 5,975 miles (see figure 9.3). On top, I throw in another 3,400 miles worth of feeder canals that collect the water generated along the coastlines from the hundreds of plants and feed it to the main lines traveling into the interior. This comes to a total of 9,375 miles (≈ 15,000 km).

The total power requirements to move the water would come to about 172 TWh. Adding this to the desalination power requirements would get us a grand total of 678 TWh to both desalinate the water and deliver it throughout the land.

Now, on to that part of the ad where it says that it comes free of charge, that is, free of the need to add power plants expressly for desalinating water. The desalination scheme would require energy all right, but it runs entirely off excess power.

Substantial excess in capacity has to be built into solar-energy generation. The reason was discussed in the seventh chapter ("Energy"). In short, in order to come up on top and provide for winter days we ended up with a generation capacity suited for the winter and over built for the summer. Some of the excess during summer days is diverted to hydrogen production. But there is a limit on how much hydrogen we can inject into underground reservoirs. Once we exceed this cap, once we have in a given hour more than 598 GWh, once the reserves are full, well, this is when desalinating seawater comes into the picture. Whatever cannot be diverted to hydrogen generation goes to the desalination plants.

A simulation of typical weather and electrical needs for every hour of the year indicates we can divert a total of 4,237 TWh to desalination plants. This is about six times more power than the desalination scheme requires. In other words, the excess power we have available is more than enough to provide for all our water needs using ocean water as a source. It is more than enough once you factor in all the things noted here. It is more than enough once you factor in the fine details that were left out. It is more than enough once you factor in all the things that I don't even know that need to be factored in. It also means we will be able to take showers whenever we feel like it, for however long we feel like it—without guilt.

If a lone day in January had excess power that could be routed to desalination operations, it is not enough of a reason for mass mobilization of the numerous employees of those plants. It has to be consistent power excess, day after day, week after week. Based on the simulation results, the desalination plants would be in operation 4,980 consecutive hours out of the total 8,760 hours of the year. They would run from the beginning of April through the end of October. Within the block of 4,980 hours, 4,351 hours provide enough power to run every desalination facility and move the water along every canal (151 GWh hourly). The other 620 hours within this block of time offered little or no power to speak of. While, as stated, a total of 4,237 TWh can be diverted to desalination, during the months of actual desalination operation, a total of 3,975 TWh could be diverted. It comes to almost the same amount, in fact.

Those seven months of operation will have to do the job of an entire year. The plants would both dispense water where it is needed and channel some of the water to reservoirs, to be tapped during the

FIGURE 9.4

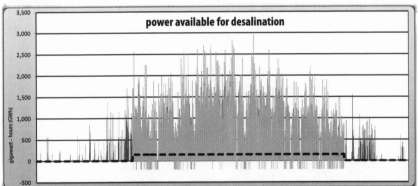

Note: The chart indicates how many gigawatt-hours (GWh) are available at each hour of the year for desalination. The light gray lines fall within the seven-month block in which the desalination plants and pumping of water in the canals would actually take place. The short gray lines pointing downward indicate the hours within this block of time in which no sufficient power is available to run the operation. This comes to about 12.5% of the time. The dotted horizontal line indicates the desalination power requirement, which is 151 GWh for each hour.

five months of the year in which the desalination plants won't operate.

The desalination operation will run for seven months of the year. Hence, we have to overbuild and overstock freshwater to compensate for the 40 percent of the time the desalination operation won't run. For good measure, we will assume plants with two and a half times the needed capacity, had they run every hour of the year. That is, we will assume a required capacity of 128 billion cubic meters annually. The *Jebel Ali* desalination plant in United Arab Emirates produces 300 million cubic meters a year. Using it as a basis suggests the need for 427 plants, spread thin over the six thousand miles of US shoreline.

The desalination infrastructure would have a real environmental footprint, no doubt. All told, though, it has a vastly lower impact than the existing practices of wringing the land dry and disrupting the natural water flow pattern of rivers.

This desalination scheme offers the potential to have fresh water wherever we need it, however much we need, and whenever we need it. Perpetually.

A lot of electricity is still left in excess, for which I have one last application—to be discussed in the eleventh chapter ("Drawing Down Carbon").

10

NATURE RESTORATION

≫≪

bringing back the wilderness

FROM A WELLSPRING OF LIFE, Earth is gradually being transformed into a glorified trailer park.

There are both practical and self-serving reasons why we need to halt the unraveling of the biosphere and reverse course. But ultimately, this line of reasoning misses the point.

Large tracts of wilderness must be restored and henceforth remain whole and intact as an end in itself. Life does not require a justification for its existence. Nature is not derived from a moral code; it is at the basis of one. Much as in the movie *The Matrix,* nature is the Source. It starts and it ends there for each of us. It is the source of all that is inspiring, beautiful, sexual, nurturing, and dangerous.

We may survive without much of it, much as crickets in a glass jar. But without everything that gives our lives context, we lose our humanity, our chance for self-actualization as a race.

With the exception of judicious mining activities, from this point forward no further degradation of nature ought to take place. Fishing activities and beach degradation ought to cease, deforestation come to a stop, unsustainable hunting end, and so-called development of new tracts of land halt. Fifty years from now, the present era ought to be regarded as the low point of recorded history. There would be exceptions to these policies; there always are. Yet, broadly, that is what is called for.

Moving forward under these precepts is good, but more is called for. We must do what is needed and relinquish some of the land we have appropriated. What's more, we ought to take an active part in the restoration and regeneration of ecological communities.

We ought to bring back the wilderness.

———— • ————

Imagine conditions somewhere in the Jordanian desert, about two kilometers from the hyper-salty Dead Sea. It's a land of dust, flies, and intense heat. The soil is salinized through and through; the aftermath of thousands of years of abuse. Long ago, goats chowed down on the land, reducing it to inert rubble.

The Australian Geoff Lawton and his team took on the seemingly impossible task of breathing life into a small piece of this land, a ten acre parcel (about four hectares).

They set up a large rain-harvesting ditch system that worked with the contours of the land. When filled, it fed the land with one million liters of freshwater. The ditches were filled a number of times over the course of a winter. After setting up the system, the team padded the ditches with a half-meter thick mulch made of crop residue from a nearby field. A drip irrigation rig was set underneath the mulch.

Above the trench, on the hill side, they planted very hardy pioneer desert trees. These provided shade, reduced evaporation, and held the soil together, enriching it with nitrogen. On the lower side of the trench, they planted date palms, figs, pomegranates, guavas, mulberries, and some citrus trees.

A web of fungi took hold and spread throughout the mulch. The fungi filaments generated a waxy substance that repelled the salt away from the area. The decomposition process locked up the salt, making it insoluble. Mushrooms started to grow in the ditch, and little insects burgeoned in the mulch. Within four months, the fig trees yielded their first crop. The land was coming to life.

If Geoff Lawton and his team can do this, it means we can take on the salinized, barren regions of the world and restore them to life.

Slash-and-burn agriculture left a swath of 2,000 hectares (about 4,900 acres) in East Borneo barren but for patches of alang alang grass. The region was practically dead, without even the sound of insects.

Willie Smits and his people planted pineapple trees with beans and ginger in between them to fertilize the soil and provide it with nitrogen. Emulating the structure of a rainforest biome, the team planted fast-growing grass underneath the slower growing primary forest trees.

They planted soil-forming pioneer trees; they planted the drought-resistant sungkai (*Peronema canescens*); and they planted legumes such as black wattle (*Acacia mangium*), which fixes nitrogen through symbiotic bacteria in its root nodules. In all, Smits and his people brought in and planted over one thousand different species. The diversity is what has made it all work: cycling the nutrients and making sure that at every stage of the forest growth there were crops that contributed to the process.

As a result of this regenerative undertaking, air temperature in the area went down 3°C–5°C. Humidity went up by 10 percent, along with an increased rainfall and cloud-cover. The newly-formed patch of rainforest is now creating its own rain, literally. It is those chemicals coming out of the leaves of the trees that initiate the raindrops. Four years into the restoration, and there are no more floods or fires. This forest is now inhabited by 137 bird species and 30 reptile species. If this can be done, it means we can bring back the forests, too.

CORRIDORS AND CORE AREAS

Tens of thousands of designated nature reserves cover, in total, seventeen million square kilometers of land—and it is not enough. Biodiversity is in steep decline. A comprehensive study affirmed that the strategy of setting aside land is insufficient to stem the ongoing biodiversity loss.[1]

Reserves of intact habitats, such as national parks in the United States, have to be big enough to survive natural disruptions such as floods, wildfires, and hurricanes. Yet, as big as those reserves ought to be in order to weather natural disaster, more is required.

Yellowstone National Park is not large enough to sustain grizzlies. Nor is Yosemite large enough for the wolverines.[2] Whether because of inbreeding or for other reasons, the isolated populations within these parks are in danger of local extinction.[3] It seems that even something the size of the Greater Yellowstone Ecosystem cannot sustain sufficient populations to have resiliency and genetic fitness for animals such as wolverines and grizzly bears.[4]

Existing reserves are but islands of relatively intact habitat surrounded by manmade landscape. Once this becomes so, the first things to go are the larger, wide-ranging animals, particularly the carnivores. Once these disappear, the ecosystem destabilizes.

Much as humans venture out of their cities to fulfill some needs, some of the wildlife needs to venture afar: to find a mate and breed. This keeps genetic variability strong. Additionally, wildlife may need to migrate between winter and summer habitats.

Thus, there is a need for large core areas, reserves, in which human activity is severely curtailed. In addition, there is a need to designate tracts of land that facilitate the movement of wildlife species between areas of core habitats.[5] These are essentially corridors that facilitate the movement of animals and the propagation of plants. These corridors make possible genetic exchange and unimpeded flow of water to and from the core areas. A modest case in point is the elephant-corridor linking two reserves in Karnataka, Southern India.

Specially designed passageways across the occasional road, both overpasses and underpasses, have been designed for wildlife. A notable example is the series of overpass and underpass animal crossings in Banff National Park, Canada. Between 1996 and 2001, researchers documented over 32,000 crossings of animals, which included wolves, coyote, elk, deer, black bear, and bighorn sheep. In some locales it makes sense to also have a game-proof fence along both sides of the road that serves to direct wildlife toward the nearest animal crossing.

Organizations that promote or undertake the formation of habitat corridors include *Yellowstone to Yukon Conservative Initiative* in North America, *Great Eastern Ranges Initiative* in Australia, *Terai Arc Landscape Initiative* in the Indian Subcontinent, and *Cantabric-Pyrennes-Alps Great Mountain Corridor* in Europe.

Let's get more specific now.

In North America, the Wildlands Network calls for the creation of four continental migration corridors: along the East Coast, along the Rocky Mountains, along the Pacific coast, and across the boreal forest in the far north.[6]

The East Coast corridor would reach from the Everglades of Florida, through the forests of Alabama, along the Appalachian Mountains, to the boreal forests and Maritime Provinces of Canada. This continental-scale wildlife corridor would skirt around towns and communities and connect working landscapes and private conservation lands to large public parks and preserves. This continental corridor traverses a wide array of ecoregions. It spans climates from arctic to tropical. The species diversity is great, from predators such as wolves, martens, and cougars, to prey such as moose and deer.

The proposed Rocky-Mountain corridor is five thousand miles long. The Wildlands Network envisions it as protecting, connecting, and restoring a contiguous network of private and public lands along the spine of the Rocky Mountains and associated ranges, basins, plateaus, and deserts from Alaska's Brooks Range to the Mexican Sierra Madre Occidental.

The Pacific corridor is to run from Baja, Mexico, to Prince William Sound, Alaska. Mostly it traverses an unbroken series of mountain ranges teeming with diverse plant life and inhabited by the most high-profile wildlife of North America. In northern British Columbia, the Rocky Mountains and the Coast Ranges come together, creating one of the most rugged and extensive expanses of true wilderness remaining on the planet.

The last of the four suggested continental corridors proposed by Wildlands Network is the Boreal corridor, running east–west from Alaska to the Canadian Maritimes across the boreal forests of North America.

Beyond designating reserves and migratory habitats, a few measures need to be set in place.

Nonindigenous species have severely altered ecological communities in some places. Thus, in a context of restoration, there is a need for the removal, partial or complete, of aggressive nonindigenous species. These may be plants, animals, fungi, or microorganisms.[7] In many cases, the prevalence of a nonindigenous species can be brought down to acceptable levels.

Periodic disturbances of fires and floods are just what the doctor ordered. For instance, longleaf pine forests require frequent, low-intensity ground fire for their well-being. Fire should be allowed to play a natural role in those ecosystems that are dependent on it for their ecological health. The suppression of fire in these ecosystems often results in reduced biodiversity and increased vegetation density, often increasing risks of catastrophic fire over time. Fire-dependent ecoregions, notably savannah and temperate coniferous forests, cover 53 percent of global terrestrial area.[8]

Last but not least is the need to reestablish the full complement of native species, both plants and animals. These will restore the full functionality of a given ecosystem, along with its built-in redundancies. A full gamut of species is what gives the system its stability and assures its long-term existence.

REWILDING

Core areas connected to each other via a series of migratory corridors form the framework, the container. As far as what to fill them with, I will discuss here one biome, grasslands. The specifics pertaining to the other biomes could be surmised.

In dry subtropical grasslands, bare ground may develop a hard crust, not allowing water and oxygen to penetrate the soil. This increases water runoff and the danger of floods. It also inhibits the growth of new plants. Herds of large herbivores break the soil's crust with their hooves. Without this, soil in brittle environments becomes more and more open to wind, sun, and rain. Bare ground between plants expands, ecological succession founders, and in extreme cases desertification follows.[9] An estimated 38 percent of the world's land area is in danger of desertification.[10]

The goal is to move from a land with high soil-exposure and overly-low moisture to that with denser grassy vegetation and higher moisture-retention. Evidence suggests that this can take place via the rewilding of grasslands, that is, repopulating the grasslands with herds of mega-herbivores.

The outcome cannot be anything other than that. Grasslands without herds of herbivores are much like fields of flowers without pollinating bees and flowers. Those large animals do not just fit into the grassland biome. Along with plants, the big animals *make up* the grassland biome.

Through the numerous cycles of glacial and interglacial periods, Siberia was a grassy steppe.[11] Mammoths, woolly rhinoceroses, bison, horses, caribou, musk oxen, elk, moose, and yaks grazed under watchful cave lions and tigers.

As the population of the great herbivore herds crashed, about thirteen thousand years ago, the vast northern grassland region atrophied along with them. Unproductive shrubs and mosses became dominant. This brings us to the present time and Siberia as we know it.

In the high latitudes, the soil is too cold to foster decomposition. A large herbivore population is an essential part of that ecosystem. Their stomachs decompose the organic matter and their dunging deposits it in the grassland soil.[12] In the past, massive herds of millions of herbivores maintained the vast steppe grasslands of the world. They fueled plant productivity by eating the grassland vegetation generated during the rainy season and by returning the nutrients to the soil through their manure. They also trampled down shrubs, preventing these plants from gaining a foothold.[13] Furthermore, as small fruits are intermingled with leaves and incidentally consumed, the herbivore herds also dispersed their seeds.[14] Last but not least, with the herds consuming large portions of the biomass, little fuel was left for possible wildfires.

With the dramatic expansion of human habitats across the globe, the population of the very large herbivores of Earth collapsed. It was a major extinction event. It took place around 45,000 years ago in Australia and most recently about 1,000 years ago in New Zealand. Some tracts of Africa and Asia make up the last remaining functional grasslands on Earth.

Until these great dieoffs, some of the fruited plants had circumvented the trade-off between seed size and dispersal by relying on massive herbivores to disperse heavy seed loads over long distances. The *piquiá* and *cupuaçu* have been faithfully producing their fruits, but their ecological partners are gone. The herds of elephants, the camelids, and the giant sloths of South America are dead. And the fruits just fall to the ground, rotten, the seeds undispersed.[15] In Costa Rica, the large palm tree *Scheelea rastrata* still drops thousands of yellow egg-sized fruits, but the mastodon-like *Cuvieronius* are not there to consume the fleshy fruits and then disperse their seeds through their dunging.[16] Year after year, the Australian Acacia has been investing its resources in producing defensive spines to protect itself from tall browsers that have gone extinct.[17] And some of the plants in Madagascar grow wide-angled branches, small leaves, and thin, wiry branches as a defense from the giant elephant birds, which were hunted to extinction.[18]

As the decomposition of grass stopped, nutrient levels went down. Seed dispersal has plummeted. Some plants that co-evolved and thrived under heavy grazing followed the mega herbivores to oblivion.[19] Once the teeming herds of mega herbivores were gone, it was but several hundred years before some of the rich steppe ecosystems across the planet disappeared and were replaced by wetlands, tundra, and shrub land.[20]

In the far north, nonproductive shrubs and mosses got their lucky break, freed at last from the tramping of heavy animals and able to out-compete in the emerging low-nutrient environment. Unlike the grasses, the mosses and shrubs that took over in the northern reaches do not transpire enough and cannot keep the soil from being water-logged. Water saturated soils inhibit decomposition of biomass and therefore the availability of nutrients to sustain grass growth. These emerging low-nutrient conditions were conducive to the ascendency of moss in the far northern reaches of our world.

When mosses are destroyed in such soils, the site becomes over-grown with grasses within one to two years. The grasses then dry out the soil through their high transpiration rates, creating a steppe-like ecosystem. However, with low herbivore populations, grass produc-tivity begins to decrease within a few years, and it's back to shrubs and mosses.

If we are intent on restoration, we need to artificially multiply the number of animals at least for a period of time sufficient for pasture development. Animals would trample all vegetation including shrubs, trees, and moss. By fertilizing the soil they would amplify the rate of biogeochemical cycling. More nutrients would accumulate in the soil, allowing higher grass productivity. Higher transpiration would keep soils dry. Under continuous pasture pressure and fertilization, grasses would once again become the dominant vegetation, and in combination with various steppe animal species they would form a lasting steppe ecosystem.

In northern Siberia, the Russian scientist Sergey Zimov and his team have commenced such a restoration project on an area spanning 16,000 hectares (about 40,000 acres). Aptly enough, they call it Pleistocene Park.

The first step in this undertaking was to gather the surviving megafauna members of the historical Siberian grassland: bison, musk oxen, moose, Yakutian horses, and reindeer. Their next step will be to increase the herbivore density in numbers sufficient to influence the vegetation and soil. As animal densities go up, the fenced boundary will be expanded with an eventual introduction of Siberian tigers.

It is befitting in this context to bring back the majestic mammoth. This may be possible. The distance to dinosaurs is about one hundred million years; the distance to mammoths is but ten thousand years—and more to the point, some of their carcasses have been found buried under layers of snow.

Akira Iritani, a Japanese scientist, and his team may be able to clone a woolly mammoth. The team plans on taking mammoth DNA extracted from preserved corpses found in Siberia. Next, they would insert it into African elephant eggs that have had their DNA removed. The researchers already did manage to extract the nuclei of a mammoth egg without damaging it. This is based on a technique that successfully cloned a mouse from tissue that had been frozen for sixteen years. The egg will then be zapped with electricity to trick it into growing and dividing, like a normal embryo. It will be allowed to mature in the lab for a few days before being inserted into the womb of an elephant that will act as a surrogate mother, in the hope that she will eventually give birth to a baby mammoth. A team at the Scripps Research Institute has induced normal adult cells to become stem cells in the northern white rhino, which is on the verge of extinction.[21] And a team at Monash University succeeded in producing embryonic stem-like cells from tissue samples taken from adult snow leopards.[22] From there it would lead to differentiating the cells into either eggs or sperm cells, and from there the process would be comparable to that described above. This technology may bring back the mammoths and also, for that matter, bring back the northern white rhino and others from the edge of extinction.

Mosses and low-productive soil belong in the Northern reaches much as a barren desert belongs in Northern Africa. Which is to say, neither do. This is not what the interglacial summers have been like.[23] The Siberia of the present is a vastly diminished ecosystem. A ghost ecosystem.

Nothing may be more pressing than the ecological restoration of the permafrost belt, which contains vast stores of frozen carbon. The chief architect of Pleistocene Park, Sergey Zimov, argues that the permafrost temperature strongly depends on snow depth and density. By digging under snow in search of food, the animals effectively trample it down, exposing the ground to colder temperatures. At the height of winter, the ground-surface temperature in the Pleistocene Park is 20°C lower than that of the ground outside the Park, which means it is less likely that the carbon will thaw and outgas to the atmosphere.[24] The next best thing to having frozen soil is to have dry soil, which grazing animals of the Park create. What's more, without wet soil, in the case of a thaw, the possibility of carbon outgassing in the form of methane is far reduced. Carbon dioxide emissions are bad enough, but the same carbon released in the form of methane packs twenty-five times the amount of greenhouse punch. Revival of the grasslands in Siberia will do away with the waterlogged grounds.

Restoration of the biosphere to its former breathtaking diversity, smells, and vast wilderness expanses is the goal we ought to undertake for the twenty-first century. Among all the far-reaching measures outlined in this book, this is likely to be the crown jewel. And the people of Great Britain and Madagascar can once again know a sense of awe and pride in the return of hippos to their respective islands.

As a case in point, I will discuss in some detail the suggested rebirth of the Great Plains of North America.

The land has been abased. It is time to claim its lost heritage. It is high time to restore the North American heartland. Bring back its vast herds of bison, horses, and pronghorns. Bring back its giant tortoise, its camels, and its hairless mammoths.

Prior to the arrival of the first settlers, for millions of years, mammoths traversed the prairie, from the Rocky Mountains to the Missouri River and beyond—both through the winter (glacial) eras and the summer (interglacial) eras. More than any other species perhaps, mammoths defined what the American grasslands are truly about. They were in the Great Plains long before the bison migrated to North America a few hundred thousand years ago.

It is anyone's guess what the early Californian settlers thought when the last dwarf mammoth of Santa Rosa Island died off the coast and the realization dawned that there were no more. They were all dead. Thousands of years later, about the time Sobekhotep the Third took the reins of power in Egypt around 1700 BCE, far to the North, the last of the mammoths on Earth died in their remaining holdout, Wrangler Island off the coast of Siberia.

One hundred and thirty centuries is not *that* long ago. The Columbian Mammoths and Western Camels living in southern California could very well have browsed on the leaves of the very colony of palmer oaks (*Quercus palmeri*) in Riverside, California that is still alive today. Many of the animals have been driven to extinction, but the key plant species are still around.[25]

We were instrumental in emasculating the land from the first. It falls on us to man up, shoulder the responsibility, and do right by this land. This is a rallying cry to make the grassland ecosystems functional again—in North America, South America, Australia, and Asia. It is time to oust the rats and the cheatgrasses of the world and concurrently resume seed-dispersal ecology and nutrient cycling with the aid of the large herbivore herds. It is time to fill in the land's vacant ecological niches and with them the full breadth of species interactions and ecological processes. It is time to have the camels browse on woody plants of the desert, restoring in the process some of the desert grasslands.[26] It is time to repopulate the land under the Big Sky.

The surviving species of horses, lions, hippos, and camels of the world cannot quite replace the galaxy of majestic creatures that crowned the open grasslands of the world. They cannot bring back the sloths that could have used tree branches for toothpicks, or the armadillos that looked like compact armored vehicles, or the epic condor-like birds that flew unchallenged in the big skies with wingspans seven-meters long (about 23 feet). What's done is done.

However, this is not about trying to re-create the historical grasslands per se, but using the pre-contact setting as a benchmark for what a functional grassland ecosystem in a given region is and what ecological niches need to be filled in. Ultimately, how close the rehabilitated grassland biomes of the world resemble those of previous interglacial periods is ecologically immaterial.

Thus, although many of the original cast members are no longer around, a second season of the show can still air.

In the case of rewilding North America's open country, the Bolson tortoise could be brought back to the Southwest and the plains bison and pronghorn antelope and horses can roam once again in their old stomping grounds. The cougars, wolves, and bears are already in the region in small number; they can once again take prominent ecological roles.

When it comes to other members, ecologists suggest that we can make do with ecological proxies.[27] Those proxies would either be species descended from those that were driven to extinction or those that play a similar ecological role. We did it in Bermuda, introducing the yellow-crowned night heron for a native counterpart, which had gone extinct.

Although originated in North America, the species of camels and llamas that roamed its open prairie are extinct. Yet, the Bactrian camels and guanaco are fair proxies. We cannot bring back the hairless Columbian mammoth, but Asian elephants can stand in for them. The Mountain tapir can take the place of the extinct tapirs of North America, and the onager and zebra can occupy the ecological niches of some other extinct species. The African cheetah can be the modern proxy for the extinct American cheetah (*Acinonyx trumani*).

Yes, carnivores do come with the package. As a matter of fact, the cheetah is the reason why the North American pronghorns developed the ability to run at speeds of up to 100 kph (about 60 mph).[28] Predators matter. A recent major study concluded that the loss of the top predators is arguably humankind's most pervasive influence on the natural world.[29]

When a hydroelectric dam flooded thousands of acres in Venezuela, the water created dozens of islands. Those were too small to support the top predators: the jaguars, pumas, and harpy eagles. No longer kept in check by predation, the populations of monkeys, leaf-cutter ants, and tapirs exploded and consequently, densities of seedlings and saplings of canopy trees were severely reduced.[30]

Predators at the top of the food chain hold in check the population of herbivores.[31] Without the top predators, there is a disproportionate surge in the populations of herbivores and mid-level predators, which in turn overtax the population of plants and small animals which are their food source.[32] In such a manner, an overabundance of foxes and opossums (mid-level predators) would cause severe declines in many songbirds and other small prey animals.

The recent loss of gray wolves in much of North America affected an increase in the population of elks and altered their foraging behavior. With lurking wolves, the elk would remain in the open country. With no possibility of being caught, the elk just went to the fridge for the really tasty stuff, allowing themselves to browse highly palatable aspen shoots or cottonwood seedlings and shoots. This has prevented any juvenile organisms from being added to the forest.[33] Comparable dynamics were observed with moose and their effect on the balsam fir's population and how historically that has been controlled indirectly by the presence of wolves.[34] In another instance, elk over-browsed riparian vegetation, leaving nothing for the beavers to eat. In turn, the beavers left, and this resulted in an impoverished ecosystem.[35]

Top predators make carrion available for scavengers who otherwise wouldn't be able to thrive.[36] In some ecosystems, the absence of top predators has cascaded to birds, amphibians, lizards, butterflies, and fish. It has also affected processes such as stream erosion.[37]

Thus, side by side with the reintroduction of large herbivores, top predators have to assume their traditional role. The gray wolf will once again hunt camels in the Colorado Plateau; cheetahs will resume their pivotal role in regulating populations in the Great Plains.

Bears, cougars, mountain lions, wolves, bobcats, coyotes, and African cheetahs—all of these, with the exception of the cheetah, already exist in limited numbers in the Great Plains. These predators are unlikely to molest humans in a normal course of events. People have learned to live with the grizzly in Montana; people in Idaho know of the gray wolf in their midst; and those living in Oregon may have encountered the cougars.

It is not possible to determine the size and distribution of the wilderness core regions within the Great Plains without ascertaining how much land the new agriculture and energy infrastructures would necessitate. Most importantly, we will need to find out what the required areas for livestock grazing would be. This will have a lot to do with pasture cropping—whether we would be able to successfully double up agriculture with pasture grazing on the same land.

I do hope that the wilderness core areas of the Great Plains would be able to encompass significant portions of unpopulated regions of Canada along with non-populous regions of North and South Dakota, Wyoming, Montana, Nevada, smaller tracts of land in Arizona, New Mexico, Texas, and the Chihuahua desert in Mexico—all to be connected via migratory corridors.

It is essential for some of the core areas to be in the warm southern deserts to offer temperate winter habitats to the Asian elephants and zebras.

The reserved wild country would be unfenced to the outside and surrounded by mixed-utility zones—areas that are semi natural, such as wood plantations with low-impact human activity. These deliberately placed mixed-use areas would function as buffer zones, discouraging the wildlife from reaching the agricultural, industrial, and urban centers that lay beyond. Having said that, there is little doubt that some of the animals, such as the coyote or the cheetah will prey on livestock. Watch dogs and makeshift corrals for the night would help, no doubt. Yet, we might just get over it from the start and assume a small portion of the cattle will be lost to predators. In the end, the bounty of nature is to be shared, at least to some degree, with others. Yes, this will take a bit of getting used to.

A comprehensive field study collected records over the last 20 years from 109 conservation areas in Africa. They found that the presence of law enforcement guards had a clear and measurable impact in the conservation of the animals.[38] Protection measures do deter poachers.

We will do what is needed to reduce poaching to insignificant levels. Every roadway leading to wildlife areas would have a permanent checkpoint, inspecting the contents of each incoming and outgoing vehicle. Beyond that, patrols and satellite monitoring could be considered. Within the sizeable wildlife core areas would be a series of two types of enclaves, hemmed in with electrified game-proof fences.

The first type of enclave will not have any top predators. This zone will be ecologically regulated by the baddest penny of them all: humans. It will be a bonanza for all the avid hunters. It will also be a genuine source of food for us, which, with billions of people on board, *is* a consideration, one must admit. I assume that those game reserves will also help reduce any poaching attempts in wilderness areas of other designations.

Essentially, this is to be a big game hunting reserve whose ecology takes a paramount consideration, and in accordance, the hunting of bushmeat would be strictly regulated, assuring sustainable rates.

This also means that if the lions or tigers could have brought an animal down, it is fair game—as this is exactly how this ecosystem operates. However, the same rationale rules out the shooting down of an adult elephant. First and foremost, our role in this zone is to provide the ecological services of the big cats. In accordance, some of the kill would deliberately be left at the field, rendering scavenger services comparable to those provided by top predators. I confess that I am offended by the prospect of a person bringing down a zebra, a pronghorn, or an elephant calf. Making it that much emotionally harder to shake this sentiment off is the awareness of our repugnant history of indiscriminate hunting. Yet, for the sake of the grasslands, the teeming population of herbivores must be kept on the move, adopt wary behavior patterns, and have their numbers held in check. This is the only zone in which hunting would be allowed.[39]

The other designated fenced enclaves would be populated with top predators. In fact, this zone would be identical to that of the broader reserve outside the fenced area with only one difference: it would have African lions as well, standing in for the larger, extinct American Lions (*Panthera atrox*).

While the lions pose an unacceptable danger level for humans much as do the alligators in South Florida, this is not about us. It is about accommodating and co-inhabiting this Earth with other dangerous species. In this case, it is about the preservation of the African lion, who is very likely to acquire a pivotal role in parts of the prairie of the Great Plains as its brethren once did.

With the exception of the African lion, any of the animals can move freely in and out of the enclaves with a series of checkpoints, which can temporarily be closed to prevent the lions from coming out.

The Asian elephants, African lions, and cheetahs are not as adapted to the colder climate of the steppe as their extinct brethren. It is not to say that these animals do not survive in the zoos located in the frigid climates of the world, but it is to say that for their benefit we need to build shelters for the cold months throughout the reserves with suitable provisions of feed. Alternatively, we may find that before the winter arrives, the elephants and Zebras may work their way to the warm desert regions in the Southwest.

——————— · ——————

In the heart of the Nevada desert, there is a knoll made of crushed rock a few hundred feet tall. The rock contained low-quality gold ore, which was extracted using cyanide. After a few years of pouring cyanide onto this mound, the mining company decided it was time to move on. There sat this hill, inert, void of any organic material, and growing nothing. Tony and Jerrie Tipton were assigned the task of restoring it.

They spread a mixture of native seeds, some organic fertilizer rich with microbes, some hay and some straw. The Tiptons then brought in the cows. These ate up the hay and straw. After a few days of dunging, urinating, and trampling of the organic matter, the cows left. After six months, the area was covered in native plants, growing well and establishing themselves. Within a few years, a community of wildlife had moved in and made themselves at home.[40]

The Africa Center for Holistic Management spans 2,600 hectares of parched, degraded grassland south of Victoria Falls in Zimbabwe. With the aid of a few hundred goats and head of cattle and under a deliberate grazing regimen, the land was transformed into a lush pasture replete with ponds and flowing streams even during periods of drought. The Dimbangombe River was flowing again in that area. Livestock have been used as a tool for improving soil aeration, water penetration, seed germination, and increasing species diversity and productivity.

I see the restoration of the Great Plain ecosystem conducted in three broad phases.

In the first phase, we would jolt the languid, degraded grassland regions back to life. This we would do by populating the land with livestock under intensive, rotational grazing practices,[41] coinciding with the deliberate introduction of a native, diverse array of seeds.

Without the massive deployment of livestock animals, this essential first step in revitalizing the land may prove unattainable. We don't have millions of elephants, camels, or bison ready out of the gate. And even if we did, they may graze unsustainably without packs of predators following closely behind and assuring that they would remain bunched. No, the first step of bringing the grasses to anything like their former density and diversity has to be done in a controlled fashion, by employing cows, goats, and sheep.

The Great Plains region starts in Alberta, Saskatchewan, and Manitoba and extends south all the way to Texas and Mexico. In total, I assume a restoration area of 700,000 square miles (about 1,800,000 sq km).

I assume that the cold-temperate grassland and scrubland of the Great Plains can sustain the grazing of forty cows per acre over a period of one day each year. Each day, a herd of forty cows will be moved to another acre sized paddock, cycling through 365 acres in all, and then, the next year, the cycle repeats itself. In other words, forty cows can sustainably live off 365 acres of Great Plain terrain. Forty cows are not quite enough to produce the needed herd effect; it is just a basis for calculation. In actuality, herds of one thousand or ten thousand head of cattle are closer to the mark.

Forty cows per acre per day means that close to fifty million cows would graze under carefully controlled herd-like practices and be able to revive the 700,000 square miles. This number of cows makes up about 45 percent of the existing inventory of cattle in North America.

I am talking about going in force to the grasslands of North America headed by numerous graziers herding tens of millions of head of cattle in the Great Plains. There is nothing remarkable about this statement; our livestock need to graze and be herded. It is just that that half would be in the rangelands of the Great Plains while the other half will be in the more humid, eastern regions, which can sustain perhaps ten times the number of cattle per acre. The intent is to herd the livestock with the high intensity, rotational grazing practice as described in the ninth chapter ("Land Use") while subordinating this to the existing wild animals of the North American grassland, such as those may exist.

The intent is to do it for a decade or so and then slowly restrict and eliminate the presence of livestock in the Great Plains as much as possible.

We still need to provide for the grazing needs of the existing inventory of cattle in North America, the equivalent of 123 million head of cattle.[42] We can take care of their needs with about 115,000 square miles of the more fertile grassland in the Eastern regions,[43] that is, an area somewhat under the size of Kansas and Nebraska combined. Or we can do it on the less fertile, more arid rangeland in the west with over fifteen times the area size. Or any combination of the two. The more we move the cattle to the domesticated regions of the east the more space wildlife would have in the west.

Coinciding with the gradual reduction of livestock from the Great Plains would be the introduction of elephants, camels, horses, pronghorns, bison, asses, zebras, moose, and elk.

Once these wildlife populations reach a critical, viable mass the third phase would come, the introduction of large carnivores.

In truth, all three processes will happen concurrently to various degrees with us having much less control over the process than may seem to be suggested here.

Beyond making the grassland biomes functional again, it would pull back from the brink of extinction the remnant population of elephants, lions, rhinos, cheetahs, Bactrian camels, tigers, and large tortoise, which under the current trajectory are headed to join their mammalian brethren and depart this Earth.

11

DRAWING DOWN CARBON

⫻

the necessity to go beyond a carbon-neutral economy

THE POSSIBLE MAGNITUDE OF GLOBAL WARMING

CURRENTLY, WE ARE AT 0.9°C mean global warming.[1] Reports are coming in on outbreaks of pine beetles in British Columbia, on King crabs appearing in the Antarctic Peninsula, on ominous methane bubbling in the tundra, and on melting of the Arctic sea ice.[2] At present, many key climate indicators are moving beyond the patterns of natural variability within which contemporary society and economy have developed and thrived.[3] Using North America as a case in point, these indicators include soaring temperatures, increasing precipitation intensity, the worst drought in recorded history, steep declines in Colorado River reservoir storage, widespread vegetation mortality, and sharp increases in the frequency of large wildfires.[4]

Currently, we are at 0.9°C mean global warming. We have experienced growing aridity in the southern United States, the Mediterranean region, Australia, and parts of Africa. Alpine glaciers are in near-global retreat.[5]

Currently, we are at 0.9°C mean global warming. Fatal coral bleaching on a wide scale started when atmospheric CO_2 levels exceeded 320 ppm. When CO_2 levels reached 340 ppm, sporadic but highly destructive mass-bleaching occurred in most reefs worldwide. Allowing a lag-time of 10 years for sea temperatures to respond, at present CO_2

concentrations most of the world's reefs are condemned to an irreversible decline.[6] Coral reefs occupy a key position on Earth. The imminent failure of reefs will affect reef-associated ecosystems:[7] seagrass beds,[8] mangroves,[9] marine reptiles,[10] seabirds,[11] marine mammals,[12] pelagic ecosystems,[13] and estuarine habitats.[14]

One study concluded that we would reach, within a few millennia, a 20 to 30 meter sea rise—provided we stabilize at 387 ppm of CO_2,[15] which we have already exceeded.

Once we get to about 1.5°C to 2.0°C warming, we are likely to see an escalation in the frequency and intensity of heat waves, floods, droughts, wildfires, and tropical cyclones in many parts of the world. These are likely to result in heightened water stress. Once we get to about 1.5°C to 2°C warming, we are likely to see irreversible losses of unique and threatened systems, such as tropical glaciers, endangered species, unique ecosystems, biodiversity hotspots, small island states, and indigenous communities.

And once we hit 2°C to 3°C warming, there is a real chance that about a fourth of known plant and animal species are likely to be at an increased risk of extinction.[16] And once we go past 3°C, over 600 land bird species may go extinct, mostly tropical.[17]

Stopping all emissions today is the equivalent of letting go of the gas pedal—without applying the brakes. We are still destined to fall off the edge.[18]

The transition to a carbon-neutral, low-impact technology would require enormous resources and the labor of many millions of people. The land-use changeover coinciding with massive rewilding efforts goes beyond, requiring a change in our lifestyle that has not been asked of us since World War II. Yet, more is called for. It may be that the most daunting technological mission has been left unstated: putting back in the genie bottle the hundreds of billions of tons of carbon we have inadvertently unleashed for the last two hundred years.

Our collective future will rise or fall on the amount of carbon we draw down from the air and subsequently sequester. The battle over the future of the world as we know it will be decided in the planetary carbon fields.

Had this blueprint been written but twenty or thirty years ago, the content of this chapter would have been markedly different. The situation as of this writing requires us to bring to bear every resource we have and every ounce of ingenuity we can muster. It also demands that we make decisions that may have dire repercussions. We don't have decades to get it right. We don't have ten years to get it right. Much of what I've stated in earlier chapters should in fact be accomplished by 2015, not 2027. We have only a few years to stop all emissions—that is, let go of the gas pedal—and proceed to draw down carbon—that is, brake hard before we hit a possible point of no return.

It is time to describe what I have not stated so far: two natural mechanisms which, once triggered, will take the matter of climate change out of our hands. These two processes have the power to push Earth into a new, inhospitable climate. If they come to pass, we will not be able to counteract this. It is time I explain why a carbon-neutral economy is no longer an option; we must get this titanic ship into a reverse gear.

———————

Under cold, high-pressure conditions, water and methane combine into a crystalline structure referred to as *methane hydrate*. Most of these hydrates are located along the ocean shelves of the world. The hydrates could be vulnerable to melting with a deep-ocean warming of a few degrees Celsius.[19] Increasing temperatures or lowering pressure can cause gas hydrates to break down, releasing their methane. And so it has happened; researchers have reported an upsurge in the methane being released in the East Siberian Arctic Shelf.[20]

The global inventory of methane in ocean sediments is estimated at 1,600–2,000 billion tons.[21] A preliminary assessment reckons that on a time frame of millennia, anthropogenic warming may melt a significant fraction of the methane hydrates in the ocean.[22] It is not clear how much of the methane would escape into the air. It is possible for the gas to migrate up to a stable zone and re-form as hydrates. And even if the gas does reach the atmosphere, it is still unclear how much of it would first be oxidized in the ocean.[23] Releasing carbon dioxide into the air is a different proposition than releasing methane into the air.

Our best, state-of-the art model is likely to prove in retrospect to be but a progress report, an early study. For all it's worth, it shows a likely outgassing range from 35 to 940 billion tons of carbon.[24] This is likely to take place over the course of many human lifetimes. This volume of outgassing ranges from the fairly inconsequential to the catastrophic. Provided it is first all oxidized, the output would range from 128 to 3,450 billion tons of CO_2. Provided all of it reaches the atmosphere as methane, the output would range from an equivalent of 875 to 23,500 billion tons of CO_2. The current combined anthropogenic GHG emissions are about 44 billion tons annually of CO_2 equivalent.

Above and beyond the inventory of hydrates in ocean sediments are the 400 billion tons of methane hydrates found in the permafrost. There is no study that models their possible outgassing. It stands to reason that the considerable warming projected for the permafrost will have a detrimental effect on the stability of the hydrates. As the permafrost thaws, the thermal shock of this warming may destabilize gas trapped underneath.[25] If and when this methane is released in full, it will be the equivalent of 10,000 billion tons of CO_2. The current combined anthropogenic GHG emissions are about 44 billion tons annually of CO_2 equivalent.

To put these numbers further in context, one more thing needs to be stated: the more a given greenhouse gas is added to the atmosphere, the less its greenhouse effect is. Thus, if at first, the addition of 280 ppm atmospheric concentration of CO_2 may cause 3°C warming; to produce an additional 3°C may require a 560 ppm CO_2 concentration; and to produce further warming of 3°C may take 1,120 ppm, and so on.

Beyond the hydrates is the carbon locked up in the permafrost region of the world. In total, the northern permafrost region is assessed to contain approximately 1,672 billion tons of organic carbon, of which approximately 1,466 billion tons occur in perennially frozen soils and deposits.[26] It is this inventory of carbon that may live in infamy for future generations. This underground deposit is largely made of dead roots that have been gradually buried under for the past tens of thousands of years. And the only thing that stands between these deposits and the outgassing of their carbon content is their permanently frozen status. Once they thaw, they will decompose with the help of microbes and subsequently outgas. This subterranean carbon has been out of circulation. And now it's coming back.

As air temperature has risen in recent times, the permafrost temperature has also climbed. Once the unfrozen layer expands beyond a critical thickness, the soil column will thaw rapidly, with rates often exceeding 100 centimeters per decade. Researchers have recently detected changes to the hydrologic cycle and vegetation composition that appear linked to permafrost degradation.[27] Recent observations indicate widespread permafrost degradation in the Northern Hemisphere.[28] In some places, permafrost temperatures at a depth of 20 meters increased 2°C–3°C in the past two decades.[29] In some areas of the permafrost, thaw lakes have appeared.[30]

So far, the thawing has not reached the frozen carbon layers.

Models project that the degradation of the permafrost will worsen, driven by surface warming. By 2100, under an emission-intensive scenario,[31] Lawrence and Slater project 90% thawing.[32] Eliseev and his team project 80% thawing.[33] Regardless if 80% or 90% will thaw by 2100, under sustained warming, within a few hundred years the entire permafrost will thaw and the bulk of the carbon inventory will subsequently outgas. Perhaps something like 15% of the carbon is likely to stay in the soil.[34] Thus, the outgassing may be estimated at 1,421 billion tons of carbon. Well, to date, our *entire*, cumulative carbon emissions since pre-industrial times have been about 520 billion tons.

The all-prevailing wetlands and lakes in permafrost regions assure us that some portion of carbon emitted to the atmosphere will be released as methane.[35] How much of it will be released as carbon dioxide and how much of it in the form of methane will primarily depend on the presence or the lack of oxygen. We don't have an answer to this very consequential question; guesses and anecdotal evidence are all we have to work with at this stage. However, it would have been grossly irresponsible to assign a default value of zero methane emissions due to the vast uncertainties surrounding this matter. During a 2011 expert gathering in Seattle, opinions voiced in private ranged from 1% to 30%, while an average seem to center around 7% methane content.

If we assume 7% methane content, this suggests that, cumulatively within the next few hundred years—within a few lifetimes—we will have 1,321 billion tons of carbon released in the form of carbon dioxide and 100 billion tons as methane. Combined, this amounts to a net emission equivalent of 7,348 billion tons of carbon dioxide.[36] The current combined anthropogenic GHG emissions are about 44 billion tons annually of CO_2 equivalent.

In the first chapter ("Climate Change"), I stated that based on existing models, we will experience a global annual warming of 5.5°C in relation to pre-industrial levels under the business-as-usual, fossil-intensive scenario. This is not what is going to happen. Neither the possible outgassing of hydrates nor the likely outgassing of carbon from the permafrost and their subsequent warming effect have been accounted for in the models listed above.[37]

Under business as usual, it is virtually certain that the atmosphere will experience a massive injection of carbon and methane from the northern reaches of our planet. Thus, under current trajectories, by the end of this century it is probable that the annual global warming will be higher than 5.5°C. And when the outgassing process is completed, be it a few centuries away or earlier, it is probable that the annual global warming will be *significantly* higher than 5.5°C.

This will be even more pronounced once the melting process of the glaciers runs its course, and there will be much less ice to reflect the sun's rays back to space.[38] Beyond that, if we continue with fossil-fuel emissions and land-erosion practices into the twenty-second century and beyond, it will add even more fuel to the fire. The warming oceans cannot keep taking in from the air the same portion of CO_2 as they have been until recently. We are already witnessing the uptake rates of the subtropical North Atlantic slowing down.[39] In other words, a bigger portion of the carbon is likely to remain in the atmosphere.

I wish to impress upon us the importance of changing the current trajectory.

Connecting the dots of all that has been stated, I will produce an assessment of the possible temperature patterns of Earth in the short term, in the span of a few human lifetimes. I will take uncertainty atop uncertainty atop other uncertainties, producing something with a very low-level of confidence. Regard it as a provisional, first-stage assessment. Yet, is a big sign reading "bridge collapse imminent; travel at your own peril" any less valuable because the authors cannot quantify with a high level of confidence the vehicle weight and speed that may precipitate this?

Under the business-as-usual, fossil-intensive scenario, it is projected we will reach a CO_2 concentration of 900 ppm by 2100.

As just discussed, above and beyond this 900 ppm, we may assume in the next few centuries a total of 7,348 billion tons of CO_2 equivalent from the permafrost. This would translate to 472 ppm.[40] Coupled with the anthropogenic emissions, this would bring us to the equivalent of 1,370 ppm of CO_2 equivalent. On top of this, we ought to add an unknown amount of emission coming from hydrates in the next few centuries. This amount, as stated, can range from the insignificant to the calamitous. The hydrates would tack on anything from 8 ppm to 2,153 ppm to the running total of 1,370 ppm of CO_2 equivalent.[41] The grand total comes to a possible carbon dioxide concentration equivalent in the following centuries that ranges from 1,378 ppm to 3,523 ppm. When was the last time we had CO_2 concentrations in this range?

A recent, extensive study reviewed 370 estimates of the CO_2 concentrations for the past 65 million years. The readings we have from phytoplankton, from palaeosols, from somata, from liverworts were averaged, and they do indicate that the Early Eocene, 50 million years ago, is likely to have had about 1,100 ppm of CO_2.[42]

So from one side we have the historical levels from 50 million years ago that may have been 1,100 ppm. From the other side we have projected levels slightly higher or noticeably higher than that within a few centuries. As explained, 3,523 ppm does not translate to three times more warming than 1,100 ppm. Figure 11.1 is more on target.

Not everything is equal, though. For the time being, we have expanses of ice at the Polar Regions, reflecting some sun rays back into space. Fifty million years ago, there were none. So this counts for 1 or 2°C in our favor.[43] At the same time, we did not account here for any anthropogenic carbon emissions beyond 2100. In addition, the level of CO_2 assumed to remain in the atmosphere does not account for the ocean's diminished ability to take in CO_2. And we did not count on tropical ancient carbon sinks possibly releasing their carbon content under increased temperatures.[44] In other words, we could end up with more CO_2 in the air than has been suggested.

FIGURE 11.1

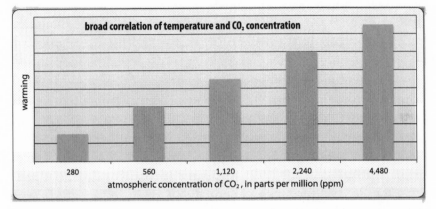

Further studies may prove this entire analysis to be off the mark, one way or the other. But this analysis cannot be *that* off the mark. That is to say, we must allow the possibility that under the business-as-usual course of action we would arrive in the next few centuries to a climate whose temperature levels correspond to the Early Eocene, around 50 million years ago.

While this means that Earth experienced these levels of warming long ago, there is nothing reassuring about that fact. A recent study modeled the temperature characteristics of the Early Eocene.[45] Compared to present day, annual mean temperatures were 30°C (≈54°F) higher in the latitudes of Northern Canada and Siberia; 15°C (≈27°F) higher in the latitudes of Europe and the United States; and about 10°C (≈18°F) higher in the tropics. Under such a climate, Paris would have temperatures akin to those currently present at the heart of the Sahara desert.

This warming range is off the charts of any existing impact studies. The closest thing we've got is one impact study that explores the outer reaches of Hell.[46] One paper does explore the wet-bulb temperatures of the world under 12°C mean average global warming. This is the level of warming that we had 40–45 million years ago (Middle Eocene). Back then, it was not quite as warm as Early Eocene times, though.

It is often assumed that humans would be able to adapt to any possible warming. That is not the case. A sustained exposure to a wet-bulb temperature, a combination of humidity and temperature, of 35°C (≈95°F) or higher would cause the body to reach lethal heat levels as dissipation of heat becomes impossible. The highest present-day values are about 30°C–31°C (≈ 86°F–88°F), during the worst heat-humidity events in India, the Amazon, and a few other very humid places. Even though the hottest temperatures occur in subtropical deserts, relative humidity there is so low that wet-bulb temperatures are no higher than in the deep tropics.

With annual global warming of 12°C, borderline-lethal wet-bulb temperatures would engulf the some of the existing population centers. During at least one day of the year, parts of the South American interior, most of the Australian interior, the Indian subcontinent, a large portion of the population centers of China, and the Mid-Atlantic and Deep South regions of the United States will become potential death zones. A corresponding map can be accessed at www.danielrirdan.com (map data courtesy of Steven C. Sherwood).

With the prospect of carbon emissions from the northern reaches of the planet coupled with manmade emissions, we are faced with the prospect of the biggest calamity that has ever faced mankind. We have mobilized ourselves into high gear and given our lives for far less. Within a century or so, all of those who are alive today and reading these words will be dead. It is not every century that a generation gets to choose the words engraved on their collective tombstone. This is one of those times. The question is not if humans as a race would survive the impending calamity. We would, much as we would have survived a Nazi Germany taking over the world. The question is what manner of human culture would emerge in the aftermath; the cost our great-grandchildren would pay for our apathy; and what sort of a planet they would inherit.

———— ——

In the long run, it would matter little if the entire release of carbon from the permafrost would occur in one lifetime or three. Either way, it is going to happen in the short term. However, in the context of trying to prevent the bulk from ever outgassing, it does matter how much and how fast.

As of 2011, we don't know enough to predict that. What we have is one initial model published in 2011 by Kevin Shaefer and his team that may be our best guidepost.[47] However, the rates are likely to be higher than modeled. To arrive at a more realistic outcome, we need to factor in augmented warming due to the thawing. We also need to account for the added carbon release due to river and coastal erosion. Finally, under a business-as-usual trajectory, we need to assume that the Northern reaches would be under greater warming than assumed in the model.[48]

What is of most interest about this model is its projection that the permafrost carbon will start outgassing in earnest by around 2026. In fact, this modeling work may give us a pretty good idea how it may unfold. It will start off with about 5 billion tons of CO_2 equivalent annually around 2026–28. The rate will quickly rise, reaching 31 billion tons CO_2 equivalent a year by 2040 and sustaining average levels of 36 billion tons from around 2070 to 2120, and subsequently it will start to decline.[49] The current combined anthropogenic GHG emissions are about 44 billion tons annually of CO_2 equivalent.

Any efforts to draw down existing carbon will be offset and negated with added emissions coming from the northern reaches of the world. We must bring down the planet's temperatures before the outgassing gets into high gear, around 2040–50.

The thawing of the land has already started; it just hasn't yet reached to the deeper layers, where the carbon is. It is a carbon bomb with a fifteen-year-long fuse. And the fuse is already lit. Even if somehow we affect an immediate cooling, it will be decades until the permafrost re-freezes. In other words, we *are* going to witness the thaw of the carbon. This is inevitable. It is now a matter of how much. A relatively small amount might be manageable. A lot may render our mitigation measures futile. The global warming process will be wrenched out of our hands.

We cannot afford that. The ecosystems of the world cannot afford that. And so now we come to setting relevant goals.

ESTABLISHING CARBON TARGETS

First, it is important to understand that things at this moment are likely to be worse than they appear. So far, the air has been losing much of its heat energy to the vast bodies of water on the planet. This is but a temporary condition, as the ocean is catching up and gaining equilibrium with the ambient air temperature.[50] Thus, beyond the 0.9°C we already experience, an additional 0.4°C to 0.6°C are in the pipeline, committing our 2027 projected emissions to a total of around 1.4°C rise above 1950s levels.[51] Yet, this does not include some possible long-term reinforcing feedback processes that further warm the climate, such as shrinkage of the ice cover or changes in the distribution of vegetation. Once these slow feedbacks are accounted for, 2015 levels might commit us in the long run to 3°C above pre-industrial levels.[52]

Then there are aerosols. There is good news and bad news about them. The good news is that as long as we keep spewing massive amounts of particulate matter and soot into the air, more of the sun's rays are scattered back to space, the reflectivity of clouds increases overall, and the net effect is to cool the Earth's surface. The bad news is that once we stop polluting, stop running all the diesel engines and the coal plants of the world, and the soot finally settles down, the real state of affairs will be unveiled within weeks.

And no one really knows what that means—beyond an additional, unspecified amount of additional warming. Additional warming is bad; the large uncertainty surrounding its extent is downright worrysome.

Thus, it is not good enough to stop all emissions by 2015; it is not good enough to have a carbon-neutral economy. In fact, it is not even *close* to being good enough. A carbon-neutral economy at this late stage is an unmitigated disaster.

Hell, what has been portrayed in the previous chapters is not even a carbon-neutral economy. Not really. The plan calls on massive concrete and steel work. And those come with a GHG-emissions surcharge.

Thus, the question is not *whether* in addition to halting all the possible GHG emissions, we should also draw down and lock away some of carbon in the air. The question is how *much* carbon should we sequester from the air, and what should be the rate of withdrawal.

There is a need for a carbon-negative economy. Essentially, it means that we have not only to stop emitting, to the technological extent possible, all greenhouse gases, but also capture much of the crap we have already outgassed and lock it down. Moreover, we have to capture the crap that will be generated by the construction of the lower-impact technologies. And once we do the above, the ocean will burp its excess gas, which has come from fossil fuels in the first place. So we will have to draw down and lock up that carbon, too. We have taken fossil fuel and released its content; now we have to do it in reverse—hundreds of billions of tons of that stuff.

I put the concentration target at 325 ppm of CO_2. All the way from 399 ppm, projected for 2015, down to 325 ppm. Bringing the CO_2 concentrations down to 325 ppm will not bring us to historical levels as far back in time as one may assume. It was but 1970 that we had 325 ppm. But wait, there is more to come.

In 1970, the levels of CO_2 were 325 ppm; however, in 1970 the methane levels were lower too, and so were the concentrations of nitrous oxide, and chlorofluorocarbon (CFC) gases. A concentration of 325 ppm while assuming 2015 concentration levels for the secondary greenhouse gases will not bring us back to 1970 conditions. It will bring us back to the combined greenhouse levels akin to that of 1982. Yet, even 1982 was not all that it could have been—aerosols had a huge masking influence. If we talk about a GHG effect comparable to that of 1982 *without* the existing effect of aerosols and without the black carbon on the snow, it is worse yet.

Let's talk numbers. Total anthropogenic forcing projected for 2015 is 2.30 w/m². If we replace the 2015 CO_2 value with its 1970 value, this gets us to a total anthropogenic forcing of 1.17 w/m², which we last saw in 1982.[53] Once we further get rid of the aerosols and black carbon on snow, we may be very well be worse off than in 2011.[54] The picture is more complex than suggested here, but the bottom line may be about the same.[55]

Incidentally, even if 325 ppm concentration could have brought us back to warming levels of 1970, would it have been enough? The oceans have been warming in earnest for over 100 years now.[56] In other words, under CO_2 concentrations of around 295–300ppm, the planet has already started working its way to a new climate regime.

Indeed, there is nothing remotely conservative or cautious about 325 ppm target levels in the context of what else is going on in the atmosphere at the moment. The 325 ppm target is also markedly higher than historical CO_2 levels of 259–265 ppm during the first half of this interglacial period,[57] a period in the history before we got our industrial pursuits under way. Considering the inevitable halt of cooling aerosols under carbon-neutral technology along with the lack of our ability to neutralize some of the other anthropogenic greenhouse gases in the atmosphere, 325 ppm may prove to be but a milestone, not a final destination.

If we assume that 2015 would be the year we halt most of our GHG emissions, CO_2 levels will hover around 399 ppm. About 157 billion tons of carbon must be brought down to achieve 325 ppm.

FIGURE 11.2

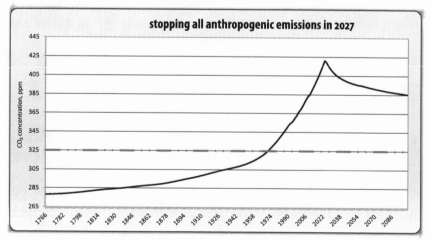

Source: See endnote 58.

Note: The dashed line is the 325 ppm target.

FIGURE 11.3

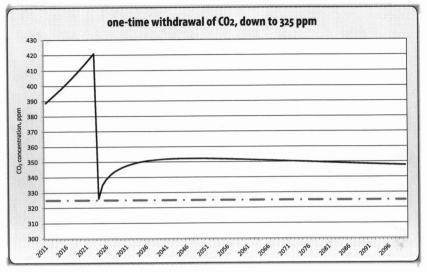

one-time withdrawal of CO₂, down to 325 ppm

Source: See endnote 58.
Note: The dashed line is the 325 ppm target.

However, this is when the ocean will start emitting its excess of carbon into the air. While this is what we want it to do, the atmospheric concentration of CO_2 will start to go back up again, as is shown in figure 11.3. This represents a model of what is likely to take place.

In order to hold the needle steady at 325 ppm through to 2100, there is a need to draw down an additional 65 billion tons of carbon, for a total of 222 billion tons between now and 2100.[58]

On top of this, we have to account for some inevitable emissions from the permafrost that will kick in sometime around 2020–2030 and will go on for a few decades before the permafrost will hopefully re-freeze and we put a lid on it. Using Shaefer's emission schedule described earlier as a basis, I peg the number at 60 billion tons of carbon to be added to the air from the permafrost, assuming that half of it will settle fairly quickly in the oceans and land. Finally, I will throw in another 20 billion tons—a result of our reduced emissions between now and 2027, when we will get to a fully carbon neutral economy. This is a grand total of 302 billion tons of carbon that need to be brought down.

This is the goal then: halt the vast majority of greenhouse gas emissions by 2015 and subsequently draw down and sequester 302 billion tons of carbon between 2015 and 2100 plus any possible added carbon that may emanate from the hydrates within ocean sediments.[59]

This carbon target is not conservative but somewhat rosy—as it does not account for the additional GHG emissions that will be incurred during the colossal construction process of the new energy grid.

TABLE 11.1

CO_2 concentration target	325 ppm
total to be drawn down by 2100	302 billon tons of carbon (1,108 billion tons of CO_2 equivalent)
commencement year	2015

Depending on the context, at times it makes sense to discuss things in terms of carbon and at other times in terms of CO_2. It comes down to the same thing. The target can either be discussed in terms of 302 billion tons of carbon or 1,108 billion tons of CO_2.

We've got eighty-five years to get down all this carbon, but we cannot pace ourselves. Until the permafrost refreezes, we have to draw down substantial chunks of the 302 billion tons and then we can slowly ease up.

Furthermore, given how late the hour is, we don't have enough field data to provide a high-level of confidence about any aspect of what needs to be done. This will have to do, and risks will be inherent to the plan; there is simply no more time. It is what it is. Very little time is left for ponderous conferences and committees. We will have to pull off the seemingly impossible.

What follows is a portfolio of carbon capture and sequestration technologies. I present their potential, taking each of them for a drive as far they go. The exact mix of technologies depends on further study. Even one year with tens of thousands of researchers employed may give us the needed minimal confidence in determining the portfolio breakdown. As it stands, we know little about strategies to be described.

On the ground there are few research teams with supporting models or demonstration units.

Beyond the technologies presented below, there are a dozen others even less tested and with even higher levels of uncertainty. Two of these are covered in Appendix B.

FLUE-GAS CARBON CAPTURE

Worldwide, there are currently about 8,100 large CO_2 point sources: fossil-driven power plants, steel mills, cement kilns, chemical plants, and oil and gas production and refining facilities. They account for more than 60 percent of all anthropogenic CO_2 emissions, which is to say, annually, upward of 19 billion tons of CO_2 out of a total of 32 billion tons of CO_2 derived from fossil fuel and cement.[60]

As discussed when dealing with energy, within fifteen years the bulk of these plants are to be phased out. However, within three years all of them will require retrofitting: their emitted CO_2 is to be captured and sequestered to the maximum extent possible.

For reasons which have nothing to do with climate change, CO_2 has been captured from some coal power plants since the late 1970s and from some natural gas power plants since the early 1990s.

One of the larger CO_2 capture systems applicable to power generation facilities can be found in a recently constructed power plant in Alabama. It is a tiny, 25 MW coal plant run by Southern Company. The system captures 90 percent of the CO_2 fed to it within the plant. In total, it captures 150,000 tons of CO_2 annually. The CO_2 capture-unit utilizes the KM CDR Process, developed by Mitsubishi Heavy Industries. Mitsubishi Heavy Industries has stated that they would be able to fabricate larger units, capable of capturing ten times the volume of CO_2 captured in the Alabama plant. Be that as it may, the consensus among experts is that there is considerable uncertainty associated with the process of scaling up the carbon-capture technology.

There is nothing trivial about retrofitting existing plants with carbon-capture units. First, carbon capture infrastructure requires several acres of space and many plants do not have this land area available. Second, other pollution control devices will need to be installed for plants that do not already have them. For example, current MEA systems are unable to process flue gas streams containing modest amounts of sulfur dioxide. Prior to attempting to capture CO_2, these plants will have to be fitted with state-of-the-art pollution control systems. The difficulties don't end there. Current capture technology requires 40 percent more water for operation. Other challenges to retrofitting existing units include large modifications of existing plants, proximity to storage options, and maintaining expected generation throughout the construction process.[61]

In addition, CO_2 capture-and-compression systems also require significant amounts of energy for their operation. It is estimated by analyses done by US Department of Energy that the total power of capture and compression of the gas could equal 20 to 30 percent of the existing coal plant's energy output.[62]

The amount of energy penalty heavily depends on the efficiency of the existing plant. And it also depends on the specific capture technology, the degree of carbon capture, and the fuel burned in the power plant (coal or natural gas). Relatively new efficient coal plants (called supercritical units), if fully fitted with a modern 90% capture system, would see 20%–25% power reduction. Older, inefficient plants would see much larger reductions. For modern gas-burning plants the penalty is smaller, about 15%.[63]

Some plants, notably older ones, may just shut down and be replaced with their zero-emissions, lower-impact counterparts. Others would be retrofitted with carbon capture installations.

Tomorrow we may have a better carbon capture technology.[64] However, then is then and now is now. So let us be clear; there is nothing warm and fuzzy about the chemicals or machinery used to capture carbon at the flue. They fit right in with the coal and gas power plants. However, until the complete transition to carbon-neutral power generation within fifteen years, this is the *only* viable course of action.

Carbon dioxide is not the only thing that requires an immediate capture.

The global-warming effects of black carbon are possibly second only to carbon dioxide.[65] The chief anthropogenic sources of these soot particles are the burning of wood and dung for cooking, clearing crop residue with fire, and the exhaust from diesel engines, such as from trucks, ships and construction equipment. Unlike greenhouse gases, the soot particles linger in the atmosphere for just a few weeks.[66]

Side by side with the CO_2 capture, the exhaust of all suitable diesel engines ought to be equipped with particulate traps for land vehicles and onboard scrubbers for ships—until they would all be replaced with or retrofitted with lower-impact technologies as discussed in the fifth chapter ("Transportation").

INJECTING SULFUR INTO THE ATMOSPHERE

As the bulk of the emissions cease, the full heat of the sun will bear down on our planet. The cooling veil of aerosols will be lifted. We have three years to decide if there is more to gain than to lose by deploying a cooling sulfur haze in the upper atmosphere in the next few decades. This is a last resort strategy—which is exactly where we are at. However, from where I stand, it seems the lesser of two evils. If it turns out to have net benign effects, as I suggest, it would be indeed a notable exception in an endless stream of what have turned out to be very bad ideas in our various attempts to tinker with the natural environment.

The idea is to inject into the upper reaches of the atmosphere sulfur dioxide, where it would eventually form small particles of sulfate that reflect sunlight back into space. This is not unlike what happens when a volcano erupts. One paper suggests that five million tons of sulfur dioxide would have a substantial influence.[67] As a point of reference, the eruption of Mount Pinatubo injected about 20 million tons of sulfur into the stratosphere. Models do concur that injecting sulfur dioxide into the atmosphere would reduce surface-air temperatures worldwide.[68] Yet, at least one modeling study suggests that heightened sulfur dioxide levels won't be able to completely offset changes in atmospheric and oceanic circulation under high atmospheric CO_2 concentration.[69]

Now, let's turn our attention to the downsides. It is likely that the addition of sulfate to the stratosphere would result in a loss of ozone and consequently allow more ultraviolet radiation through.[70] Indeed, the eruption of Mount Pinatubo in 1991 had just such an effect, even as it cooled the climate. That is not all. Cutting down solar radiation is apt to reduce rainfall, creating a risk of widespread drought and reduced freshwater resources.[71] The rates of cooling can exceed the capacity of plants and animals to adapt.[72] Other possible adverse effects may include reduction in direct sunlight reaching the Earth, higher levels of acid in the rain, and whatever other effects we don't know about.

This geoenineering exercise will do nothing about the alarming ocean acidification and effects of high CO_2 concentration on the nervous system of the marine animals. It may though, stop the global warming clock in its tracks, albeit with a climatic and ecological surcharge. It gives us borrowed time as long as we keep administering sulfur injections, say, for a period of twenty years. And borrowed time is exactly what we need as we draw down and sequester colossal amounts of carbon from the air without the sheltering aerosols.

The most straightforward method of deployment would be to develop an aircraft capable of flying at altitudes of 20–25 kilometers, distributing ten tons of sulfur per flight. Sulfur aerosols are relatively short lived, so we would have to keep applying them into the atmosphere. About eighty such planes would allow the delivery of sufficient acid to the stratosphere every year. Another idea is to use a sort of hybrid plane-blimp along the lines of Lockheed's experimental P-791, which generates lift through both buoyancy and aerodynamics.[73]

LAND TAKING IN SOME OF THE CARBON

I will start at the end. Our practices have caused wholesale deforestation and land erosion. In turn this led to the soil losing large amounts of carbon to the atmosphere. Land regeneration offers the potential to replenish most if not all of the carbon thus lost.

Having said that, while the soil has lost carbon, it could not have lost *that* much. Since 1850 the soil is estimated to have lost a combined amount of about 70 to 90 billion tons of carbon.[74] For comparison, about 440 billion tons were outgassed from the burning of fossil fuels. It seems that most of the excess carbon out there has not come from the soil but from fossil deposits. In fact, we know it to be so, as much of the carbon sampled from the air bears an unmistakable signature identifying its ancient, fossil origins. We may very well be able to help the soil restore its lost carbon. However, we have to be skeptical about the prospect that the soil could store carbon above and beyond its pristine, pre-historical carbon levels, taking in the extra carbon that was never emitted from it in the first place. In truth, we don't know; it may go either way.

In fact, it is downright astounding how little we do know on the matter of carbon in the soil.

If we had the time, numerous soil tests at many sites over a few decades would have given us a good idea of the global soil sequestration potential. Let's work with what we think we do know.

Forests were cut down, and annuals replaced perennial plants. The shift from natural ecosystems to agriculture has released substantial quantities of carbon into the atmosphere. Conventional agricultural practices have resulted in the loss of as much as 60% of the soil carbon content in temperate regions and 75% or more in cultivated soils of the tropics.[75]

Time to reverse the process. Not only must we prevent further greenhouse gas emissions from the soils, we also have to convince the soil to sequester some of the carbon that is in the air. Carbon is in a constant flux between the soil and the air. However, a certain amount of carbon is resistant to decomposition and remains in a stable form within the soil. This is the humus, what makes up the carbon store within the soil.

There have been at least 39 published comparative studies on organic and non-organic farming, studying the carbon levels stored in various soils. In at least 15 of these studies, organic farms averaged 20 percent higher soil carbon levels.[76] In fact, those studies are likely to have seriously understated the case for organic farming, as they have just measured the amount of carbon in the topsoil. All the action is in the subsoil, where the bulk of the stable carbon may reside. A synthesis study concludes that under worldwide organic farming we would be able to sequester about 1 ton of carbon per hectare annually for a total of 1.5 billion tons of carbon per year worldwide.[77] To put this number in context, our target is 302 billion tons by 2100, with the majority being sequestered in the first few decades.

The synthesis study mentioned above speculates that the amount of carbon that could be stored may end up being higher, or even much higher, once we apply a host of beneficial practices to organic farming. There is a general consensus that some farming practices would enhance the storage of carbon in the soil: no tillage, cover crops, mixed cropping systems, agroforestry, and the addition of organic matter such as crop residue and manure. As suggested in the ninth chapter ("Land Use"), these practices are expected to be the de facto standard in the future.

Rattan Lal is regarded in academic circles as an authority on the matter of carbon sequestration in the soil. In one of his papers, Lal lends another view. He estimates the rate of sequestration under the above mentioned practices to range from around 0.1 carbon tons per hectare a year in dry and warm regions to 1 ton per hectare in humid and cool climates. Those rates would be sustainable for 20 to 50 years, or until the soil is filled to its absorption capacity. In total, this would amount to about 30 to 40 billion tons of carbon.[78] As a reminder, our target is to draw down 302 billion tons of carbon.

Lal suggests that an increase of carbon in degraded cropland soil boosts crop yields markedly.[79] It is a win-win proposition. It is generally agreed that boosting the levels of soil organic matter (which contains carbon) improves soil's water holding capacity, soil fertility, and water quality.

Beyond croplands are the billions of hectares of grassland throughout our planet. The substantial stocks of carbon sequestered in temperate grassland ecosystems reside largely below ground in roots in the soil. The single greatest potential lies with the shift from annual to perennial species in pasture lands.[80] In fact, at ten farms in the northern Great Plains, this very changeabout from annuals to perennials has resulted in carbon sequestration of 3.8 tons of carbon per hectare annually.[81] Breeding perennials—with their long-lived, two-meterlong roots—could double the amount of CO_2 captured from the air.[82]

While we cannot convert the many hundreds of millions of hectares of cropland from annuals-based to perennials-based as of yet, we may achieve comparable results if we switch to a polyculture that incorporates perennials.

For one, Christine Jones, founder of Amazing Carbon, has mea-sured the carbon levels of land under pasture crop cultivation. Jones is a soil ecologist gone rogue for personal reasons, and those who know her vouch for her integrity. If her readings are to be taken at face value, over a twenty-year period, the soil under pasture cropping sequestered a total of forty-six tons of carbon per hectare. In the last two years of the study, 2008–2010, the rate was an astounding nine tons of carbon per hectare annually.[83] A confirmation and reproduc-tion of these results is in order. Pity we don't have twenty years to spare for that.

Jones maintains that when carbon enters the soil ecosystem as plant material, it decomposes and returns to the atmosphere as CO_2. Enter mycorrhizal fungi, and the picture changes radically. Jones states that these fungi must form an association with living plants. The mycor-rhizal fungi acquire their energy in a liquid form, as dissolved organic carbon, siphoned directly from actively growing roots of the host plants. These fungi cannot obtain energy in any other way.[84] Now we are getting to the interesting part of the story. Provided appropriate land management systems are in place, the major portion of soluble carbon is transferred from the symbiotic plant roots to mycorrhizal fungi and is subsequently converted into stable humus. It is a process in which simple forms of carbon are resynthesized into highly complex polymers. The resultant humus is a stable, inseparable part of the soil matrix that can remain intact for hundreds of years.

The important role of mycorrhizal fungi has been generally over-looked, as they are present in low numbers in annuals-based, conven-tionally managed agricultural landscapes.

The soil environment required for these fungi and for humus gen-eration is commonly found in association with year-round green farm-ing practices such as pasture cropping. Photosynthesis occurs for a much greater portion of the year in perennial pastures. Furthermore, the permanent presence of a living host provides a reliable supply of soluble carbon and a suitable habitat for colonization by mycorrhizal fungi.[85] Jones claims that the practice of pasture cropping, where an annual crop is grown out-of-phase with perennial pasture, can result in even higher rates of soil building than under a true perennial pasture. This, she speculates, may be due to year-round transfer of soluble carbon to the root-zone and formation of humus during the non-growth period of the perennial plants.

The carbon sequestration rates reportedly achieved under pasture cropping management are most substantial and offer the potential to sequester much of the runaway carbon in the air, and that is one more way cattle can play an important role in the overall plan. There is no doubt that a thriving grassland ecosystem spells more carbon capture and carbon retention. Rotational mob grazing of large herbivores, described in the previous chapters, seems to prove vital in the restoration of the vast ranges of grassland that are on the verge of desertification and are losing carbon due to bare soil.

Last but not least are peatlands.

Peatlands are wetlands whose soil consists primarily of decomposed organic matter. Not surprisingly, they store enormous amounts of carbon.

In the California Delta, a freshwater tidal marsh once formed carbon-rich peat soil sixty-feet deep in places. In the 1870s, farmers began to build dikes, drain the marsh, and farm the peat soil, which has oxidized rapidly upon exposure to air.

The US Geological Survey began a small experiment to see if this loss due to oxidation could be reversed. They flooded about six hectares, and put in some clumps of tules and cattails. As the plants grew, they raised the water level. After ten years, this experiment has built two feet of peat soil that you can stand on. The plants drew carbon dioxide from the air and built complex carbon compounds. Protected from further oxidation by careful management of the near-stagnant water, the partially decayed dead stalks and roots became organic matter: peat. More carbon went in than went out. On average, the experiment yielded 10 carbon tons per hectare annually. This is a respectable amount.

Large tracts of peatlands have been drained worldwide, followed by the outgassing of massive amounts of carbon. The historical drainage of peatlands ought to be reversed and the wetlands restored.

BIOCHAR

Each year, as a part of their photosynthesis process, terrestrial plants and trees of the world capture about 440 billion tons of CO_2 from the atmosphere.[86] Due to respiration, combustion, and biomass decomposition, almost all of it goes back into the atmosphere.

What if we could hold down some of it?

Through photosynthesis, plants capture carbon dioxide from air as they grow. If the plants are left to rot, the carbon and oxygen within the plants combine again as carbon dioxide, which is released back into the air. However, if the dead plants are heated in the absence of oxygen, in a process called pyrolysis, charcoal is formed. If the charcoal is buried, it will be a long while until its carbon content will join hands with oxygen. This is what the biochar sequestration is all about.

Charcoal, or biochar as it is referred to in this context, is mostly indigestible to microorganisms. That spells stability. It is generally accepted that biochar is a highly stable form of carbon, and thus, its burial sequesters carbon. Estimates for the turnover time of biochar in the soil vary from hundreds to thousands of years, as is attested by the dark soils present in parts of the Amazon region. Seven thousand years ago, people buried charcoal deep down, and it is still very much intact.

In addition to its ability to sequester carbon, biochar can be a beneficial soil amendment.

Charred plant-material retains the porous structure of the plants. This is why one gram of biochar may hold tens to hundreds of square meters of surface area. Biochar' s high internal surface area and its ability to absorb organic matter and inorganic nutrients provide a highly suitable habitat for microbes and mycorrhizal fungi.[87]

The structure of the biochar is ideal for trapping nutrients, and it provides surface on which beneficial microorganisms can attach themselves. Overall, biochar augments the soil's water and nutrient holding capabilities. In a nutshell, this is biochar' s claim to fame as a soil amendment substance. In one study, rice and sorghum showed little response to the application of biochar alone. However, in comparing plots that only had fertilizer applied to them, plots that received fertilizer in combination with biochar experienced a nine-fold increase in yield.[88]

The agronomic effects of biochar application are somewhat mixed, though. In many trials, dramatic yield improvements have been observed;[89] in other trials, yields have not changed; in still others, biochar has exhibited inhibitory effects.[90] It does seem that the high alkalinity of the biochar had a net adverse effect in soils that were alkaline to begin with. We may choose to rule out the application of biochar in alkaline soils. But then again, maybe not. A study suggests that biochar can possess almost any pH in between 4 and 12, and its pH value can decrease to 2.5 after a four-month incubation period.[91] As a matter of course, in overly acidic soils, biochar has value as a liming material by reducing acidity.[92]

However, biochar also has downsides. The heat inherent to the process of producing charcoal causes some of the minerals to outgas.[93] Moreover, the nutrients left within the biochar are not very available to plants. It is like having meat in the fridge, and the fridge being locked. Plants will not be able to live off biochar by itself.

The biochar itself is not a crop nutrient except for its ash content. This is the crux of the problem of large-scale biochar application. All organic matter, from manure to food waste to crop residue, needs to return to or remain in the soil. Returning organic matter to the soil as charcoal might not do. Converting billions of tons of food waste, crop residue, and excreta to biochar is a lot better than landfilling or incinerating them, but this is not much of a point. It is a given that we will do away with the decadent practices of the present.

If we apply biochar to the top soil, as with spreaders or by hand, we need not bother. Not only may the carbon dust spread far and wide (with adverse effects on health), but the black cover will sharply decrease the reflectivity of the land surface, resulting in warming of the surface air. This warming can more than offset any cooling due to the reduction of carbon dioxide in the air. The only way the application of biochar can be made to work is by integrating the biochar dust into a compost application.

Finally, if we use it as a soil amendment, biochar may come with an expiration date. It may be more in the category of buying the biosphere a few thousand years—and then we are back to where we started.

This all translates to biochar being a carbon sequestration technique under some conditions and under some prescriptions. There are both pros and cons to the use of biochar. Up to a point, biochar may be able to make up for the low bioavailability of its nutrients by making more available nutrients of other soil sources and for prolonged periods. Hence, a combination of crop residue and biochar is likely to have a net positive affect or at the very least no net adverse effect. This suggests that a portion of crop residue may be converted to biochar, provided that another portion is converted to compost.

Under optimized heat, for every 100 kilograms of biomass used as a feedstock, we are likely to net about 20 kilograms of carbon embedded in the resultant biochar, or the equivalent of sequestering just over 73 kilograms of CO_2.

A facility that may be a prototype for the future production of biochar is a design by Pacific Pyrolysis in Australia, with its *PyroChar 4000* design that will process 96 tons of biomass a day. In goes the stock feed and out goes biochar and clean syngas, after a portion of it has been diverted to run the process of pyrolysis. The scaled down demonstration unit of Pacific Pyrolysis has attained proof of concept. The apparatus can process almost any type of organic matter. It can also work with a wide gamut of moisture content and a large range of particle sizes.

To process anything in the vicinity of, say, three billion tons of biomass annually would require the mass production and deployment of somewhere under 100,000 *PyroChar 4000* units. This is good in itself, as a wide deployment of biochar processing units would cut down on transportation of feedstock.

While about half of the feedstock ends up converted to biochar, the other half is converted into oil and gas. With 100 kg of biomass and the addition of water throughout the conversion process, the total yield came to 126.6 kg of syngas in one study. About 56% of the syngas was used to drive the pyrolysis and steam reforming process.[94] This means not only that biochar generation requires no outside energy, but 100 kg of feedstock can yield upward of 56 kg worth of syngas that could be applied elsewhere.[95]

Syngas could be processed and made into methanol. The methanol can then be converted to formaldehyde, which is a building block for numerous resins used in plastics, plywood, and the manufacture of automobiles. There are numerous materials made from methanol such as synthetic fibers, paints, adhesives, and insulation.

Let's talk now about possible feedstock sources. Biochar can be made from any organic material, whether it is wood, leaves, straw, food scraps, forest thinning, rice husks, groundnut shells, or paper sludge. The biochar scheme would be as effective as the amount of organic matter available to work with.

The heavy metals found in biochar from paper sludge or tannery waste and the like cannot be used as a soil amendment.[96] While biochar from industrial sources cannot be used as a soil amendment, it is still of value in the context of carbon sequestration, and it can be hauled away to a coal mine. One way or another, the volume involved is too small to make a dent. This also holds true for sawdust. In contrast, crop residue has enough of a volume to matter, a bit.

Annually harvested crop residue is about 3.63 billion tons, and unused residue is about 1.85 billion tons, or a combined 5.47 billion tons of residue.[97] If we collect and run 35% of this mass through pyrolysis,[98] it will yield somewhere along the lines of 0.37 billion tons of stable carbon each year. This will not get us all that far in our intent to sequester 302 billion tons by 2100.

To the extent that it will prove to be ecologically sound, biochar will play an important role once we bring down a big chunk of the carbon from the air and reduce the rate of capture to a few billion tons of carbon each year.

DIRECT-AIR CAPTURE MACHINES

For the first few decades, it is likely that more carbon has to be drawn down than is ecologically possible by tapping into the photosynthesis process. This is where the heavy hitters come into play: structures expressly devised for the capture of carbon from the atmosphere.

For decades, scrubbers have removed CO_2 from the air breathed inside submarines and spaceships. In recent years, a handful of academic teams have endeavored to make this technology work on a far bigger scale, one suitable to tackle global warming. All the designs are variations on the same theme: as air passes through a device, it comes in contact with an absorbing material that chemically binds the CO_2 while the other elements drift away.

A direct-air capture technology has been developed by a team at Columbia University and is being commercialized now by Kilimanjaro Energy.

The absorbing material has been molded into large, flat panels akin to furnace air filters. The array of upright filter-panels revolves slowly in a carousel-like setup. The filters are exposed to the air and absorb CO_2. The carousel of filters sits atop a large cabinet, which is housed within a 40-foot shipping container, literally.

The people at Kilimanjaro Energy have come up with a special formulation of carbonate resin. Filters made from this resin are exposed to the wind, soaking up carbon dioxide until the resin reaches the saturation point. It takes about an hour for a given panel to get saturated with CO_2. At that point, a conveyer belt moves it off the rack and lowers it into a regeneration chamber inside the shipping container. One of the unique properties of the absorbent filters is their reaction to water, which causes a rapid release of their CO_2 content. As it turns out, even water vapor would suffice to drive most of the CO_2 out of the resin. Since the resin only comes in contact with water vapor, it would be possible to use sea water in a pinch. The moist resin releases the CO2, which is to be pumped out and compressed into liquid. Compression also forces any residual water vapor to condense into pure water, to be reused. Once the resin dries, it is ready to begin absorbing carbon dioxide again. The refreshed filter rises back to the carousel-like structure above to draw more CO_2.

It is not a coincidence that this contraption can fit within a cargo shipping container. This capitalizes on existing manufacturing lines and on the hundreds of millions of containers already in existence. As we need to deploy rapidly, these are good things.

One collector system should have the potential to absorb 10 tons of CO_2 per day. This means that we will need just over 8 million of these collector units in order to capture 30 billion tons of CO_2 each year.[99] This number of units is but a tenth of the number of cars we manufacture each year. With each unit 12 × 2.5 meters in size, this also means that their collective land footprint would be only 246 square kilometers. Their real land footprint lies elsewhere, with the generation of the needed electricity to run them.

The entire process of collecting 1 kilogram of carbon dioxide requires 0.305 kWh of electricity.[100] This includes the power needed for the removal of air from the regeneration chamber, compression of the extracted CO_2, and compression of the water vapor in order to condense it. This means that for 30 billion tons of CO_2, we need upward of 9 million GWh worth of electricity. To put this number in context, the world production of electricity in 2008 was just over twice that amount. This is to say, it will take an *astounding* amount of electricity to capture and subsequently compress the called-for amounts of carbon under this technology.

The timing could not have been better for that.

For one period in history, under this plan, we would have double if not triple our needed power-generation capacity, as the lower-impact technologies (e.g., solar power towers) would be coming online while the old fossil guzzler power-plants still remain in operation. The excess in generation capacity will produce more than enough juice to power the carbon-capturing carousels for fifteen years. Past that period, we will dismantle the carbon-capture machines and throw the switch off on the fossil power plants.

I also have an even better answer.

In the seventh chapter ("Energy"), using North America as a case in point, we were left in the end with vast amounts of excess energy. After diverting all that we could toward water desalination, we were still left with close to 3.5 million GWh. Now, if we scale it to the entire world generation of electricity, the excess would come to 13.9 million GWh, which is more than needed for the yearly capture of 30 billion tons of CO_2. In other words, if we run the carbon capture machines for seven months in each given year, we can work exclusively off the excess of the world electrical generation. It would work. It is just that with a carbon capture machine out of work on average 40 percent of the time (5 months), there is a need to deploy more machines to compensate. However, the footprint of having more machines would be negligible in comparison to the footprint of power generation facilities dedicated to capture carbon.

FIGURE 11.4

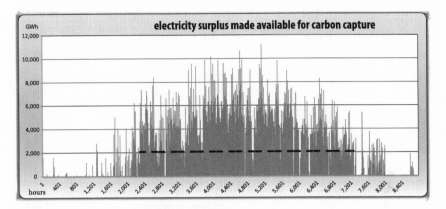

THE SEQUESTRATION OF CO_2

Next, we need to take out of circulation the hundreds of billions of tons of CO_2 that would be captured at the flue and by direct-air capture machines. This we do by injecting the CO_2 into deep geological formations or into depleted oil and gas fields or coal seams.

Under high pressure, the CO_2 would be injected into rocks bearing countless, tiny pores that trap natural fluids. In fact, some of those rock formations have trapped and held onto CO_2 for millions of years. The deeper the injection, the higher the pressure and the more dense and compact the CO_2 becomes. Suitable formations are regarded as those eight hundred meters or more below the surface. What makes them suitable is the increased pressure due to the depth, which keeps the CO_2 in a super dense form, and hence the pressurized CO_2 takes much less space.

When CO_2 is injected into a porous rock formation, multiple physical phenomena allow it to remain trapped in the rock. The rock into which the CO_2 is injected must be porous and able to store the CO_2, and there must be a layer of impermeable rock on top of the formation to ensure that the CO_2 does not rise through the rock layers and escape to the surface. Incidentally, under such conditions, the chances of leaks to the surface with consequent lethal exposure to CO_2 have been reckoned to be one in 100 million.[101]

A 2005 report by the International Panel on Climate Change (IPCC) stated that less than 1 percent of the amount initially injected underground will seep out over a one thousand year period.[102] In fact, the vast majority of the CO_2 will gradually be immobilized and thus could be retained for millions of years. A more recent study has arrived at a similar conclusion.[103] The chance of leakage is very small to begin with. On the off-chance that something does leak, it is expected to be of an insignificant amount, and the leakage—such as it may be—is not expected to occur in the first one thousand years.

This procedure is not something novel; injection of CO_2 into geological formations has been practiced for many years in the context of coaxing oil out of the ground. Injection wells are similar to oil and gas wells and are drilled and completed using the same types of rigs and construction methods. It is a mature technology.

The total capacity of suitable geological storage is highly uncertain. The IPCC report concluded that it is likely there is a technical potential for the storage of upward of 2,000 billion tons of CO_2 worldwide. The US Department of Energy projects that the United States has a combined storage capacity between 1,000 and 3,500 billion tons of CO_2 in saline formations and oil and gas reservoirs.[104] In total, we need to draw down 1,108 billion tons of CO_2. If these estimates hold true, the worldwide storage capacity would suffice. However, a recent study casts a doubt. The study concluded that CO_2 can occupy no more than 1 percent of the pore volume into the rock into which the gas would be injected and likely as much as 100 times less. The net result is that it takes 500 times more volume to sequester carbon dioxide than was originally occupied by coal.[105]

The most straightforward part of the sequestration is likely to be the transport of carbon. We have over 3,400 miles of CO_2 pipelines in the United States. In other words, we know how to handle this stuff. Naturally, we will need a lot more pipes than we have at the moment in order to transport the captured CO_2 to locations where the compressed gas will be injected deep under. One paper seems to sum it up by stating that there are significant challenges to scale up the carbon sequestration technology but no insurmountable obstacles.[106]

At some stage, we may bring online a second, possibly superior method of sequestration. Instead of compressing the carbon dioxide into liquid-like form, we will solidify it through mineralization. The idea is to mimic the naturally occurring weathering process. The CO_2 may react with calcium within basalt rock and form calcite. At the Pacific Northwest National Laboratory, Pete McGrail has shown that exposing basalt to CO_2-saturated water yielded calcium carbonate mineral formation in four to six weeks and extensive mineralization within several months.[107] The potential is there. The implications are significant.

SUMMARY

This is the proposed plan of action: In the first three years, a number of things are to be done concurrently. We spray sulfur dioxide to give us temporary relief from climatic warming. This we keep up for about twenty years. Once this anesthesia is in place, we immediately get to work.

From day one, we capture all the main sources of greenhouse gas emissions—those discharging from the thousands of power plants, steel mills, and cement plants. Each of them gets a gold watch or an upgrade, which is to say, either an early retirement or an adjoining facility that captures the vast majority of the emitted CO_2. This would serve to stop most of the emissions. Once that's done, we get the big ship into a reverse gear.

In the forefront of this campaign would be a vast network of pipes and injection wells along with direct-air capture machines consuming large portions of our generated power. The incredible amount of power would be met by the creation of a new energy network, described in the seventh chapter ("Energy"). For a certain period of time, both the old and the lower-impact power-generation systems will exist side by side, making it possible to provide for our needs and support the carbon-capture installations. Alternatively, it may be possible to run most or all of the direct-air capture devices by working off the power excess of the new grid during the warmer periods of the year.

It is unlikely that biochar and new farming and land-use practices will be any more than sidekicks in the first few decades; the required rate of carbon capture appears too high for them to take on by themselves. However, once we can afford to slow down the rate of carbon capture and dismantle the energy-hungry CO_2 collectors, holistic farming and biochar applications would be the chief technologies that will keep drawing down carbon decade after decade.

We partied like it was 1999. The blowout fossil party was fun while it lasted, at least for many of us, as not everyone got an invitation. The 1,000 billion tons of CO_2 is the tab we have to pick up as the night is drawing to a close. It sounds about right. The party lasted for about 150 years and that is about the amount of time it will take us to clean it all up.

12

❊

CONSUMPTION
overpopulation, consumption habits, industry

OVERPOPULATION

THERE ARE TWO WAYS we ought to go about reducing our ecological footprint: through modifying our consumption habits and by restricting the number of new club members.

A global population of a few tens of millions of people on Earth may be a stretch—when it comes to having this population fully integrated into the overall biosphere. A more sustainable upper figure is likely to be that of a few million, namely a few thousand population centers, each with a few thousand or tens of thousands of people. Such concentrations may be more robust and can harvest the surrounding fauna and flora sustainably, that is, for an indefinite period. In some regions, a concentration of one thousand people may be too much. Other areas can support a town of a few tens of thousands of people. The prehistory of many regions attests to this. At the same time, the fate of mammoths, bison, and moa is a testimony to how such numbers can develop unsustainable practices—and lead to mass extinctions. Be that as it may, we have destroyed most of the ecosystems, so the ability of the biosphere to sustain in that fashion even a few million people is unlikely.

Anyway, as of this writing, we don't have a population of a few million but that of 7,000 million people.

It boggles the mind to realize that about one thousand million (one billion) people were added in just over a ten year period. There were but six billion of us in 1999.

Every year, there are about seventy-eight million more people walking the streets and frequenting the food stores. Every week, a population the size of San Diego is coming online—once those who die are accounted for. Every four years, the equivalent of the entire United States population is added on.

That is to say, every four years, we add about twenty-five billion kilograms of human biomass, which in turn displaces a comparable— in truth, far larger—biomass of other living things.[1] This cannot work out any other way.

The planetary ecosystem has only so much energy to go around. The net amount of plant matter produced each year via photosynthesis is the basis for the existence of all higher life forms. Given the fixed surface area of Earth and the fixed amount of solar radiation reaching it in any given moment, the volume of plant matter is capped. In turn, this sets a cap on the amount of animals that graze on the plant matter and in turn on the number of predators that feed on the herbivores. The more people who are placed on the game board, the less there is room for other animals competing for the same resources. A model by Brian Maurer projects that if we reach a population of fifteen billion, we will consume two-thirds of the global primary production. Consequently, we will experience a drastic decline in biodiversity.[2] At a theoretically impossible twenty billion people strong, the human family would consume all energy for itself with nothing left over.[3] This is how you get to add Soylent Green to the menu.

Each day, as about two hundred thousand people are added to the human family, they inevitably appropriate energy from the global ecosystem.

Under the current trajectory, we are destined to reach over ten billion people by the end of this century.[4] Thus, we continue our march to assimilate the rest of the Earth's biomass into our Borg-like collective: a never-ending expanse of fields of genetically identical corn and soya punctuated by bustling cities and populated with cattle feedlots, pigs, and people—alongside growing numbers of sparrows, mice, dogs, and cockroaches. The shining beacon and source of inspiration for the cancerous organisms that we are.

The human situation circa 2100 is what we've got coming to us in the next few decades, plus three billion people stacked on top of it, each with his or her inalienable right to life, hamburgers, and a dwelling place.

Within a living system that has finite production capacities, it has always been a matter of *what* we indirectly force out of existence and of *when* much is too much. Do we really want to win the human race—a human habitat from sea to shining sea standing at the finish line with an unraveled, simplified planetary ecosystem tumbling about our collective ears? Under the already heavily degraded planet, short of reducing our numbers to under a few million, the question would become not whether our presence impinges on the planetary ecosystem but what is an acceptable degree of degradation.

A 2009 study concluded that the potential carbon savings from reduced reproduction are huge compared to the carbon savings that can be achieved by changes in lifestyle. The study calculated the footprint of a hypothetical woman in the United States who increases car fuel-economy from 20 to 30 mpg, reduces her driving from 231 to 155 miles per week, and replaces single-glazed windows with energy-efficient windows. In addition, the study assumed that this hypothetical woman retrofits her home with efficient light bulbs, recycles aluminum cans, newspaper, magazines, glass, and plastic. In total, this woman would save about 486 tons of CO_2 emissions during her lifetime. On the other hand, if she were to have two offspring, this would eventually add nearly 40 times that amount of CO_2 to the earth's atmosphere.[5] This analysis gives one a sense of the broader ecological footprint of bringing people into this world, which can be inferred.

We have never balanced our checkbook. When we needed more food grown, we just cut down more forests. When we liked a new place, we claimed it as our own and slaughtered or drove off competing animals. When we wanted to have children, we just went ahead and had them.

It is impinging on our civil rights to be told where to cross the road, when to stop, where to walk, and where not to walk. Yet, this is a necessary set of restrictions in city streets when there are millions of us congregating together in mega population centers. By the same token, when you live on a boat that is laden with more cargo than it can handle and you invite more people aboard, it affects everyone and everything. The right for reproduction cannot be construed as a license to multiply without limits. The very existence of the planetary ecosystem is at stake. This supersedes our rights to swell our ranks further beyond our existing numbers.

I want to be really clear about this. Choosing as a race to maintain or raise existing population levels is profoundly immoral. Those who may elect to have large families are committing us all to consumption levels that cannot be supported without further degradation of the planetary ecosystem, which Earth can ill-afford. We would be bringing a generation of children into a world that can support them only at a cost of impoverishing the world for many generations to come. We can't afford to be wrong on this one. Past this point, there isn't any *undo* button to our choices.

I call on bringing our population down to one billion people and letting future generations in the next century decide if the still-astounding number of one billion humans is acceptable. After all, the first part of the nineteenth century had about that number of people. And it sure was a busy, bustling planet back then.

Yes, having six billion less passengers aboard takes the immediate, obvious pressure off. However, if we then decide to take the population numbers down further, the fun and the potential for quality life really begins to come through. Consider, really consider, the breathing space for the various ecosystems, the ratio of resources per capita, and the political possibilities of having a stable population of one hundred million people on a future Earth other than one that is post-apocalyptic.

The demise of tens of millions of large mammals in former millennia have made possible a full-blown agricultural system and with it our empire-style civilization. It is the big herds' stake on life, their share of the photosynthesis pie, that we largely appropriated in the first phase of our expansion, refashioning their share to vast wheat fields and myriad head of cattle.[6] Reducing our numbers south of one billion would let us, once again, coexist with other large mammals.

For the time being, one billion people sounds like a sweet deal. It is like rush hour traffic—with seven times fewer cars on the road.

We ought to go to one billion people tomorrow if there is a humane way to do so. Alas, that is not the case.

TABLE 12.1. Number of children per woman, existing US profile.

number of children	percentage of women
0	18
1	19
2	33
3	19
4	7
5 or more	4

Table 12.1 shows the current breakdown of kids per household. No comparable statistics have been compiled for the world at large. However, from all the countries that may have such information available, I find the United States to be the best yardstick, with its diverse ethnic and socio-economic profile. Countries whose fertility rate is beyond replacement rate would have more households with three and four children. But it does not matter here, as will become evident in a moment. All we are looking for is the percentage of people who have no children and the percentage of those have one child.

The goal is to have two children or less per family—at least until we bring our numbers down to one billion. I assume that the percentage of those choosing to have no children or have one child would not change significantly, and those that would have two children would be the largest category, as they represent those that currently choose to have two children plus those that currently elect to have a higher number of children. I assume future fertility breakdown to be as shown in table 12.2.

India and China have shown that coercive population policies backfire and work poorly. Voluntary targets such as those that were practiced in Iran point the way and are to be discussed at some length.

TABLE 12.2. Number of children per woman, assumed under a two-child policy, a global profile.

number of children	percentage of women
0	19
1	26
2	52
3	3

I took the fertility breakdown in the table above as a basis for calculating the world population of the future. To arrive at population numbers, I relied on the mortality rates of each age segment of most countries of the world.

TABLE 12.3

year	world population	year	world population
2015	6,995,445,000	2105	3,016,361,000
2020	7,096,368,000	2110	2,786,723,000
2025	7,168,685,000	2115	2,565,328,000
2030	7,205,087,000	2120	2,357,904,000
2035	7,183,761,000	2125	2,168,135,000
2040	7,074,954,000	2130	1,996,612,000
2045	6,887,635,000	2135	1,840,845,000
2050	6,646,751,000	2140	1,696,985,000
2055	6,369,353,000	2145	1,562,896,000
2060	6,062,046,000	2150	1,438,379,000
2065	5,724,675,000	2155	1,323,807,000
2070	5,359,863,000	2160	1,219,029,000
2075	4,975,589,000	2165	1,123,061,000
2080	4,586,440,000	2170	1,034,665,000
2085	4,211,473,000	2175	952,924,000
2090	3,862,908,000		
2095	3,547,635,000		
2100	3,266,839,000		

Under existing mortality rates and under fertility rates indicated in table 12.2, we will reach 3.2 billion people by 2100 and the 1 billion mark around the year 2175 (table 12.3).

The obvious question is how we would go about changing our cultural practices to reach this fertility profile.

Many women today are not using any method of family planning because of their desire for large families, because of lack of correct information, or because of social opposition. These are the things we need to turn the tables on.

It is a given that family planning centers have to be highly afford-able and accessible. This aside, I do not suggest a one-size-fits-all policy. We have to work with the intricate socio-economic-cultural mosaic of our world. In some places, it is befitting to employ a large force of female field-workers to visit women in their homes and provide highly subsidized contraceptives. In other places, it is fitting for highly subsidized contraceptives to be made available via dispensers at street corners and given out in schools without a charge. Sterilization such as vasectomies ought to be offered for free as well. In some cultural landscapes, it is suitable and necessary to have for teenagers frank sex education grounded in reality. In other places, it is fitting to get the husbands and wives to talk to each other about the use of family planning.

All of the above is important, perhaps even essential, but it does not address the heart of the matter. Aside from unintended pregnancy, people would be motivated to restrict their family size only if they saw how this may lead to some benefits, if not for themselves then at least for the world in general. These perceptions have to make up the centerpiece of the campaign to reduce fertility rates.

In some cultures, there is an obvious, direct correlation between girls' education and fertility rates. Education empowers women in the workplace, raising the value of their time. At least that's the theory. No one is sure of the cause and effect, but the correlation is there, without a doubt.

Family-planning education can reach large numbers of people very effectively through entertainment programs broadcasted on radio and television. Serial dramas are especially well suited for this, allowing audiences to watch or listen along as their favorite characters shed their traditional large-family outlook in favor of that reflecting a smaller family size.

Financial incentive to have fewer children, taxation on baby apparel, celebrity role models—we will try some things, and if they do not get us on course, we will try other things. In the end, having two children or fewer is not a far-flung social convention. If whole nations were made to feel that mutilating the clitoris of young girls has high merit, I feel confident that we can bring about a universal perception that two or less children per family is appropriate and highly desirable. Iran is an inspiring case in point as it managed to implement an effective family planning campaign in spite of having largely a traditional, devout Muslim population.

During 1989–2006, the Iranian government was intent on curbing family size. It encouraged families to space births 3–4 years between pregnancies. It discouraged pregnancies among women below the age of 18 and over the age of 35, and it sought to limit the total number of children per family to three.[7] After the third child, the government stopped paying for social welfare. Birth control classes were required before a couple could get married. Thousands of Iranian women underwent training as midwives and family-planning advocates in their respective villages. Youths got to learn about family planning and population growth issues from all media channels, mosques included. Pre-marriage counseling was required prior to the issuance of a marriage certificate. Free voluntary sterilization was provided if a couple had three or more children and the mother's age was over 30.[8]

While many other countries' high-fertility rates went down during that period, none moved as fast or dipped as low as Iran's. A fertility decline of more than 50 percent in a single decade is not only unique for a Muslim country but has never been recorded elsewhere.[9] From an average of having around five kids per mother in 1988–89, the numbers in Iran went down to just under two per mother in 2006.

We don't need to win them all. We just need to get down to four billion and then to three and eventually to one. Seven hundred years into the future, that is all that anyone will remember or care about.

In the end, either the people of the world are on board with what is at stake and what needs to be done or they are not. If they are not, the ship will take in water.

CONSUMERISM

Our economy is a sub-system of the biosphere. For its existence, the economy utterly depends on the larger system for raw materials and services. As the solar energy powering the Earth is finite and as the resources of the planetary ecosystem are finite, so is the viable extent of economic activities.

We remove copious and ever-growing quantities of raw materials and water from the ecosystems of the world. We cover ever-more surface area with buildings and roads. The process of fulfilling our wants and needs is stripping the earth of its capacity to beget life. It is a feeding frenzy by a single species that overwhelms the sky, earth, water, and fauna. Every living system is in decline.

FIGURE 12.1. Total primary energy consumption for the years 1980–2008.

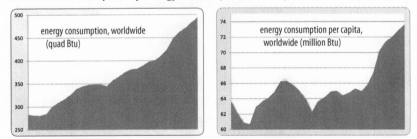

Source: US Energy Information Administration "International Energy Statistics" Database (accessed March 5, 2012) http://www.eia.gov/cfapps/ipdbproject/iedindex3.cfm

Schmidt-Bleek and his associates at Wuppertal Institute for Climate, Environment and Energy, Germany, have estimated that the amount of global resources withdrawn has to be reduced by half before sustainable coexistence of commerce and the biosphere can be expected.

The well-off strata, the 20 percent of us in the first-class cabin, rake in 80 percent of all extracted natural resources. If we wish to cut combined human consumption by half and at the same time level the field some among the haves and have-nots, it would mean that first-class passengers have to cut back their consumption by a factor of ten. This would require a fivefold decrease in material goods coupled with a twofold reduction in ecological footprint for each commodity.

What I am saying is that consumerism ought to come to an end.

Efficiency gains by themselves do not reduce our total ecological footprint. Cheaper, more efficiently produced goods and services can potentially increase economic activity, thus indirectly adding to our footprint. As a case in point, we have devised more gasoline-conserving car engines, but this was not being translated to better miles per gallon, that is, to reduced energy consumption. Rather, this made it possible to boost the vehicles' horsepower and introduce other energy-draining features such as air-conditioning. In other cases, efficiency gains may truly reduce the energy bill (e.g., a low-energy dryer), but, by itself, this freed-up money makes it possible to direct the purchasing power elsewhere, on other footprint-bearing goods and services.

A fundamental shift in our values is called for.

There are but certain material things we need for our well-being and prosperity. By and large, we need satisfactory food and water; we need protective clothing; we need a place that keeps the elements out and has humidity and temperature within comfortable ranges; we need a sanitary and easily accessible means to do our business; we need some things that help us heal or aid our failing body; we need clean air to breath. All the rest of the stuff is icing on the cake.

We need to give a different thrust to the big economic engine of the world. In the context of the well-to-do populace, it may mean a shift from banal and frivolous goods and services to these that are meaningful and matter. It may mean moving away from goods and services that are short-lived and of marginal utility to those that stir us or leave us richer in the end. We need to do so, if nothing else because many of these quality services have a negligible environmental footprint.

Consider the ecological footprint of a band playing in a piano bar, of skinny-dipping in the ocean under moonlight, or of a work of art. Think of the footprint of an aromatic massage, of a standup comedian performance, or of listening to a good bedtime story. Reflect on the footprint of a whitewater rafting trip, a dance party, or of a good conversation in the company of friends. Note the footprint of gardening, great sex, and of a neighborhood basketball game.

We can fulfill the need to have a business meeting with people overseas with a far smaller environmental footprint by replacing the flights with a telepresence video system. As Friedrich Schmidt-Bleek argues, we can look at our products as service delivery-machines and start to imagine, then fashion, novel solutions to provide the services we need. Concentrating on the benefits in utilizing a product rather than on the product itself opens up whole new dimensions of development. We ought to hone in on our needs and aim to fulfill them in ways that require the least possible amount of resources. We need to ask what are the true services needed—and go after them.

Rory Sutherland, an advertiser, gave in his public talks a few apt examples of how identifying the service needed helps in identifying solutions with a vastly reduced environmental footprint.

Twenty-four billion pounds of sterling are planned to be invested in a project to make the London–Birmingham train journey thirty minutes shorter and presumably an overall better experience—which is the point of the whole exercise. Sutherland reckons that we could retain the existing train-system and for a fraction of the money spent on shortening the trip duration, it would be possible to employ the top male and female supermodels to walk up and down the train cars with offerings of free wine. In fact, he reasons that people will ask the train to be *slowed down*. The key is to tinker with perception. Once a certain level of material needs is met, wealth is just a matter of social context and perception. This is how we get people to buy bottled tap water or convince them to get it from a remote island in the Pacific Ocean.

Sutherland gives a second example. Asked to help in the war effort against the French, the wealthy Prussians were encouraged to turn in all their jewelry. The jewelry pieces were replaced with replicas made of cast iron. For fifty years hence, the highest-status jewelry one could wear in Prussia was made of cast iron. It was a badge of honor. It said that one's family had made a great sacrifice.

I, for one, just love the lobby of the URBN Hotel Shanghai: its back hall is primarily a collage of old leather suitcases. The luxury hotel itself is a renovated factory warehouse.

As a group, we need to radically cut down our ecological footprint.

To that end, what is an obvious first step is to audit the ecological aftermath of our designs. What is necessary is an assessment of the environmental footprint of our various industrial activities: those that extract various raw materials, those that fabricate products, and those that are within the service industry. We need to assess the extent of environmental degradation and the extent of resource depletion. While inherently crude and generic, such an assessment would be close enough to the mark to be of service.

Only under a rigorous analysis could we have ascertained that, in fact, the transportation infrastructure has up to ten times higher ecological footprint than that generated by the manufacture of the vehicle itself—once we divvy up the ecological footprint among the estimated number of cars using the road.[10] Only a rigorous analysis could have found that the production of each desktop computer necessitates upturning more than fourteen tons of earth.[11]

The idea is to establish ecological footprint indices that would assess the impact throughout the lifecycle of a product along the five to six parameters listed below. For ease of discussion, I refer to those five to six parameters collectively as *footprint*.

TABLE 12.4

environmental footprint throughout the lifecycle of a product ("footprint"): six parameters
Direct Disturbance and Degradation of the Earth: acuteness of disturbance, volume of earth affected, duration of the disturbance, and the ecological sensitivity of the locale involved.
Toxins, Pollutants, Debris: effects, coverage, persistence (includes heavy metals that are not salvaged, global-warming potential, ozone potential, various forms of water contamination, acidification, harmful radiation, aerosols, poisons, and solid non-recyclable waste).
Material Scarcity: Each non-renewable commodity is assigned a scarcity value. Different values are assigned to virgin materials than to recycled ones.

environmental footprint throughout the lifecycle of a product ("footprint"): six parameters

Water Resources: amount of water used compounded by the sustainability of the source. In the case of desalination, the footprint is just that of the power required. In the case of recycled graywater, it is assigned a zero footprint value.

Biomass Used: Renewables are measured against a natural rate of regeneration. Generally, the rate of replenishment will dictate the rate at which renewable material is harvested. Beyond that, this parameter measures the amount of biomass displaced on Earth and the overall degradation for given ecoregions.

Special Considerations: effect of wind turbines on the global winds, marine noise pollution, etc.

The assessment would take into account the combined *footprint* of the creation, usage, and eventual disposal of a product. One product may require less by way of raw materials; however, if most of it cannot be recycled it may have higher combined, total *footprint* throughout its lifecycle than another product that requires a lot of raw material—all of which would be recycled in one fashion or another.

First comes the creation, and with it the raw materials that are needed.

Every commodity requires more material than is contained in its final form. For instance, in the creation of 1 kg of propylene, 1.74 kg of earth has to be disturbed. The extraction of 1 kg of gypsum causes the disturbance of 1.83 kg of earth. In the case of 1 kg of lead, 15.6 kg of earth needs to be disturbed. The extraction of 1 kg of tin requires 8,500 kg of earth disturbance. And to end up with 1 kg of gold necessitates 540,000 kg of earth disturbance.[12]

As may be noted in the matrix above, earth disturbance is only one of the parameters. In mining for the materials, the *footprint* also factors in depletion of resources involved, pollutants and toxins, and possible water consumption. In addition, it also factors in the *footprint* of the energy required for mining, transportation, and eventual distribution of the raw materials.

During the manufacturing process of the product, we would consider all the materials that go into it and the packaging that may be required. In addition, the assessment would consider the *footprint* of the manufacturing plant throughout its operational lifetime, or rather, the proportional fraction of that *footprint* that it takes to fabricate the product in question.

During the operation of the product, we need to tally the *footprint* of routine maintenance and the total energy input during its anticipated period of operation. We also need to include the *footprint* of the surface area the product may take. Then there is the matter of replacement parts. There is a big difference between a design that allows replacement of one component for another, and a design that requires the technician to toss out an entire section. Above and beyond all of these, there is the *footprint* of the infrastructure needed for the functionality of the product. Like everything else in the assessment, we are talking about the proportional fraction related or allocated or used by the product. Thus, we don't need to account for the *footprint* of an entire shelf, just the portion that the product is likely to take on it.

In order to evaluate the *footprint* during the operating stage of a product, we have to put it in context. The production of one Blu-ray Disc player may have the same *footprint* as another player. However, if its components suggest it has twice the longevity, this reduces the *footprint* score dramatically. If the product lends itself to recycling of its raw material perpetually—fashioning the same product again and again, that dramatically changes the *footprint* as well. And then there is a matter of how much bang for the *footprint* a product yields. A bus may have a larger environmental impact than a van, but a bus can carry a few times the number of passengers. By the same token, the *footprint* of providing towels at a hotel obviously improves if customers use the towels longer, and consequently the frequency of washing towels is reduced. And wool's *footprint* has to be ascertained in the context of being a byproduct of another product, meat. All of those things are to be factored in.

Sooner or later, the product is destined to be recycled. As with other phases in the lifecycle of the product, what matters here is the related *footprint*: how much energy goes into the recycling process; what materials can be recycled; and what is the combined collateral mess. When it comes to the recycled materials, what matters is how energy-intensive the recycling process is and the quality and usability of the product that comes at the back end of the recycling process. It goes without saying that if a raw material can be extracted and put back in circulation, it ought to be. No more throwing metal-rich appliances into landfills.

Ultimately, the *footprint* of a product is case-specific. Harvesting and shipping kiwis from New Zealand to Japan has a different *footprint* than kiwi grown and consumed in Japan. Ultimately, every ecoregion has its own unique set of circumstances. Some areas are more degraded than others and by and large are more vulnerable than others. The same activity performed in two very different areas of the world may have two very different *footprints*. Hence, every supplier and manufacturer would have its own code, its own *footprint* that reflects its own chain of suppliers.

Much like Wikipedia, the assessment of a product' *footprint* is to be a vast open-source database compiled and constantly updated by millions of distributors, suppliers, and manufacturers worldwide. Each of them may insert an additional piece of data, a new link between products or industrial activities. Each of their customers in turn builds upon the data of their suppliers in arriving at their own *footprint*. The creation of such a database is a cumulative process. Once the *footprint* of a ton of iron from each of the respective mines is established, the steel mill just plugs in the amount of iron delivered and its origin. The *footprint* of the iron has already been calculated by its respective mining company. A car-manufacture analyst just uses the *footprint* of the derived steel without having to trace back its origins. And thus the analyses build upon one another. All parameters are ultimately interlinked, reflecting the true dynamics of the marketplace. The *footprint* values within this colossal spreadsheet are altered continuously. A manufacturer may increase the percentage of recycled iron in its production of steel, and as its *footprint* is updated to reflect this, it causes a ripple effect throughout all the derivative products and industrial activities.

This open-source database ought to come with a modeling feature, letting vendors play with various chains of supply and to arrive at different hypothetical *footprints* for their own product or service. In attempting to raise the score, there are many other things a vendor may do. It is possible to improve the engineering of the product to last longer or to engineer it so that it would allow the replacement of individual components. It is possible to design the product to have a second life as something else. It may be also be possible to change the application of a given product and give it a broader functionality.

The incentive for fraud in this process is likely to prove irresistible—operating from within our existing economic paradigm. But then again, we won't use the existing economic engine. This is to be discussed at length in the next chapter ("Economic and Political Paradigm").

As a matter of course, the *footprint* score ought to be accessible to the general public and to prospective shoppers in particular. In fact, labels with a *footprint* summary ought to be plastered on products. Specifically, the data could be distilled into three composite indices: impact on water, depletion of resources, and direct disturbance. In addition, the label will have a radar chart (aka, polar chart or spider chart) that displays the impact of distinct toxins and pollutants.

The goal is to start in the upcoming decades a generation of products and services that have a minimal *footprint*. After the first few decades, though, we ought to slowly make the shift from products that degrade and deplete to products that, if not downright regenerative, at least have a net *footprint* that is neutral.

There is no need to conduct a rigorous lifecycle analysis in order to make some obvious observations.

On average, a power drill is used 10–20 minutes in its entire life. It makes far more ecological sense to rent one. We already extensively rent books and movies and less extensively bicycles and cars. We ought to expand it to the maximum viable extent across various product categories. Everything that bears substantial *footprint* or is only used rarely could lend itself to a rental scheme.

Public carpentry and metal shops can dot the city, where people can work on projects or rent tools they need. In fact, every large apartment complex or subdivision could rent items that cannot be mailed easily or which require immediate access: heavy machinery, boats, etc. Inventory may be adjusted automatically to reflect demand.

Other products lend themselves to online order. A case in point is www.toygaroo.com. One rents a toy and returns it to get another one. In terms of environmental bang for the buck, nothing comes close to car sharing—short of an urban environment that eliminates the need for motor vehicles, that is . In places where it makes sense, we can create an infrastructure that can guarantee a shared car to be no more than five minutes away from anyone. And with that, people may choose private cars, but there would be a viable, and far more environmentally friendly alternative. The outfit www.zimride.com offers a spin on the car-sharing model: ride sharing. In seconds one can set up a profile, book a ride in one's area, or post a ride of one's own. With Zimride profiles, one can check out interests, music tastes, and feedback before one shares a ride.

RAW MATERIALS

FIGURE 12.2

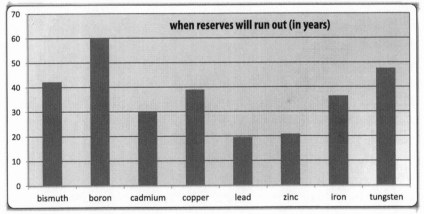

Source of data: "Commodity Statistics and Information," USGS, http://minerals.usgs. gov/minerals/pubs /commodity. I took the reserves and divided them by the rate of 2010 mining to arrive at the number of years left.

Note: At the current rate of production, using 2011 as the starting point, this is when known reserves for the respective minerals would run out—provided there are no new economic discoveries and related mining.

To a large extent, known reserves are like inventories. Economically recoverable cadmium may run out in thirty years. Or maybe it will turn out that we find there is ten times as much economically recoverable cadmium, and it will provide, say, for three hundred years. This is not all that reassuring if you stop and think about it. There is nothing sustainable about using up minerals. How *can* there be? They don't renew. It is a matter of time until we transfer all the ores to a landfill fused in myriad ways to other materials. Henceforth, we need to transform our approach to raw materials. In the short run, we ought to recycle and conserve. This is of course but a stopgap measure. In the long run, we must make a switch to renewable materials, whether materials that can be tossed after use into a vat and refabricated in a new form endlessly or materials that we can synthesize from organic material or fashion from truly abundant minerals such as silicate minerals. In accordance, we ought to launch a vast-scale research and development initiative to find a set of sustainable raw materials.

More broadly, the intent is to reduce the *footprint* in producing raw materials. This largely means reducing the related energy consumption, reducing the emission of CO_2 to negligible amounts, and reducing the mining of virgin material to a minimum. (Unless we are dealing with truly abundant material.)

What follows highlights how we can achieve some reduction of *footprint* with the main industrial processes and raw materials.

Heating limestone with some other materials such as clay to temperatures around 1,450°C, produces Portland cement. With the addition of water and some aggregates, such as gravel and sand, concrete is produced. Next to water, concrete is the most widely used material in the world. The problem is that during that process carbon dioxide is driven off the limestone, and the resultant GHG emissions are considerable—billions of tons annually. It is assessed that cement accounts for about 5% to 8% of all CO_2 anthropogenic emissions.

For starters, we can use less cement in concrete. This is achieved by substituting some of the cement with a more benign filler. Two parts cement to one part limestone is found to be adequate in some applications that do not require very high strength or resistance to chemicals. But we may very well do one better.

The STEP cement process developed by Stuart Licht and his associates at George Washington University produces quicklime (calcium oxide) without any CO_2 emissions. The carbon and oxygen are driven from the limestone all right, but they are broken down to carbon, which is a stable graphite that can be easily carted away, and to oxygen. The breakdown is done via electrolysis. The process requires 500°C, which can be provided from the molten salt in solar tower powers, and small amounts of lithium carbonate, which are used to dissolve the limestone. The lithium carbonate is not consumed during the process and can be used again and again. Electricity is required to drive the electrochemical reaction, yet it is minimal.[13]

The conversion of limestone to quicklime accounts for just over half of the total GHG emissions of cement production. Furthermore, once we feed the quicklime—instead of CO_2 packed limestone—into the kiln, there would be a need for only about half of the fuel to diffuse the quicklime with the other minerals for the production of Portland cement. In total, this is a reduction of about 75% in GHG emissions from the production of cement. Carbon sequestration of the flue gas will cut the remaining emissions to insignificant levels—or perhaps to zero if we can find a viable way to heat and maintain the kiln at 1,450°C using heat sources other than the burning of fossil fuel, such as via plasma torches.

Given how prevalent iron is, any reduction in the *footprint* of its production would be significant. A few things can be stated. To the extent possible, we ought to replace blast furnaces with electric arc furnaces. They use one tenth of the fuel, one eighth of the water, and one fifth of the air compared to conventional blast furnaces.[14] Steel mills collectively produce each year millions of tons of slag. A research team has established that these can be incorporated into commercial bricks used in construction, without any adverse performance or health effects.[15]

The production of plastic should be carbon neutral—such as is the case, at least in theory, with plastic made from corn. But this is not enough. Plastic should be discarded and then biodegraded in a short period of time. Thus, biomass is to be taken out of circulation only to return to it later, as in ashes to ashes, or in this case, biomass to biomass. Yet, this would still require a large area of Earth dedicated to the growth of biomass that would serve as feedstock for plastic. It is less than desirable, but it may have to do in the short term. One good example of this is the packaging materials made by Ecovative. They are not plastics but do act as a substitute and conform to the principles laid out here. These are made from agricultural byproducts bonded together with mycelium, which are basically the roots of mushrooms.

Better yet is the development of a synthetic material that may require a one-time sacrifice in terms of biomass followed by endless recycling of the generated plastic, which would require only minimum additions of plastic from grown biomass to compensate for inevitable losses.

This brief analysis of the prime raw materials of our world is but a rudimentary draft in this broad subject. It is an invitation to open the door wide and make progress in the extraction, processing, and usage of these materials in ways that are environmentally benign.

13
ECONOMIC & POLITICAL PARADIGM

≳⤫≲

what exists, what is called for, and getting there politically

THE EXISTING ECONOMIC AND POLITICAL SYSTEMS

DURING THE GREAT DEPRESSION, many went hungry. In my late teens, I heard of this and wondered why. Foremost, though, I wondered why instead of increasing the production of food and other material necessities, production slowed to a crawl. Many millions of adults, who could have been working on necessities, spent their days in long bread lines and looking for work that no one could offer. Observing this it became apparent to me that our economic setup was daft.

Many go to bed hungry or live in crowded, dilapidated housing units. Some find that they have to work for more hours in a day than they spend time with their children throughout the week. Others put on cheery makeup and go to work in jobs they detest. The whole arrangement does not work for us; it has never worked for us. But there is always the hope: everyone has a shot at the brass ring, striking it big. In reality, few escape the rat race. But we do not complain. It is maintained that if we do not make it, we have no one to blame but ourselves. Our economic model is popular because it has the titillating dynamics of a lottery system: some make it, most don't, but *everyone* gets a fighting chance to become rich. Many millions keep this economy in business. They feed it with their hopes of riches; finance the lottery prizes with their mortgage interest-payments; give their oath of fealty via the binding chains of credit cards and student loans. And the joke has been on us.

We could be cultivating a Heaven on Earth. Instead, we make money. Stripped down to its core, the marketplace is piles of money, each striving to get bigger. Just that. And those piles of money hire the services of the shrewdest and brightest in order to impregnate the broader culture with the essence of the current moneymaking dynamics.

The gist of the movie *Groundhog Day* reflects the theoretical best this marketplace can be.

In the movie, the protagonist wakes up every day at 6:00 a.m. And every day is Tuesday. For him, time halts and merely oscillates back and forth within the span of a single day. He lives one Tuesday over and over and over again. A second chance—forever.

The protagonist is enamored with a female co-worker and tries to win a place in her heart. He has only one day to do it; tomorrow, after all, will be today all over again and he will have to start anew. He retains a memory of the previous Tuesdays. No one else does; for them every morning is a new day.

Through countless dates, he perfects the best lines and learns to say the right things. He becomes a consummate reflection of this woman's whims and alleged preferences. She declares her love for French literature, and he learns to recite French poetry. She toasts to world peace, and he formulates the perfect comeback. But every night the date ends with her slapping him once he attempts to maneuver the date past a certain point. It is everything she wanted in a date. And yet the date doesn't quite cut it for her, evidently.

The makeup and quality of his courtship is comparable to what the moneymaking drive brings forth at its absolute, theoretical, impossible best. Anticipating and catering to the customers' wishes and whims is the highest expression of our marketplace. And yet, our marketplace doesn't quite have it—just like the date who was everything she wanted yet lacked both fundamental authenticity and something bigger or more meaningful, beyond the wish to score.

In the role of producers, we invest our time devising sixty-seven competing kinds of toilet paper to wipe the consumer's personalized arse. The customer side of our persona is a reaction to what is out there, which in turn is a response to this customer persona—ad infinitum. It is a hall filled with mirrors, bearing endless reflections of nothing worth looking at. Businesses aim to get ahead of the pack; they don't aim to elevate humanity. It is a race to win favorable attention, not an expedition to heighten the quality of life. Any business initiatives to the contrary are in spite of the marketplace dynamics, not due to them.

To the tune of thousands of commercials annually per square person, we promote the lesser path of life. Ratings-driven talk shows and formula-proven movies have been devised; inane products bearing alluring veneers have been formulated; multi-million dollar lawsuits and mutual funds have been mass-produced. Another "life-changing" self-improvement book, another "groundbreaking" electronic tablet: torrents of inane things pretending to be something.

We channel a tremendous amount of energy into safeguarding products, speculating over the monetary value of products, acting as superfluous intermediaries to products, or outdoing other people's attempts to sell products. This economy is like a Rube Goldberg machine: inefficient and full of superfluous functions and redundant parts.

Operating under the existing economic setup, the impetus is to grow and to maximize. Earnings are tied in to the growth of a business. Venture capital will be invested under the prospect of a far larger return. Companies won't stay in business long if their profit to expense ratio isn't maximized, that is, maximum profits at minimum production costs. All of these translate to more consumption and to more production. These two imperatives are real enough; if consumption and production were to slow down significantly, we would plunge into a deep recession with an eventual economic collapse.

Thus, under this economic paradigm, it is inevitable that beyond just offering products and services for sale, we will manufacture demand for them. This is apparent in a multitude of sectors and across a multitude of economic activities—as when some automobile manufacturers waged an all-out campaign to demolish public transportation in order to increase the sales of cars.

To maximize consumption, a ravenous consumer culture was contrived and is peddled and pushed on us in countless ways. Branded advertising clamors for our attention when we fill our gas tanks, turn on the TV, tune in to the radio, or go to a football game. And ever-altering fashion renders last year's shoes obsolete, making room for the purchase of new ones. Cumulatively, these things perpetuate two memes essential to the preservation of this economic entity: a financial incentive is the driving motor of the world, and our aspirations are tied to things or experiences that require money to be spent.

The foundation of our free-for-all market is consuming as many resources as possible as rapidly as possible to generate as much money as possible, which in turn is used to exploit more resources. Aside from us having to run ever faster and ever harder on the hamster wheels of the economy, this sets human activities on a collision course with the physical reality, with nature. This is where the pathology of our economic system reveals the full scope of its harm.

As long as money is made by utilizing natural resources, and as long as we have entities whose purpose of existence is to amass money, there will be a fundamental conflict of interests. The economic imperative to produce and consume as much as possible is the quickest way to suck dry the living environment.

Irrespective of the above, while we can assign a monetary value to one's labor, we cannot barter with nature and with the living. It is not for us to uproot or give parcels of nature away. Some argue that the cost of goods has to incorporate the true external cost, the accompanying degradation or depletion of nature. But this is farcical. It is not possible to offer a cloud rainforest a wad of cash in exchange for its existence.

Under the existing setup, what is the likelihood of a carbon-neutral economy if this entails shutting off the billions of dollars that are gushing from the ground in the form of black oil? For that matter, under the existing economic system, what is the likelihood of establishing an array of genuine cures for cancer when these would put an end to the giant, very lucrative radiation and chemotherapy industry—notably so if the cures can easily be replicated and cannot be patented? Where is the economic incentive to devise umbrellas that could last for decades or motherboards with architecture allowing for easy replacement of the microprocessor rather than throwing out the motherboard? In a cutthroat competitive economic paradigm, a company cannot afford to take a path other than the one with the flashing neon sign noting, "Profit—this way!"

All in all, the moneymaking economy is the nastiest arrangement in the history of our civilization. And that's saying something.

Nothing quite casts in sharp relief the nature of this economy as well as the tobacco industry. Behold the modus operandi of this economy without the veneer that gives it semblance of something humane.

In 1940, it was asserted in the closing session of the Annual Clinical Congress of the American College of Surgeons that an increase in smoking was the cause of lung cancer.[1] In 1949, it was noted in the first National Cancer Conference that the increase in cigarette smoking paralleled the increase in the incidence of lung cancer.[2] Then the 1950s rolled in and with them rigorous and comprehensive studies, which taken as a whole, have shown beyond any reasonable doubt that smoking causes cancer.

Schrek and his associates found that a relatively high percentage of cigarette smokers were found among patients with lung cancer as compared to the control group.[3] Wynder and Graham found that only 2 percent of nonsmokers develop lung cancer; the population of those who did have lung cancer was overwhelmingly comprised of smokers; and the enormous increase in the sale of cigarettes corresponded to the increase in lung cancer.[4] Wynder and his associates tested the equivalent of cigarette smoking on mice, and established tobacco tar to be carcinogenic.[5] A slew of additional studies during these years further affirmed the obvious correlation between lung cancer and smoking.[6] By that time, articles to that extent appeared in the popular press quite frequently.

By 1954, many physicians, scientists, and public health professionals were convinced of the hazards of smoking.[7] By 1954, only the most obtuse, uninformed, or addicted would have failed to note the connection between smoking and lung cancer. At the 1954 annual meeting of the American Cancer Society, the director of statistical research, Dr. E. Cuyler Hammond, urged the industry to devote research funds to investigating how to manufacture safe cigarettes.[8] Under a human-centered economic paradigm this could have been the end of smoking as we know it. But it wasn't, of course, and many millions of people have since paid for this with their lives.

Had the captains of the tobacco companies cared, it would have mattered little. Any CEO in a publicly traded company who had chosen to truly act upon the cost of human life at the expense of the company's stock would have been replaced by the boardroom. Under our economic setup, the board of directors has a fiduciary, financial responsibility toward the shareholders, who are in it for the profit. However, realizing that they were accessories to the premature deaths of millions of people, tens of thousands of employees should have resigned—from the CEO down to the person who assembled the cigarettes. Yet, virtually no one resigned. And society did not expect them to. We have come to accept that our economy is divorced from any endeavor to promote the welfare of society. The significance of this cannot be overstated.

Presenting the choice of a drug to a consenting adult may be distasteful, yet fundamentally it honors one's right to choose. A true choice; however, means an informed choice, not one based on deceit and manipulation. This is the second line the tobacco industry crossed, from the early days of the industry, in fact.

An emergency meeting was convened in December 14, 1953 by the CEOs of all the major tobacco companies. The objective of maximizing the sales of cigarettes was a given. They decided that the best strategy would be to co-opt the science. They set up an institute with scientific trappings to serve as a front, the Tobacco Industry Research Committee (TIRC). Under the pretext of doubt and skepticism, the tobacco industry engineered controversy and confusion over the effects of smoking. In the decades that followed, they were also to persist in asking for more research, and in that way, they would appear to play ball all the while continuing to profit from the sales of the drug. They succeeded in selling a big segment of the public on the notion that there was scientific controversy over the effects of smoking, much as other corporations succeeded in later decades in selling a big segment of the public on the notion that there is scientific controversy over global warming. After all, comforting perceptions have been on their side, even if these have been illusionary.

Far from offering a product to an informed consenting adult, the tobacco industry pursued a deliberate campaign of disinformation and lies, which inflated demand for the product. Since the early 1960s, the industry was aware of the presence of polonium, an intensely radioactive substance, in tobacco.[9] With smoking, polonium concentration builds up in the air passageways leading to the lungs. An internal secret report concluded that "over a 25-year period, a sufficient exposure to this radioactive material could occur to produce cancerous growth."[10] The industry assessed that over a 20-year period, someone smoking two packs of cigarette daily would have been exposed to 1,000 rems (50 rads, 10sv).[11] Annualized, this dose is 25 times the maximum yearly dose permitted for radiation workers.[12] In 1980, acid wash was found to be highly effective in removing the polonium from tobacco leaves; however, the industry would not permit it. The tobacco companies were concerned that this would ionize the nicotine, reducing its ability to be absorbed by the brain of the smoker and thus eliminating the nicotine's kick sensation.[13]

Every year, six million people die due to smoking. It is anyone's guess how many out of the six million deaths are largely due to advertising and public relations campaigns of the tobacco firms. At present, 80,000 to 100,000 individuals become new smokers each day; most of these are children and adolscents.[14] Youth are important to the tobacco companies as they are the only source of replacement smokers.[15] Therefore, notably with the Joe Camel mascot, the industry targeted children.[16] Research has ascertained that it worked: advertising has hooked a portion of the teenagers.[17]

Worldwide, every year, over 150,000 children die due to second-hand smoke after contracting lung infections, asthma, and Sudden Infant Death Syndrome.[18] This figure, incidentally, is not likely to account for thirdhand smoke: cigarette smoke that deposits on carpets and later reacts with ozone within the house to form harmful organic aerosols.[19]

The nicotine-addicted parents are responsible for their children's deaths. Yet, the blood of these children is also on the hands of all those who make this industry possible, producing or selling the five to six trillion smoking sticks each year. Their blood is on the hands of the 78,300 employees of Philip Morris International, and 95,710 employees of British American Tobacco, and 49,665 employees of Japan Tobacco. It is on the hands of the personnel of freight ships that consent to transport the stuff. It is on the millions of storeowners who are willing to peddle the drug. It is also on the countless people all around the world injecting billions of dollars into tobacco production with the purchase of shares accompanied by the implicit mandate to generate more profit from cigarette sales. More specifically, it is on the hands of those who purchased shares in *Mutual Global Discovery Z*, *Vanguard Total Stock Market Index*, or any of the other mutual funds that have traded in blood tobacco.

We do not really think about mutual funds or about our jobs quite that way. Dying children are what we see on the Discovery channel. Buying mutual funds shares or working for some company is just what we do to make a living, to pay the bills, and to put a roof over our heads.

Whether it is the cigarette industry or that of the asbestos, benzene, vinyl chloride, or private prison industries, these are the dynamics true to our economy in general, not restricted to one industry or another. These inhumane dynamics can be readily seen in many sectors and economic activities. At the end of the day, it is about the money. At times, this results in the generation of the frivolous and superfluous. At other times, this results in murderous practices.

The defects inherent in our economic setup pose insurmountable obstacles in tackling the planetary crisis. This mode of economy has a flawed design—structural insanity, really—and cannot be used to affect the needed transformative measures. Even if we overwhelmingly choose environmentally-benign products, even if we tax or boycott harmful products, even if we enforce complete transparency, it would not suffice. The big engine of the world, the moneymaking engine, is not taking us where we need to go.

Our time and ingenuity are primarily tied in to the moneymaking engine. We need these vast human resources elsewhere, routed to another path altogether. We need a new engine, an engine whose expressed purpose for existence is the well-being of the world. Nothing short of that will put a stop to our withering of the living environment and the biosphere as a whole.

The stakes have never been higher.

The only way upward and onward, first of all, is to replace the moneymaking paradigm with a different economic paradigm. This is at the heart of the next section of this chapter.

The only way upward and onward, second of all, is to anchor and integrate our cultural dynamics with the broader dynamics of nature.

We have been preoccupied in social mind-games full of sound and fury signifying nothing. Most people are disconnected from the planetary web of life eddying and flowing all about: the genuine McCoy dormant underneath our city blocks, beyond the virtual make-believe worlds of cyberspace, and outside the gray stretches of our streets. It is imperative that we hold nature in sanctity and give it a centerpiece role in our hearts and minds. This is fundamental to all that has to be undertaken.

These two elements, two prerequisites, make up the building blocks of a social edifice whose dynamics would make it possible to take on the planetary crisis.

It matters not in the least how far away this is from where our civilization is at present. This fact counts for nothing. Reality cannot be negotiated with and brought around to see our existing way of looking at things. We have tried that, and it didn't work. We cannot afford to stay saddled with the existing social constructs. We have to be realists.

Function dictates form.

In dealing with the global crisis, our national political institutions are as effective as constructing a bridge from Styrofoam cups—and with about the same results.

The existing governing bodies are not cut out to handle the global crisis; bank bailouts and running a national vaccination program are about their speed. They are too pedestrian or otherwise acting as water carriers for various powerful interest groups. At the end of the day, the heads of state are glorified public relations people. None of them possesses the breadth of vision to see what needs to be done—or the balls to back it up.

The most significant administrative division of our world is that of countries. These make up about two hundred squabbling, gun-toting administrative regions, each looking out for and holding fast onto its respective territories. In turn, each of these turfs is mired by the pull and tug of myriad interest groups. The political arena of our world is more of a Middle Eastern bazaar with armed militia patrolling the alleyways than a Zion Council of elders in the movie *The Matrix*.

Democracy as we know it is a highly representational form of government—representing anything from the oil industry to pharmaceutical interests. The more influence an entity or a group exerts, the more representation it gets. It seems that democracy, as George Bernard Shaw said, is a device that insures we shall be governed no better than we deserve. The existing political system is a different thing in different contexts. Insofar as what is relevant here, it is a giant, intricate mechanism spewing money and statutes beset by thousands of yammering interest groups, pulling and tugging any which way as they grease the wheels to get legislation, money, or tax breaks their way. Everything is negotiable; everything that the involved parties can get away with is on the table. More often than not, any possible aspirations to advance society's well-being by individual politicians are being overshadowed by the eddies of power throughout the system.

Looking at the system as a whole, there is no broader agenda, no underlying compass or sense of morality.

In the recent decade, the flurry of conferences, NGO initiatives, and government subsidies to curb our carbon emissions have netted zero results. If anything, the reverse took place; the concentration of CO_2 in the air not only increased between 2009 and 2010, but the rate of discharge increased.[20] In the 90s we emitted between 6 and 6.5 billion tons of carbon a year. Since then, the annual rate of emissions has been going up. Over nine billion tons of carbon were dumped into the atmosphere in 2010.[21] In respect to the planetary crisis, nothing substantive has been done by our governments, and nothing substantive will be done. Our illustrious governments can neither draft a vision nor chart an economic course of action for even the next century, let alone for the next millennium. In the context of the global crisis, the existing power brokers are irrelevant. Worse, they are in the way.

These ponderous systems have been set up to assure that nothing of substance would come into being beyond the ongoing redistribution of wealth and making the world safe for democracy. This stands for a reason. We are the daughters and sons of slaves, peasants, and the persecuted. The system may not lead us anywhere; it may accomplish nothing of substance; but at least it assures that no man can assume a position of unbridled power.

We don't have time to stay wedded to this ball-and-chain political system; real trouble is brewing. We have to move past the childhood traumas of our race and act as the adults that we are.

From within our existing paradigm we cannot fathom a way out. To hell with the existing paradigm, then.

A NEW ECONOMIC PARADIGM

Close to one and a half billion people live in extreme poverty. They do what the rest of us would have done in their stead: cut down anything that can be used as food and as wood fuel. It is not possible to sustain the planetary ecosystem while every seventh person is living in a state of hunger. More broadly yet, it is not possible to implement the needed all-embracing, worldwide measures from within the each-to-his-own political-economic paradigm of the present. If it could have been done from within the existing paradigm, it would have been.

Out of a sense of universal responsibility and out of identification with the greater community of life and the needs of future generations, I call on instituting a new economic paradigm.

The bedrock for the economy is Earth: the world in which we are born, live, and die. It is a finite system with a quantifiable set of natural resources, a certain level of incoming energy from the sun, fixed land and water areas, particular rates of regeneration, and certain ecological needs.

First and foremost, I maintain the objective is to manage Earth in order to assure an everlasting well-being of its biosphere and to meet the necessities of people, all people. To that end, the first order of the day is to identify and uphold the needs of the planetary ecosystem. The second order of the day is to quantify what we have left to work with: how many trees can be harvested sustainably, what is the total amount of economically-recoverable minerals, and what would be a responsible and viable rate of extraction. Furthermore, we need to determine what is to be the ecological ceiling on our various activities across a broad range of parameters—from emitted toxins to disturbed earth. Studies along those lines have been conducted, but not at the scope and thoroughness called for.

In prioritizing economic activities, we ought to make accessible to all health care, drinking water, and dignified and ecologically-sound sanitation. We need to provide universal access to electricity, transportation, family planning, and telecommunication. Moreover, we should provide economic safety nets for those who are unable to support themselves due to maternity leave, physical or mental debilitation, or circumstances beyond one's control. Along these lines, the objective is to enable all to achieve a secure and meaningful livelihood that is ecologically responsible. Hence, we are to provide universal access to education of all levels and other resources needed to procure a standard of living adequate for one's health and well-being. Within the context of the broader planet, we need to restore the biosphere to health and draw down carbon, as outlined in previous chapters.

Conversely, we must avoid industrial activities and products that are likely to inflict serious environmental harm or serious harm to our bodies. When knowledge is limited, we should apply a precautionary approach. The burden of proof is on those wishing to advance a certain practice, having to show beyond a reasonable doubt it won't cause long-term, indirect harm. This is a reversal of the existing doctrine, where a substance can be used in products or dumped into the environment until proven harmful, usually by the state.

All of these are the basics; we have been screwing around with these principles for a very long time. We need to get them handled expeditiously and move on, charting an economic paradigm that will also reach beyond that which is adequate and satisfactory. Listen, our notion of cultural achievements to date has been the recognition that women are not inferior to men or that slavery is inhumane and therefore ought to be banned. That these are to be regarded as achievements at the tail end of one hundred thousand years of human culture is *pitiful*. It is time to unleash the human potential in earnest.

This is a call to chart an uplifting future and build for posterity. It is a call to build edifices to the human spirit, not to quarterly reports. It is a call for streets that would be adorned with art installations that are an expression of the best within us, not with billboards for soft drinks. It is a call for the construction of city centers for which people would compose ballads of praise rather than the construction of strip malls, parking lots, and pawnshops.

The desire to matter and to engage in what is meaningful is bursting through the seams of our sordid, pedantic economic paradigm. It is a testimony to the human spirit that in the face of the indoctrination in the merit of competition and financial incentive, our spirit to volunteer, contribute, share, and matter will not be stamped out. From the helpful, competent public librarians to the design teams responsible for the complex dazzling world of fantasy brought onto the large screen, it is not the pursuit of riches that drives these people. It is not out of the drive for material riches that Tesla envisioned global wireless communication, Einstein formulated his groundbreaking Relativity theory, and Van Gogh drew his galvanizing paintings. It is not out of the drive for material riches that countless people from all around the world have contributed their time free of any financial compensation and as of 2011 have written and edited in concert about nineteen million articles that make up Wikipedia, bringing into being the largest general-reference body of knowledge in the history of humankind.

The goal is to set up an economic-social milieu that will empower people to have a sense of ownership over what they do and responsibility for their actions; a milieu that will honor and dignify each person; a milieu that will empower people to produce that which matters—enriching people's lives in particular and the biosphere in general. And to share and make accessible to all what has been gained and learned. Information, knowledge, and art are to be the common heritage of humanity, an open source, made freely accessible to all.

Envisioning a new operating system on which a marketplace with the above characteristics may exist is no small feat.

I call for the formation of an economic-social milieu that will empower people to produce that which matters—promote people's lives in particular and the wellbeing of the biosphere in general, and with no one left out.

The format that seems to lend itself best to bring this about is an open-source economy that retains a sense of ownership, responsibility, and pride over one's work.

These are two big, general statements, and they solicit big, general questions. I don't have some of the key answers. Looking at the past, we know what types of economies failed spectacularly. Looking at recent trends, we know what works. All in all, there are a few things that can be said with a measure of confidence. I will proceed to state what seems evident and on-target. Regard it as a first bid, which I put on the table for others to mull over and improve upon. For your consideration, here is one possible set of ideas for a viable economic and political setup.

FOOTPRINT CREDIT

In the context of advocating a resource-based economy, Peter Joseph, founder of The Zeitgeist Movement, said he often runs into people who challenge the precepts of the economic model his movement espouses. They may ask, "What if I want a fifty-room mansion in a Resource-Based Economy? Where's my freedom to have that?" Peter Joseph fires back with "'OK, well, what if I want a million-room mansion?' or perhaps 'What if I want the entire continent of Africa as my backyard?'" He goes on to ask at what point the selfish, acquisitive, spoiled interest becomes blatantly irresponsible and socially offensive. Given that we live on a finite planet, he argues, and in a society in which resources must be shared, excessive and ostentatious living is really an anti-social form of neurosis.[22] And so it is.

We have to work within the confines of the planetary ecosystem—much as pilots have to work with the fact that they've got gravity on their hands, and much for the same reasons.

The first priority of the ecologically available natural resources and activities should be applied to the creation and maintenance of infrastructure facilitating our basic needs. Insofar as the balance of the environmental footprint we can exert, it would be apportioned among all the people of the world to be used as each sees fit. People would be given a monthly allotment of *footprint* and this would hem in, define, the outer limits of their consumption activities. This quota would be determined by a global body. Every month, *footprint* credit is added to one's account, adjusted to the current state of the planetary system and the number of people alive.

Naturally, people would be wise to manage their *footprint* with care, much as they would be wise to manage their money with care. Money does not grow on trees and *footprint* even less so.

The introduction of a *footprint* allotment for each person fundamentally alters the thrust of the marketplace. The incentive would be to purchase, and therefore produce, commodities and services that have a lower *footprint* price tag. It also means that a twenty-room mansion is far more of a liability than an asset. This wouldn't just be a socially-contrived principle; this really is the case.

The ecological footprint, referred to as *footprint*, is complex and comprised of many parameters. Some of its aspects have already been described in the previous chapter ("Consumption").

TABLE 13.1

environmental footprint throughout the lifecycle of a product ("footprint"): six parameters
Direct Disturbance and Degradation of the Earth: acuteness of disturbance, volume of earth affected, duration of the disturbance, and the ecological sensitivity of the locale involved.
Toxins, Pollutants, Debris: effects, coverage, persistence (includes heavy metals that are not salvaged, global-warming potential, ozone potential, various forms of water contamination, acidification, harmful radiation, aerosols, poisons, and solid non-recyclable waste).

environmental footprint throughout the lifecycle of a product ("footprint"): six parameters
Material Scarcity: Each non-renewable commodity is assigned a scarcity value. Different values are assigned to virgin materials than recycled ones.
Water Resources: amount of water used compounded by the sustainability of the source. In the case of desalination, the footprint is just that of the power required. In the case of recycled graywater, it is assigned a zero footprint value.
Biomass Used: Renewables are measured against a natural rate of regeneration. Generally, the rate of replenishment will dictate the rate at which renewable material is harvested. Beyond that, this parameter measures the amount of biomass displaced on Earth and the overall degradation for given ecoregions.
Special Considerations: effect of wind turbines on the global winds, marine noise pollution, etc.

Every product and service would display both a price tag and a *footprint* tag. While people may waive off the monetary charge of products and services, they can never waive, unless they have a magic wand, the related *footprint* charge, whatever it may be. When a person obtains a product, a due amount of *footprint* credit is deducted from one's account alongside, if applicable, a due deduction to their currency credit. More specifically, insofar as the *footprint* credit is concerned, there will be a deduction in the allotted credit along each of the parameters that make up *footprint*; the specifics depend on the particular product or service. Beyond the cost in dollars and cents, a cell phone may come with a *footprint* overhead of 53 Direct Disturbance, 79 Toxins, 122 Material Scarcity, 39 Water Resources, and 16 Biomass Used. One's balance—both monetary and *footprint*—is something that could be accessible from any terminal, checkout counter, or hand-held communication device.

Once a person uses up his *footprint* quota, that would be it until one's tank is refueled again with more *footprint* credit at the beginning of the next month. To avoid finding oneself with zero *footprint* credit, it would be possible to set up an automated system that will keep some *footprint* in reserve, setting aside monthly savings to use as a buffer. It is also possible to have a publicly-traded loan website. Participants set up a contract with hundreds of millions of others, each agreeing to loan to each other automatically when any hit a zero balance in a given month. The system would look at other accounts and would borrow based on carefully pre-set criteria. Each person can define a threshold in which he is comfortable loaning out and can define the loan period. There would be no technical possibility for the borrower to obtain some of the lender's *footprint* and make a run for it. When the borrower is due to receive a new allocation of *footprint* credit, before it hits his account, a portion of it is automatically credited to his lender's account. The incentive for partaking in such a loan scheme is that it is to be a two-way street: today one person may loan out, but tomorrow, he may be in need of a loan. What goes around comes around. This aside, the extent of *footprint* loans would universally be capped by the system. If one was able to borrow a five-year supply of allotted *footprint,* later this could very well come to a choice between waiving off the loan and letting the reckless borrower starve—neither of which is really acceptable.

It may be evident from the example of the hypothetical cell phone that the measure of *footprint* is not as simplistic as suggested in the above paragraph. *Footprint* is comprised of at least five independent parameters. It is not only possible but also likely that one would use up all of his credit in one parameter (e.g., Toxins) before the other four at any given time. However, with billions of *footprint* accounts, this is not much of a problem.

It would be possible to trade credit in one *footprint* parameter for that of another. For instance, if a person has an abundance of Earth-Disturbing credit, he can trade with someone else who is low on that count but has an abundance of Toxin credit, which the first party happens to need. In fact, what would be called for is an automated management program that balances one's credit on the fly. The system would make millions of transactions at lightning speed every second of every day. It would all be seamless and would operate behind the scenes. Thus, if someone draws heavily on one parameter, the overwhelming chances are that the system can trade some of his credit in other parameters and even out his *footprint* credit across the five different parameters, or, better yet, study one's purchasing patterns and attempt to maintain an optimal profile of parameters at all times. Naturally, it would be possible to manually program the trading profile, if one needs to amass credit in, say, the Water Resources parameter in anticipation of a trip that would draw heavily on that parameter.

Behind the scenes, the trade system would be able to smooth out the daily operation to a large measure. Yet, we must never forsake the underlying reality. Water Usage has nothing in common with Earth Disturbance—let alone Material Scarcity, which puts value on the availability of nonrenewable resources. Thus, if many in the world draw heavily on Earth Disturbance, we may all run into the imposed ecological limits in a given month. If push comes to shove, no amount of creative accounting and trading would change that fact. Unlike the concocted notion of money that has always been just a cultural construct, *footprint* has a physical reality and very physical consequences, if overplayed.

It could be that some people would end up with more *footprint* credit then they really need in their lives. Beyond the desire to sustain the biosphere, the incentive to conserve excess *footprint* resides in the fact that at any time it would be possible to redeem some of the *footprint* credit for currency credit. *Footprint* would be worth money. However, there is no intention to allow the well-off to buy their way to eco-salvation, and in accordance, a few restrictions would be set in place. While within the automated system one can redeem some of his *footprint* for currency credits, no one can buy or sell *footprint* from anyone else. Furthermore, it is a one-way street; it is not possible to purchase *footprint* credit from the system, only to redeem it.

There would be a universal, fixed exchange rate between currency and *footprint* credit points. The money redeemed from *footprint* ought to be high enough to encourage people to trade in excess *footprint* rather than squander it on what is decadent, but low enough that it will not start to constitute a source of income, which would bring us back to people making money out of thin air.

As mentioned above, it would be possible to accumulate *footprint* credit, allowing one, for example, to take a *footprint*-draining vacation once in a while. However, there would be a ceiling on how much could be accumulated so as not to create a considerable disparity between the actual ecological footprint we exert on the environment and the credit that could be exercised—and was not. Hence, once a certain ceiling is reached, all additional *footprint* credit would be automatically converted and then deposited in one's account as currency credit.

Most everyone will have the same *footprint* credit, adjusted to the number of dependents. Yet, this *footprint* equity would exist only once a certain, minimal income level is reached. To encourage people who make USD 300 a year to conserve environmentally, their *footprint* cap is to be lower than those having one hundred times their income. Or else, we may find businesses producing ultra-high *footprint* items for the poor—knowing that they wallow in excess, well above their means to spend. There is an additional reason for a *footprint* quota adjusted down for people with ultra-low income. If there is no lower allotment for the poor, they could forego work, converting excess *footprint* to money and simply live off it. As is discussed later, the intent is to rapidly eradicate poverty. Thus, lower levels of *footprint* for the poor are only an interim measure. The long-term aim is to maintain an identical *footprint* credit for all.

The *footprint* is amortized over the lifespan of the product usage. If a person purchases a chocolate bar, the entire *footprint* is deducted immediately from one's account. If one buys a house, the *footprint* over the lifecycle of the house is amortized over the projected period the house is going to be in existence. Whether a one-time lump withdrawal or installments, the schedule of *footprint* withdrawal is set for every commodity and service to reflect its individual nature. The practice of *footprint* amortization of some products would create a disparity between actual *footprint* exerted on the planet in a given period and the *footprint* on record. In actuality, the construction phase of a house accounts for the bulk of the house's *footprint*, and the notion that we can amortize it over time in equal installments is but a necessary fiction.

The construction of a manufacturing plant needed for the fabrication of a certain commodity bears its own *footprint*, and it is to be passed on and distributed equally among all consumer goods that would be produced in that plant. This is of course also a necessary fiction. We cannot know how many products are going to be sold, thus we cannot know what is the fractional share to pass out to each individually sold commodity. When it comes to overhead *footprint* costs, creative accounting is called for. The agency governing *footprint* would forecast, or estimate, the number of units that would be sold— based on historical records, market trends, and such. This would determine the fractional share of the overhead *footprint* to be added to each sold commodity. If it is projected that one thousand units would not be sold— due to loss, returns, or lack of demand—their combined *footprint* is to be equally apportioned and tacked on to those units that are sold. In other words, unsold or recalled units are to be treated as a part of overhead *footprint* costs.

If a person gets something as a gift or inheritance, they are responsible for any *footprint* balance of that artifact. This is of course also the case when one purchases something from a private party. Along the same vein, if a man rents a table saw, he would be responsible for the fractional *footprint* of the incurred usage.

In theory, it should be possible to make and subsequently sell a product that on some or all parameters has a positive *footprint*. Put another way, the product has a net benefit to the ecology of the planet. In this eventuality, one's *footprint* credit actually increases along that given parameter. And why indeed not?

Everything else being equal, it is obvious that people living in heavily degraded, ecologically-fragile regions of the world have to tread more lightly than those who live in a relatively intact environment. Indeed, commodities and services have *footprint* that is locale- and context-dependent. See the previous chapter ("Consumption") for more on this.

Every additional layer of nuance would allow the *footprint* system to track more closely the reality of our ecological activities. Every added layer contributes.

The Toxins parameter belies the true complexity of hundreds of very different toxins, not interchangeable in any sense. So even the ostensible complexity of having five different parameters to keep tab of is a vast simplification. It is a balancing act between what is practical to manage and what still presents an effective tool to monitor and constrict our environmental impact. Having said that, what would be overwhelmingly complex for people would be but child's play for a computer. Once again, behind the scenes, every commodity may be comprised in truth of hundreds of distinct parameters, such as those indicating the level of arsenic or GHG emissions. In fact, everyone would get a *footprint* allotment that would be comprised of those hundreds of parameters. So while it would appear to us that we purchase a pen that has five Biomass Used points and this is the amount being deducted from our account, in actuality numerous, differing deductions would occur in dozens of sub categories within the Biomass Used umbrella. The system will continuously trade from within all the users' deficits and excesses of the various sub categories, compensating and adjusting. None of this may be apparent to the consumers, but in setting up the next month's *footprint* balance, the agency administering it would adjust the *footprint* points of various products to reflect our activities. If we, across the board, pushed seawater acidification to the outer edges of ecological wellbeing, the credit of the overarching category of Toxins would have lower availability in the future.

The *footprint* system tracks the physical reality close enough to be useful. But there is no need to confuse it with perfection. To various degrees, we all impact the environment outside the purchasing cycle, outside the *footprint* system discussed here. If one changes oil in his car and dumps the remains on the ground, the resultant *footprint* is not accounted for anywhere. Beyond, of course, enacting and enforcing due environmental laws, we just have to write these things off, as it is not feasible to closely monitor such activities.

In assessing our collective, ongoing ecological footprint, the responsible agency is likely to mostly rely for its data on reports from the manufacturing sector—the source—versus tracking *footprint* credit used up by the end-consumers. As a rule, this would track environmental impact more closely and better monitor our actual impact. In determining future available *footprint* credit, the agency would also look at the actual impact our activities have had on the environment—along countless locations and parameters.

OPEN-SOURCE ECONOMY: ADDITIONAL PRINCIPLES

It should no longer be feasible to make money out of money; the new economic setup would render any such practices unworkable. No longer would some people control the means of production nor control natural resources. Under the new economic setup, people could not be absentee owners of firms nor of properties. The new setup would also have no speculative financial instruments of any kind. Existing debts will be annulled along with the industry of banks and lending institutions as we know them. Broadly speaking, people will no longer be able to generate a living by the making of and trading of hot air.

While interest-bearing loans—along with other methods of making money out of money—could be a thing of the past, loans by themselves have a most legitimate use. It is enough that each of us, or each person who wants to participate in this plan, would set aside a tiny portion of their credit money as a possible loan to be made available to the general pool of billions of people. And suddenly vast amounts of money are available, to facilitate once-in-a-decade large purchases, such as a remodeling project, or to facilitate a small purchase such as an oven. Much as it is now, a person may have to qualify for a given loan and have collateral set in place. No interest would be charged on loans. The incentive to participate in the loan scheme is the same as that related to the *footprint* loan scheme: what goes around comes around.

Credit of one's work ought to be given where credit is due, and, much as in the contemporary science culture, all findings and results would be made accessible to all.[23] Thus, various outfits throughout the world would be able to build upon each other's work. Amid mutual criticism and analysis, cumulative knowledge and products would evolve. All forms of art and knowledge are to be a common heritage of humanity. Everyone will have open access to everything, from a new engine schematic, to the formula of a new substance, to all the books and songs in print. Nothing is to be patented, and in a commercial context, no technical details of a discovery or development would be held back. This is not a matter of affixing the human populace with halos and wings but of creating an economic setup devoid of any incentive to patent and otherwise hold back key details of inventions and discoveries.

A through-and-through open-source economy would be possible, and viable, only if all participants would have access to basic necessities, both via ubiquitous infrastructure and via a sufficient purchasing power, as through some sort of a credit system. If basic necessities are not met, and one is not duly compensated for one's work, we are back at patents, trade secrets, and the competition paradigm in general. We are back to bribes, sellouts, and each person out for himself.

The new economy rests on a bold premise that we live in a world that can provide for the needs of all. And once one's livelihood is not on the line, the transformation of the manmade world becomes a possibility. No one who is a part of the new economic paradigm would have to fight over the basic necessities. There would always be work to be had, and there would always be a safety net in moments of need. Both of these statements require an explanation.

First, let's get the obvious out of the way. The government would provide unemployment benefits for those who are unemployed due to maternity leave, debilitating conditions, and other circumstances beyond one's control. The proposed system would take care of those who can't work, yet it would offer no free ride to those who can but won't.

What requires more elaboration is the notion that work is always to be had.

The question is not what tens of millions of unemployed people could possibly do, but rather what is the best use of their time. After all, our capacity to dream of what is beyond the here and now is infinite, and, in accordance, the called-for labor is potentially without bound. For instance, it may be determined that there is a need to construct a bridge. The only question is how far we care to take that project. If the workforce is available, such a project can easily be opened up for a lot of people—beyond those required for the construction of a purely functional bridge. We can build a bridge with subtle artistic carvings on every square yard of it, with a hand-painted finish of glaze in myriad pastel hues. It can have an embedded hidden power system that translates the vibration of traveling vehicles to an energy source that emits a soft glow at night synchronized to the breeze and ambient temperature—a duet of soft light and the passage of time. And the bridge's glow in the autumn would be straw-gold.

No, there would neither be a need nor a justification for anyone to experience long-term unemployment.

Each person ought to be financially compensated for his work—whether via royalties or a salary. A system of wages has obvious merits. Wages encourage acquiring goods in a judicious manner; they help to retain a sense of ownership over things one obtains; and they hold one accountable to one's purchasing choices.

The currency used is proposed to be generated as it is generated now: from thin air. But it will not be tied in to debt, issued by banks, profited by credit card companies, controlled and manipulated via exchange rates by the various central banks. It would be a globally uniform unit and with a virtual existence only, much as the envisioned *footprint* credit. More than anything else, money in this context would function as a bookkeeping method, a glorified ledger.

The global workforce pie comes to about 6,000 billion labor hours a year. If one person works about 2,000 hours a year, it makes no sense for them to be able to lay claim to the fruits of labor worth 60,000 hours of others. In fact, that would be impossible to execute, unless many others would lay claim, say, for only 200 hours, to compensate. At any rate, one need not overstate the extent of contribution of some individuals over those of others. All our creations and projects are tightly interrelated, building on prior insights, inventions, and work. Edison made great strides in the architecture of the light bulb, but it was based on prior light bulb designs. Furthermore, Edison did not accomplish this single-handedly; he advanced the light bulb design in collaboration with a hard-working team of tinkerers. In conclusion, there is a merit for salaries to vary from person to person, but within reason.

If an expert gets twice or even three times the salary of a trainee, it makes sense. By the same logic, a rocket scientist, who undergoes years of study, deserves to get more than a bus driver. If a miner, a lumberjack, or a firefighter gets two or three times the pay of a barber, a tour guide, or a curator, it makes sense—given the danger and hardship involved in the former vocations. If a portable toilet cleaner or someone handling entrails and offal gets two or three times the salary of a coffee shop manager or a real-estate broker, it makes sense; it is due compensation and incentive for a job that may be essentially of little intrinsic appeal or physical comfort. A director of a big budget movie or an air-traffic controller is likely to have a higher level of investment of one's self than that of a janitor or cashier and ought to be duly compensated. It also makes sense that someone who works more hours will be compensated in accordance.

The disparity in income ought to be within healthy social bounds, such as up to three times from the lowest to the highest wage within the economy. Vast disparity puts pressure on societies; they tend to go haywire. Inequality weakens the social fabric, damaging health and increasing crime rates. Everything else being equal, it seems that relatively egalitarian societies are more socially cohesive. They have a stronger community life and suffer less from the corrosiveness that afflicts societies with a high income disparity.[24]

To maintain relatively equitable income bounds, all wages, universally, would be subject to the same filter, a mathematical function, where up to a certain income level, each currency point entered is one currency point credited to one's account. Once an income is beyond a certain bracket, it is subject to an exponential curve, where an incoming 1.25 units are credited into one's account as 1.00. At higher income levels yet, each incoming 2.5 units are credited as 1.0. And if, say, fifty million units come in, the conversion ratio would be in the stratosphere, to offset what is basically a sound economic principle that has run amuck. There are potential problems with this mathematical-function model, but it can be made to perform.

Let's now move from the consumer end of things, with its wages, to the production end of things. Let's start by describing sectors that essentially work well within their current mode of operation.

The United States Postal Service, which relies on revenues to sustain itself, seems to work well enough—once we preclude its downsizing due to an overall shift to electronic mail, which is immaterial to our discussion here. The Postal Service balances its checkbook, and there is no agenda but to provide the best service it can. The public library is an example of something else that works well. Much as with the postal service, there is no intent to make money; however, its business model is different. Libraries work within their budgetary constraints to the best of their abilities without any revenues to speak of from their patrons. For that matter, the existing business model of an iron-smelting plant or a producer of traffic cones seems to be sound, as they don't lend themselves to a money-feeding frenzy. In contrast to these examples, the creation of a movie is a mixed bag. The art, makeup, or sound departments make the movie the best it can be; however, the choice of what movie to produce is dictated by the drive to minimize risks and maximize profit and, consequently, oftentimes panders to the baser elements within us, working within bounds of well-worn winning formulas.

All in all, a host of economic activities would be served well within the public sector framework: the transportation infrastructure and public transportation; municipal sanitation and sewers; universal health care, which would include clinics, hospitals, pharmaceutical industry and medical equipment manufacture; telecommunication, energy infrastructure and power generation; science research; environmental management; and postal services.

There is no need to levy taxes to fund the public sector. Instead, about 30 percent of the total man-hours of the global workforce may be declared the basis for the government budget. This would be the basis for generating money credit for public-sector jobs. As is apparent, money credit in that context would be nothing but a bookkeeping measure.

On a slightly different note, many millions of young people around the world can join or have it be a part of their curriculum to go and partake in any of the colossal public projects that are needed—from a new power grid to a desalination infrastructure to rewilding initiatives. They can be organized along the lines of boy and girl scouts clubs.

The goal is to have a dynamic, vibrant, innovative marketplace committed to producing commodities made to last and to promote wellbeing. Keeping these objectives in mind, many economic activities do not lend themselves to the public sector model. Even a cursory examination of the Soviet Union or China at the height of the Cold

War makes this obvious—as they produced commodities that oftentimes were insipid and lacking any spirit. But then again, these economic activities also do not flourish under the existing, profit-driven free-market—as this market produces commodities that oftentimes are inane or shoddily made.

Thus, much as it is now, many economic activities ought to be in the realm of the free market. Yet, a free market with different dynamics than those we have had.

The design flaw of the existing economic setup has been discussed at length. It revolves around moneymaking. Like any other social construct, moneymaking generates justification for its existence as such. It is an entity by its own right. It creates a need for its prominence, a self-serving need, a manufactured need. Much as with salaries, money in the private sector would aid with maintaining discretion, a sense of ownership, and accountability. Yet, historically, these merits have been completely overshadowed by the pursuit of money as an end in itself. The target is to dissociate the accounting aspects from the moneymaking drive. Or if this proves impossible, dispense altogether with money in the private sector of the industry. For one, it is not like financial incentive is the only driving force of competitive spirit—as can be noted in other human endeavors, as in science or in amateur sport. If necessary, it should be possible to decouple the purchasing credit of the consumer from that of currency to be had by the entity providing the goods or services. When one acquires a hair dryer, the price would reflect the labor invested and would be deducted from one's account. However, this does not necessitate a comparable sum to be credited to the business from which it was acquired.

For all we know, money—much as school grades—started as a good idea, an inconspicuous tool of bookkeeping and monitoring. But then it took on life of its own. It is a problem, one that is not only specific to money but to a broader social phenomenon.

We note the reflections, the indicators, of successful life, dub them "results" and then say, "Let's go after the 'results'." Success or sex appeal are no longer regarded as indicators but are becoming a deliberate subject of pursuit. They take on a life of their own. And as we regard a byproduct as a deliberate target, something happens: it undermines the real thing. Thus, the pursuit of good school grades—that is, the favorable byproducts of learning—eclipses the joy of learning and the thrill of exploration and gives a paramount role to state tests. Thus, preoccupation with the *appearance* of partner dancing alters the nature and context of partner dancing, from that of an intimate interplay to that of a performance, moving it from the realm of intimate relatedness to the limelight of dance competitions. Thus, the market's fixation with the so-called 'bottom line' corrupts and sabotages meaningful progress.

This is the dilemma. It is a problem of a complex-system behavior—the marketplace—and it does not lend itself to a theoretical fix drafted on paper. What we need are dozens of teams staffed by those excelling in systems game design and economic models to get working on devising various economic systems: a series of open-source models addressing various aspects of any possible, future economy. In the end, one way or another, we need half a dozen or a dozen really promising and differing market models and one final model that delivers.

For an initial test run, a 3D, online virtual world environment would be set, programmed with natural resources. Such a virtual world ought to handle concurrently hundreds of thousands of volunteer players. People would form detailed avatars and interact with others. Much as in *Second Life*, we would need modeling software that would allow people to create objects, be it buildings, consumer goods, or food products. In short, it is to be a detailed simulation running under the economic roles set in one proposed economic model or another. The virtual world will offer opportunity for the modeling groups to tweak and adjust various features. Yet, the usefulness of these simulations can only go so far. Very little is at stake in a virtual world. There is no possibility to either feel the pangs of possible hunger nor to work at manual labor in the baking sun. Face-to-face interaction with a real human being is absent. And risking one's children in a virtual world ultimately means nothing; it is just a gesture. Thus, the final, bona fide test would be conducted later in the real world, opening up the economic model to large groups of people, one million people here, half a million people there. These large groups would live for six or twelve months under a new economic paradigm while the world watches.

In the end, specific principles of a new, viable open-source economy would become clear and subsequently be established throughout by a planetary-wide agency. More on this agency later.

It should be clear by now that a technological centerpiece of the economy is a ubiquitous access to online data. Every point of purchase and service ought to have a biometric ID system, which would facilitate transactions by updating one's account—both the currency credit and *footprint* credit.

While we are at it, we might as well consolidate all of one's records into a master account that can be accessed in a highly selective fashion by a potential employing outfit, or by a physician accessing one's medical record, or by the department of motor vehicles. This will obviate the need to fill out countless forms throughout one's lifetime, and for a small army of clerks to file and update them in myriad offices as is done now. And with a biometric ID, there is no need to carry any form of identification beyond one's retina, fingers and such. I do wish to acknowledge that there is more than a touch of hysteria surrounding the issue of privacy. As this centralized records scheme is beneficial but not essential, and given the emotional stir it may generate among some, joining it ought to be a matter of personal choice.

A last point to be made pertains to advertising. It is said that advertising is the art of convincing people to spend money they don't have for something they don't need. This is said to be done largely via a judicious administration of dissatisfactions into one's life. The advertising message keeps many in a perpetual state of infantile self gratification subject to the acquisition of goods.

Gratefully, an economic paradigm with neither the incentive nor ability to turn a profit offers little reason for advertising that is manipulative in intent. Bernays got the gist of it when he stated that in a public relations campaign, one must never assume a notion that puts his duty to the client above his duty to society. In the future, any professional association for publicists and advertisers should monitor and revoke one's license if a practitioner is found guilty of breaking his oath.

No more half-truths and lies by omission. A price tag on a printing device needs to display the upfront cost and also display the costs of ink cartridges throughout the anticipated lifetime of the product. There is something...liberating about conducting business in a world where we don't feel the need to manipulate and to overstate the case for our products and services.

As a matter of course, it ought not be mandatory to participate in the new economic paradigm. Anyone should have the right to take hundreds of cans of SPAM, make it to the hills, and build a bunker. Much as the Amish choose to lead life outside the predominant economy of today, groups of like-minded people can have settlements where they can enact the finer points of Grand Theft Auto or Monopoly games, that is, as long as they don't coerce or dupe multitude others into demeaning, subservient, or hard physical labor so the players can assume the more glamorous parts of the money-amassment game. Those groups can play games to their hearts' content, yet, it would be within physical bounds that are to be shared by all people. Each of the merry bands of capitalists would be subject to the same *footprint* limits as everyone else. In other words, products and services everywhere and sold by anyone would be mandated to carry a *footprint* surcharge.

THE WORKPLACE ENVIRONMENT

The Soviet economy took out of the equation the drive to make money as an end in itself. However, with the sense that everything belongs to everyone and thus to no one, there was little sense of ownership or that anything one did mattered. This sense of apathy was reinforced even more so by domineering, dictatorial central planning. Alas, the modern day, Western workplace has its own mechanisms that suffocate and disempower. I refer to detailed company policies, insurance dictates, and the myriad of existing laws. The bulk of these things have to go to the incinerator, along with the blizzard of paperwork they have been generating.

Laws and codes are inherently blind to context and nuances; inherently generic and superficial; inherently encouraging power lust, greed, and validation of the victim life view. Their sheer presence deters personal, direct handling of situations. Their absolute nature and simplicity lure people to trade a perception of the dynamic quality of the world for an uninvolved illusion of certainty. Laws are but an approximation; they seek to point to some fundamental truths and values. And they should come as a companion to reason and judgment call—never as a substitute. Laws are to function as a set of guidelines, not as verses in a holy book of some secular, modern religion.

We are enmeshed in a web of codes and regulations dictating every interaction and aspect of living—the magnitude of which never existed in the history of humanity. We can't take a crap without a permit, let alone try to deal with the pile of radioactive waste without a lawsuit. Moreover, be it practicing physicians or school teachers, the fear of litigation is just as restrictive as any law, acting as a very strong damper over what people would say and do in the workplace. As Philip Howard said, what people *can* sue for ends up defining the limits of freedom. It does not matter if a lawsuit wins or loses; no one wants to practice or introduce something with the potential for a lawsuit.[25] Our social reality is suffused with impotence and control. It is oppressive. And the spirit of initiatives is wrung out of us, repeatedly. We no longer trust human judgment to do what is right; even judges have to follow a prescribed schedule of punishments. We no longer look within and deem what is right or wrong; we look instead at what the law says we should do. When laws are absolute, no judicious, context-specific choices are possible. When there is nothing but laws, as Philip Howard argues, it drives people away from the subconscious, the deep well of instincts, experience, and good judgment. It drives people to the thin veneer of conscious logic.[26]

We need to restore our trust in ourselves and in our capacity to make judgment.

The potential exists to transform the climate of the workplace from that mostly of indifference in performing a job—oftentimes demeaning or inane—to a climate that instills in people a strong sense of ownership, initiative, and pride. To that end, a few sweeping measures are to be introduced.

The existing colossal body of statutes should be replaced by Common Law: a broad and relatively simple set of guidelines and principles. With the implementation of comprehensive universal health care, the health-insurance industry would pass away and with it its numerous constrictions and dictates. I urge to can the adversarial system, with attorneys hired to bend the truth to benefit their clients. I suggest it would be succeeded by a tribunal panel of a few expert judges joined by research assistants and detectives, dispassionately committed to uncover the truth and particulars of a given case. The judges would have the authority to evaluate each case on its own merits, attributes, and circumstances—using the Common Law as a guiding principle. What's more, it is to be the mission statement of the courts to provide justice that is not only sound but also carried through in an expeditious manner.

Under this proposed setup, the courts are to remain open to all who seek in good faith to invoke the protection of law. However, baseless litigations, whether out of malice or greed, ought to be harshly dealt with. We need to uproot and stamp out the litigious trend of recent decades. We will go a very long way toward doing so with a wholesale shift from verdicts that award dizzying amounts of money to those meting out Community Service sentences—whenever possible.

In order to give people a strong sense of ownership, responsibility, initiative, and pride in the workplace, additional far-reaching measures need to be set in place.

Owned by no one in particular and created for the sole purpose of generating profit to millions of anonymous people, the corporate entity should be phased out entirely. This also goes for any other kind of absentee ownership over any means of production. In their stead, the people who run the business would be the owners, the sole owners. I refer to those who currently manage, and also to those who operate the machinery, and to those who design, and to those who engineer. As the functions and attributes of currency in the private sector are yet to be determined, it is yet to be seen what ownership would exactly mean under the new economic paradigm. Yet, it obviously would entail a measure of control and participation in decision-making. Thus, there would be no more employees in the traditional, neo-feudal sense of the word. It also means that owners of companies, including by definition all staff members, are personally accountable. In a liability case, however else they would be made to compensate, they would have to compensate also with a portion of their time on Earth.

By and large, the logical transition to the new outfits is to have the employees of the existing public corporations assume ownership and continue to run them. The fundamental differences between now and then would be the thrust and mission statement of the new business entity, and the fact that ownership is in the hands of the people who actually run and operate the business.

With an economic paradigm that upholds humans' dignity, we should see the demise of modern-day sweatshops and the serf class as a whole. Without starving people compelled to do anything and everything to secure their next meal, prostitution in all of its forms is likely to go away, and jobs that are debilitating and gruesome are likely to acquire humane work conditions via the use of automation or otherwise. At present, many tens of millions of children are pressed into bonded labor, toiling from dawn until the dead of the night. They never know childhood so that others can have sparkling diamonds or inexpensive basketballs. With serfdom gone, not only would these youngsters have the opportunity to experience childhood, but the rest of us could regain our humanity, not having to be callous nor having to convince ourselves that pressing children into bondage is the inevitable, albeit regrettable, order of things.

HUMAN RESOURCES

If we extrapolate from US labor data, we can assume that we have, or would have if the unemployed held a job, about 2.7 billion full-time workers and 0.6 billion part-time workers. This comes to a total of 6,000 billion work hours to go around in every year, or more if we assume a crisis footing and move beyond a forty-hour work week.

Whether we are creating a universal sewage system or desalinated water infrastructure, the goals are lofty. Where would the man-hours come from?

As it turns out, this is not going to be a problem. There are industries that either entirely or almost entirely will be rendered superfluous and irrelevant under the new economic system. While every country has its own distinct labor profile, table 13.2, pertaining to the United States, should give us a sense of the labor potential. It lists the current economic sectors that either entirely or in the majority of cases contribute nothing of true substance. In this table, fully 23 percent of the workforce would become available.

TABLE 13.2

potentially available workforce	
existing, largely superfluous economic sectors	number of people (USA, 2011)
department of defense, civilians	700,000
defense industry	564,000
military, active personnel	1,477,000
credit intermediation and related activities	2,542,000
finance and insurance	15,571,000
securities, commodity contracts	809,000
unemployed	14,000,000
lawyers	554,000
telemarketers	345,000
advertising and insurance sales agents	489,000
IRS	106,000
total	**37,157,000**

Sources: Bureau of Labor Statistics, under Databases & Tools, One Screen, "Employment and Wages from Occupational Employment Statistics (OES) Survey, http://bls.gov/data; Mary Meghan Ryan ed., *Handbook of U.S. Labor Statistics 2011: Employment, Earnings, Prices, Productivity, and Other Labor Data* (Lanham, MA: Bernan Press, 2011); "Active Duty Military Personnel Strengths by Regional Area and by Country," US Department of Defense, http://siadapp. dmdc.osd.mil/personnel/MILITARY /history/hst1109.pdf.

Consider what is possible instead. The United States comprises about 5 percent of the world's population. In line with that, let's assume that the number of capable adults who contribute nothing constructive around the world is ten times the above figure, roughly 370 million.

Let's assume that about 30 people are involved in building a house (framers, brick masons, electricians, plumbers, excavators, etc.). Furthermore, let's assume that it takes about 4 months to build a house, that they work in teams of 3 or 4 members, that the part of each team in a given house is 2 weeks long, and that then they move on to work on another house. Assuming 253 work days in a year, had they been trained to build houses, all of those who currently contribute nothing of real value would be able to construct in one year about 310 million houses at an American standard of living. Within one year we could have easily taken care of everyone who has no adequate shelter (if we assume a few people per household). This calculation is crude and simplistic, yet it gives a sense of what is possible. In truth, there is also a need to account for all those people who manufacture and transport the raw materials required for the construction. But then again, there are many more than the speculated 370 million people whose jobs are of little merit, as other industry sectors are permeated with the superfluous and the inane. In reference to this example, there was no intention to suggest that anyone would be drafted to construct houses, willingly or otherwise.

PROPERTY, LAND, AND FREEDOM OF MOVEMENT

Earth is owned by no one and is home to all. In accordance, everyone ought to have unrestricted rights of passage, residence, and work throughout the planet.

These three rights are ends in themselves, requiring no justification for their existence. Yet, there is a utilitarian reason as to why we must insist. We cannot hope to bring about the needed transformation in values while hanging on to the age-old segregation of humanity within fenced territories. Even if all would be granted the same opportunities, rights, and range of salaries, it won't work. Been there, done that, with the separate-but-equal doctrine. The us-versus-them paradigm underlies the more colorful and sordid aspects of our history. No, given the nature of the planetary crisis, it seems that we will rise or fall together.

Yet, if we brought down all borders today, tens of millions of people might descend on San Diego from Haiti, Burundi, and Niger, forming hundreds of favelas around it, à la Rio de Janeiro. Thus, given the vast socio-economic gulf currently existing between various regions, a transition period is called for to level the field some.

In accordance, in the first phase under this proposal, the world is to be divided into nine broad blocks made up of countries of comparable socio-economic profiles: the equivalent of nine regions each with an internal setup akin on the whole to the European Union's. Within each region, the inhabitants would have the right of unrestricted residence, work, and movement. Those wishing to emigrate or move to countries outside their region, may or may not do so, subject to the immigration policies of each country, much as it is at the present time. International travel may be subject to health-related requirements and possible restrictions, and it may entail things such as vaccinations and medical exams.

The proposed transitional division is indicated in the following table:

FIGURE 13.3

transitional phase: division of the world to nine socio-economic blocks with unrestricted rights of passage, residence, and work within each
Afghanistan, Armenia, Georgia, Kyrgyzstan, Mongolia, Tajikistan, Turkmenistan, Uzbekistan
Albania, Azerbaijan, Belarus, Bosnia and Herzegovina, Croatia, Iran, Kazakhstan, Macedonia, Moldova, Montenegro, Northern Cyprus, Russia, Serbia, Turkey, Ukraine
Algeria, Chad, Djibouti, Egypt, Eritrea, Ethiopia, Gambia, Gaza Strip, Guinea, Iraq, Jordan, Lebanon, Libya, Mali, Mauritania, Morocco, Niger, Senegal, Sudan, Syria, Tunisia, West Bank, Yemen
Angola, Benin, Botswana, Burkina Faso, Burundi, Cameroon, Cape Verde, Central African Republic, Comoros, Rep. of the Congo, Dem Rep. of the Congo, Côte d'Ivoire, Equatorial Guinea, Gabon, Ghana, Guinea-Bissau, Haiti, Kenya, Lesotho, Liberia, Madagascar, Malawi, Mauritius, Mozambique, Namibia, Nigeria, Rwanda, Sao Tome and Principe, Seychelles, Sierra Leone, South Africa, Swaziland, Tanzania, Togo, Uganda, Zambia, Zimbabwe
Fiji, Kiribati, Marshall Islands, Federated States of Micronesia, Nauru, Palau, Papua New Guinea, Samoa, American Samoa, Solomon Islands, Tonga, Tuvalu, Vanuatu

transitional phase: division of the world to nine socio-economic blocks with unrestricted rights of passage, residence, and work within each

Antigua and Barbuda, Argentina, Barbados, Belize, Bolivia, Brazil, Chile, Colombia, Costa Rica, Cuba, Dominica, Dominican Republic, Ecuador, El Salvador, Grenada, Guatemala, Guyana, Honduras, Jamaica, Mexico, Nicaragua, Panama, Paraguay, Peru, Saint Kitts and Nevis, Saint Lucia, Saint Vincent and the Grenadines, Suriname, Trinidad and Tobago, Uruguay

Australia, Austria, Bahamas, Bahrain, Belgium, Brunei, Bulgaria, Canada, Cyprus, Czech Republic, Denmark, Estonia, France (all six parts), Finland, Germany, Greece, Hong Kong, Hungary, Iceland, Ireland, Israel, Italy, Japan, South Korea, Kuwait, Vatican City, Latvia, Liechtenstein, Lithuania, Luxembourg, Malta, Monaco, Netherlands, New Zealand, Norway, Oman, Poland, Portugal, Qatar, Romania, San Marino, Saudi Arabia, Singapore, Slovakia, Slovenia, Spain, Sweden, Switzerland, Taiwan, United Arab Emirates, United Kingdom, United States

Bangladesh, Bhutan, India, Maldives, Nepal, Pakistan, Sri Lanka

Cambodia, China, Indonesia, Myanmar, North Korea, Laos, Malaysia, Philippines, Thailand, Timor-Leste, Viet Nam

Note: A color-coded map illustrating this is available at www.danielrirdan.com

This interim division of the world seems to make sense for the reason stated above. Yet, perhaps not to the extent one may fancy. The majority of the world's poor (72 percent) live in middle-income countries, that is, countries with an average yearly wage of between about USD 1,000 and USD 12,000. Middle-income countries have twice as many people living in grinding poverty than those in low-income countries.[27]

Interim measures have a way of becoming permanent, and this won't do here. Hence, collectively we have to do what it takes to empower those in poverty-stricken regions because, ready or not, the regional borders ought to come tumbling down fifteen years from the time they would be set in place. After fifteen years, the last of the corrals segregating us would be dismantled. And a few decades afterwards, a new generation would come to its own, likely to find the concept of restrictions on one's personal freedom of movement and residence…incomprehensible.

Earth does not belong to anyone in particular, and no one can lay claim to own parts of it. This is one of the premises of this plan. What one *can* lay a claim to is one's personal possessions, one's place of residence and the grounds around it. A person may lay claim to the fruits of his labor on a given land, but one cannot otherwise assume that a tract of nature is his personal fiefdom.

Any residential house and the grounds around it would remain in the possession of the current owner. Everything else is to come under the stewardship of one government agency or another. Non-degraded land may acquire a status akin to the contemporary US designation of a National Wildlife Refuge. Degraded land would be leased out free of charge, as to those applying to use it in a model comparable to community gardens. That is to say, the stewardship of the land in question is passed on to an individual or a group for an indefinite period under broad guidelines for its use. Subsequently, those who lease the land could maintain stewardship for this land as long as its utility and practices stay within prescribed guidelines. The intent is to accommodate, whenever feasible, existing land-use claims over land tracts—whether by farmers or by manufacturing outfits.

The new economic setup would do away with the whole class of institutions and people obtaining riches by shuffling around piles of money as they contract people to build and then rent out the units to tenants that have no access to piles of money. It would do away with banks generating money from thin air, loaning it, and then charging interest for decades to come. In fact, in the context of moving into a housing unit, there shouldn't be a need for a loan. One just needs to make the monthly payments.

The rent and mortgage are proposed to be replaced with something that incorporates a bit of both. A person would make monthly payments. If they stick with it, the housing unit becomes theirs, free and clear. If they move out and otherwise stop paying, they retain a fractional share in the unit. Say they paid $7,000 on a $70,000 home. Their fractional share is 10 percent. Like now, houses can be bought and sold subject to supply and demand pressures. Once the property is sold, they would get 10 percent of the proceeds.

One apartment unit may cost $50,000 to construct, and assuming it has a 100-year lifespan, the monthly fractional portion of payment is $42 a month. In the case of a house, if the construction is $75,000, the house could be repaid within fifteen years with monthly payments of $625. What accounts for the relatively-low monthly payments hypothesized here is the lack of interest tacked on top of the loan and the cost of the land itself.

EDUCATION AND ACCESS TO INFORMATION

For the most part, our high schools are a cross between daycare and detention centers; while the major triumph of our baccalaureate program in universities is in extending adolescent, carefree living of the middle-class for another four years.

From reading to nanotech engineering, education and training at all levels will be available freely to all. Furthermore, it is not to be restricted to any age group. This means that anyone can learn to become a physician or get trained to cultivate the land without incurring a tuition fee. In that regard, Finland comes to mind. Its students have been turning in some of the highest test scores in the world, while no institution of learning in Finland—from kindergarten to graduate programs at university—is charging tuition. It is telling that the Finnish school system has eschewed national, standardized tests and has placed far more trust in the judgment of educators and principals. It is also telling that it may be harder in Finland to get a teaching position in an elementary school than to be accepted into a medical program.

Education ought to be integrated into the workplace and society and to be immersed in the working of the natural world. We may find training integrated with work in the fields and construction projects, in the hospitals and railroad track installations, in the architectural offices and movie studios, and engineering outfits. Forms of study will range from on-site, apprentice type arrangements to becoming a junior member of a work team to listening online to university lectures, in the style of those currently offered by Open Yale Courses.

At the ages of twelve through fourteen, young people would be expected to acquire some life skills through hands-on learning. Table 13.4 contains a nascent, rudimentary compilation of a few topics that fit the bill.

TABLE 13.4

strand	topics
land	ecology, permaculture, herding, soil chemistry, restoration
preservation & survival	extensive first aid, nursing, herbalism, cooking, nutrition, health and sex education, family planning, outdoor survival skills, food preservation (e.g., salting, smoking, drying), car maintenance
analysis & planning	inventive thinking, decision making processes (groups & individuals), management (time, goals, people), product lifecycle management
engineering	thermal (heating & cooling), electrical, civil (structural, transportation, construction), manufacturing
misc.	English language, knitting, road grading, electrical wiring, rain harvesting
computers	desktop publishing, programming, keyboarding, Internet know-how
building construction	architecture, carpentry, masonry, electrical work, plumbing
industrial machines	industrial design, machining (both manual and via computer aided manufacturing process involving both CAD and CAM)

Everyone ought to be but a few strokes away from every song, article, and book. Everyone ought to be a few strokes away from every schematic of every machine or invention and from countless lectures of the top people in every field.

Devices such as e-readers, netbooks, and computer tablets offer access to the entire heritage of humanity. Those should be subsidized and made truly ubiquitous. For instance, the Aakash computer tablet, launched in the Autumn of 2011, is sold by the government of India for USD 50. The quality of this first-generation product is questionable, but it does lead the way for having Internet access at a relatively low cost.

Thousands of languages may thrive and be sanctioned in the various regions. Yet, uniformly, a second language ought to be spoken in schools throughout. Universal fluency in one language would allow us to talk to one another and meaningfully connect. English is the obvious choice: It is the most used language in science, and it is the most widely used second language. Along the same vein, every youngster may engage in some public work in a country outside their socio-economic block. Thus, a young person from New Zealand may spend six months in Russia, someone from China may live for a while in Botswana, and someone from Armenia may travel to Kiribati. What better way to dissolve bigotry and to discount outmoded fenced borders from a bygone era?

ERADICATING POVERTY

TABLE 13.5

poverty-related world statistics	
people without electricity	1 of every 5
people without access to improved water services	1 of every 8
people without access to improved sanitation services	1 of every 2.5
people below income poverty line (under USD 1.25 a day)	1 of every 5
people who are undernourished	1 in every 7
child laborers (ages 5–14)	1 in every 50

Source: Computations based on data found in *Human Development Report 2010* (UNDP, 2010), 137–211.

Power, sewage, potable water, and health care would be a part of an infrastructure that various governments would make universally accessible.

This aside, the source of all material progress is rooted in four things: a certain set of attitudes, know-how, land area, and natural resources. If those exist in satisfactory measure, then a group of people can construct a flourishing community around them. One by one, in broad brushstrokes, what follows are the highlights of how we may bring these about in the poverty-stricken regions of the world.

Let's begin with the first of the four: a certain set of attitudes. Charity and handouts demean. They leave the recipient with neither a sense of accomplishment nor a sense of ownership over their lives. We all have to own our lives and our circumstances. Treating people as somewhat diminished does them no favors. In this context, the single most meaningful contribution one may make is to empower those in need.

In the 70s and 80s, the Institute of Cultural Affairs (ICA) put into practice this principle, empowering villages in various regions of the world. One case in point is the settlement Maliwada in the state of Maharashtra, India. Within a period of three years from the time ICA got involved, this small rural town with two hundred households had in place extensive new housing, electricity, upgraded roads and drains, clinics, and school for all ages.

Volunteers taught the villagers systematic planning, how to set up community organization, and how to gain access to agricultural training. To train people to become active participants in their own destiny, the very training had to reflect this spirit. At the end of the process, a two-year plan was formulated. In turn, this led to the formation of specific task teams: doctor, farmer, teacher, trader, and builder teams. Each community member was a part of one task team or another in accordance to their interests. Special training programs were scheduled for the village residents on a regular basis. The doctors' task team led a ten-day session on preventive health measures. The teachers' task team conducted weekly sessions, preparing curriculum and training village residents to become teachers.

Once a week, the population gathered for a community workday. Cumulatively, during those days, roads were built, trees were planted, debris was cleared, and children's playgrounds were created, giving rise to a community spirit. Monthly town meetings were opportunities for the total settlement to meet and learn of progress made. Music underscored the celebratory spirit of the meetings.

In the beginning, the small town was mired by factions and tensions. While tensions did not entirely dissipate, they were sufficiently overcome to allow the community to coalesce. Under the plan, the town was arranged into five neighborhoods. This assured the inclusion of the less developed sections of the settlement, who got to have their representative in the village leaders' meeting.

Maliwada was transformed from a dusty and neglected settlement, with donkeys wandering about, into a clean and attractive place. New and remodeled houses were seen in all parts of the town. Waste and debris were kept to a minimum. The residents acquired electricity, piped water, and effective transportation. They underwent changes in their attitudes and self-image. There came about a willingness to plan for the future, to work together, and to take business initiatives. At the end of the three-year period, economic ventures within the town became profitable and patterns of open and informed participatory decision-making were established.[28]

Spirit or attitude is a battle half won. But of course having relevant technical know-how is essential.

Open Source Ecology is a network of farmers, engineers, and supporters that has been designing what they think are the fifty most important machines that it takes for modern life to exist. In the spirit of an open-source economy, the network subsequently makes publicly available the respective designs, schematics, and instructional materials.

The people behind Open Source Ecology have designed an inexpensive cement mixer. They have come up with an inexpensive design for a sawmill. And they have designed an inexpensive compressed-earth block press, which, as its name suggests, produces earth blocks. A local enterprise can download the schematics and then fabricate these brick presses en masse. Locals may check out an earth brick press, much as one would check out books from a library—or may purchase a brick press outright. With 15 people loading up the machine with sand, clay, and silt, it is possible to crank out 2,000 bricks a day.

It is easy enough to envision the potential of a network such as the Open Source Ecology network. Open source technology can encompass talents and ideas from throughout the planet. Such an open source technological database with countless participating industrial designers, engineers, physicists, chemists, and permaculturalists who expressly bring to bear their knowledge and talents can transform the technologies available to those with little purchasing power. It is not just the influx of designs and blueprints that people can upload and make available for the entire world to access, it is the processes of collaboration and on-going improvements that can take place. It also means that part of a job description of an expert physician is to make themselves available. Thus, at any time, myriad expert physicians in every field would be accessible for video conferencing to consult and diagnose health problems. If a general practitioner or a village health-care giver is unable to diagnose a health problem, another physician can sit in on it, virtually speaking, regardless of where they are physically in the world.

Then there is the matter of granting access to land for building homes or growing food.

In the landslide election of 1951, Arbenz was elected president of Guatemala. The feudal legacy of Guatemala left the country with about 2 percent of the landowners controlling 70 percent of the arable land. The centerpiece of Arbenz's program was land reform: the expropriation of large tracts of land, which was to be distributed to approximately 100,000 landless peasants. However, much of that land was in the hands of the multinational corporation United Fruit, which commandeered much of the land for the growth of bananas for its foreign clients. It goes without saying that the United Fruit Company did not care to see its land holdings slip away. With the help of the strongman in a neighboring state, the CIA, and a small army of mercenaries, the United Fruit company saw the land of Guatemala liberated, Communistic forces vanquished, and the feudal system restored.

With the phasing out of corporations and the claim of any group over pieces of the earth, neo-colonialism would come to pass. With all due respect to bananas or coffee beans for the well-to-do, the foremost priority of the local administrations would be to provide people access to land for housing and subsistence.

The last issue is that of access to natural resources. This would still be served best by the supply and demand of the marketplace. Clay or straw are readily available. Copper, steel, aluminum and other key metals can be subsidized at first and thus be more readily available for the poor.

Another source of considerable help may come from microfinancing, or more specifically peer-to-peer microfinancing, a business model
that Zidisha, a nonprofit organization, promotes. People put out
their money, say $25, and choose who it goes to. When the loan has
fulfilled its function, it is repaid. Subsequently, lenders can redeem
the money or put it somewhere else with another entrepreneur. Love
combined with money is a powerful combination, as is proclaimed
by Kiva, another type of microfinancing organization.

GOVERNMENT

In the 1770s, about half of the provinces of Great Britain in North
America seceded.[29] In 1783, with the Treaty of Paris, Great Britain
recognized thirteen of its former provinces as sovereign, independent states. Concurrently, those thirteen new states came together
in an alliance, a confederation. It had many of the trappings of the
present-day European Union with a governing body reminiscent of
the United Nations. The Continental Congress could enforce nothing
and therefore do nothing of substance. Six years later, in 1789, the
country of the United States came into existence with the formation
of a federal government.

Once again in the course of human events it becomes necessary
to consolidate and bring about a broader, unifying government. This
time, we require something that represents the interest of the biosphere and that of humanity as a whole, transcending national dis-
tinctions and interests.

We need a transnational body formed expressly to deal with the
planetary exigencies. To that end, this agency is to have the executive
and legislative powers necessary for carrying out its mandate. This
government is to be independent of and have supremacy over all other
governing bodies. The planetary government should be a permanent
institution. Once we bring humanity together as a whole, we are
likely to leave the divisive old social paradigm for good. And it is not
a moment too soon.

Henceforth, I refer to this new entity as world government.

This proposed world government will set specific environmental-
footprint objectives and accordingly will hem in the various industrial

practices. The world government will coordinate the called-for public projects. It will assume responsibility over the deployment and makeup of energy generation. It will set concrete population fertility goals and undertake related public-relations campaigns.

I do not maintain that this world government will bask the planet in celestial, radiant well-being. I do not even maintain that having this world government may usher in a global milieu of relative peace and prosperity, although it may. I *do* maintain that, by and large, its establishment would be an improvement over the existing political status quo—enough of an improvement to avert a global collapse.

The proposed world government would not rule in lieu of the thousands of existing local governments but in addition to them. The creation of the envisioned world government would not substitute the need for us to discard our juvenile mindset; rather, it is *predicated* on such personal initiatives by the majority.

The proposed world government will be made up of twelve ministries listed in table 13.6 and from other agencies of more limited scope.[30]

TABLE 13.6

ministry	a non-exclusive list of overseeing functions and reach
Biosphere Well-Being	rewilding initiatives, natural parks, well-being of ecoregions
Food Production	agriculture and livestock
Goods & Services	environmental footprint on economic activities and industry
Nutrient Recycling and Water Usage	sewage system, water system
Climate Action	carbon negative economy (carbon capture, sequestration, etc.)
Fertility Management	two-child policy campaign
Energy	power generation and related infrastructure
Natural Resource Management	inventory of natural resources, footprint allotments
Economy	mechanics of the economy

ministry	a non-exclusive list of overseeing functions and reach
Education & Information Dissemination	environment, science, energy, engineering
Labor, Health, and Social Security	universal healthcare, welfare services, employment services
Transportation	infrastructure, public transportation, construction standards, vehicle registration

The heart of each ministry will be a dynamic data set, filled with schematics, blueprints and time tables. This ever-evolving body of information will span the gamut from the global to the local. It will be the product of countless private citizens, thousands of research teams, task teams, local governments, and enterprises throughout the planet. And in a sense, the first cursory draft of these schemes has been drawn and makes up this book.

There is no need to wait for anything else to take place; the expansion and improvement of the blueprints could and ought to commence immediately. The more concrete and technologically viable the plans are, the easier it would be for the public to cast aside the existing state of affairs and elect the new ticket.

The intent is to have the draft that makes up this book be elaborated and improved upon, open-source style, as numerous professionals make their contributions. Some of the schemes laid out here may end up being implemented largely as they are envisioned here; others may be changed considerably. Nothing here merits being cast in stone—except the objectives and stated principles. Once world ministries come into being, they would assume a leading, administrative charge over the respective blueprints.

Each of the twelve ministries would be led by a team of three ministers, working as a tightly knit team and arriving at decisions consensus-style. Thus, altogether twelve teams of three ministers each for a total of thirty-six ministers. There is no need for any separate, additional coordinating body. As needed, the ministers would get together on matters that require coordination, collaboration, or budget prioritization across the ministry spectrum. Each leadership team of three ministers may structure their ministry as they see fit.

And we need ministers who are leaders: those who stand tall enough to peer into the distance, sagacious enough to comprehend possible ramifications, comfortable enough to surf rather than pretend to control the chaotic, uncertain currents of reality, and courageous enough to make the tough decisions.

A written charter would define the basic nature of all governing bodies and set relevant bounds on the world government in particular and on all governments in general. It is to be comprised of three tenets. The first tenet asserts the ecological integrity of the Earth and its life diversity to be paramount and made secure in perpetuity. All living things are henceforth to be treated with heartfelt consideration. The second tenet is a mandate to promote the wellbeing of people and to provide or make accessible life's material necessities to all. The third tenet recognizes and honors the innate dignity of each individual and their inalienable right to be free from oppression.

This charter notwithstanding, one may wonder what would hem in the powers of the world government and prevent it from becoming overbearing. Controlling the consumption of marijuana in one's home would have nothing to do with the vested powers of the envisioned world government—any more than it has to do with the powers vested within the US federal government. And we see how well *that* worked out in the United States. In fact, the expanding power of the US federal government and its charter, the Constitution, are a good case in point.

The US federal government was mandated to establish and regulate currency and the postal system; it was mandated to fend off or maintain relations with foreign powers; and it was mandated to regulate interstate and international commerce—creating a cohesive market economy. In sum, the US government was formed as an overarching apparatus to bind together the federated states in an effective union. The federal government was not assigned other duties; local functions and discretionary powers have been reserved for the individual states or people, whichever the case may be. By the basic charter that defines it, the federal government could not take upon itself additional duties short of a due constitutional amendment. While the US Constitution was never amended in that regard, through the years the federal government broadened its authority all the same. The attempt to set in place a constitution as a bulwark against the expansion of the US federal government has failed.

There is no reason to think that the charter would manage to hold in check the world government any more than the US Constitution hemmed in the federal government in the United States—but for four things that are arguably different enough to matter. First, the proposed world government would not have the equivalent of a legislature. When you have hundreds of people whose only job is to sit around and come up with new laws, it is just a matter of time until they legislate everything that moves. Second, there would be a combination of radical transparency in government proceedings coupled with a speedy, straightforward procedure to send a minister packing if they exceed their mandate. This is to be discussed a bit later. Third, without mega corporations, whose reason for existence is to amass money, and without select individuals commanding ungodly amounts of resources and power, much of politics as we have known it would not exist. Fourth, the election process of world-government ministers, which will be described shortly, is profoundly different from popular elections as we know them. Among other things, this election process is likely to result in candidates with integrity on an entirely different scale from that of contemporary office holders.

And it still may not suffice. But then again, do consider the alternative of sticking to our current trajectory. In all, we have much more to gain than to lose by unifying the world in the manner described.

Currently, there are tens of thousands of governments throughout the world, from the smallest municipality to the largest of federal governments. For the most part, they would remain in existence in one capacity or another.

Municipal and county-size governments will play a similar role to that they have now: maintaining local roads, zoning, fire protection, sewer services and managing local water supply, building permits, and such. The local governments would be, much as they are now, the stewards of the land within their jurisdictions, while larger tracts like continental migratory corridors and full-scale rewilding initiatives, would be spear-headed by the world government.

Insofar as a national and ethnic identity is concerned, countries have always been lousy containers. At present, there are close to two hundred sovereign countries, but there are thousands of nations. If anything, dismantling fences and overbearing regimes would allow many of the downtrodden nations and ethnic groups to acquire their cultural independence. This would also allow those who wish to practice alternative lifestyles to do so without fear of persecution. The new political paradigm may very well usher in a renaissance of cultures, lifestyles, and nations. Under the new setup, many territorial disputes will cease to have any real relevancy. The Kurds in Syria would stop being harassed and could reconnect, for that matter, with their brethren in the territory of Turkey. And the people of Tibet would live free of the Chinese yoke.

One of the important roles of a regional government would be the preservation of the local heritage, customs, and flavor of laws specific to a given territory—at least as long as it does not include stoning adulteresses to death. The local government would maintain the official language and determine the school-curriculum subjects that by their nature are specific to the given territory such as literature and history. In the territories of Japan or the United States, cultural preservation is likely to be largely in the hands of the national government. In the territory of Canada, it is likely that the Quebec government will have cultural jurisdiction over Quebec. In Papua New Guinea, this task is likely to befall local tribes.

Global measures such as new power grids or the deployment of carbon-capture machines would be orchestrated and done in coordination by government at all levels. In the case of a power grid, fourteen regional governments would be formed, each in charge of the particular makeup of each of the transcontinental grids. The regional administration, such as that of the Greater Horn of Africa area (see a corresponding map in the seventh chapter ("Energy") will in turn involve and work in cooperation with the many hundreds of localized governments through which the power lines traverse. In this particular example, state governments may or may not have any role.

Those have been the highlights of all functions of the governments. As is obvious, state (country) government would get the biggest makeover. They would have far fewer functions to perform than they do at present. For one, they wouldn't be in charge of money supply nor would they maintain a military. Universal health care may be managed by the newly formed administrations overseeing the nine socio-economic blocks described earlier in this chapter, or it may be managed by each national government for their respective territories.

In order to take on the complex planetary issues at hand, all governments, from the local to the planetary government, ought to be re-forged. They would need to be highly participatory, lug about relatively few statutes, and be propelled by firsthand thinking.

All government data will be publicly made available online and be updated continuously. Radical transparency in all government proceedings is a given, but this would only be a starting point for inviting the public to be active collaborators.

Instead of relying on relatively few government employees, millions can be the eyes and ears of one agency or another, texting to designated websites about anything from a pothole they just passed to an overflow of a creek in a place they were hiking. In a wiki style, many government documents would be living documents on which people would be called upon to comment, improve, or work. We do need leadership and final decision makers, yet, millions can and would be expected to lend their insights and expertise. Most sessions would be broadcast live on the web. The public can chime in, help develop a given issue, resolve problems, and make their voices heard in real time. In that context, I will mention the *Deliberatorium*, a software tool developed by The MIT Center for Collective Intelligence. Using such a tool, the public would be invited to insert their input into a tree-like network of posts, each representing a single issue—which can be a question, an idea, or an argument. Crowdsourcing works.

Having said that, it cannot come instead of leadership and decisiveness. It cannot be counted on for decisions that are emotionally arduous, multi-generational in ramifications, or entail in-depth consideration that only years of work can muster.

More will be required from everyone: more from the public, more from the business community, more from the office holders.

GETTING THERE POLITICALLY

I call on the formation of a volunteer, grassroots organization (www.getreal.info). The organization would have two objectives: further develop and expand upon the technical schemes laid out here, and initiate the process of forming a world government, which would start on implementing them.

Anyone who has followed a reality television show, such as *America's Got Talent*, may note that discriminating judges in collaboration with the public are shown to be astute in identifying the most talented individuals from a big pool of candidates. In some ways, these reality shows have been a source of inspiration for the election process to be delineated. There are endless ways to appoint elected officials. What is outlined here is one route, as good as any and better than most.

The months-long or years-long selection process of the ministers is to be a rallying point for the entire worldwide transformative thrust. It will be as much of a media event as anything else.

At the beginning of the process, all applicants for the positions of ministers will go through a very rigorous background check ascertaining, to the extent possible, the integrity of each applicant.

Those who make the first cut will go through a series of hearings in front of a screening committee. Members of the screening committee will not elect anyone to office; the public will. Rather, the committee would conduct the initial screening of candidates. Much as with the United States' Supreme Court nominees, each candidate would come in front of the committee and go through a series of publicly-viewed hearings that will last as many hours as is deemed necessary, with the committee grilling the candidate to a great depth on a wide range of pertinent topics. Platitudes and a full mane of hair would count for nothing; the candidates will have to deliver. At the end of the hearings, the relatively few who make the cut will advance to the next stage and be eligible for public voting.

The grassroots organization will elect members of the screening committee, nine members in all. They are to be very highly accomplished individuals, collectively bringing to the table a wide array of disciplines and experiences. To maintain a level of credibility and trust, the potential committee members will have to meet any one of certain internationally-respected achievements and standards.

Those who are or have been supreme-court judges would be eligible to be members of the screening committee. Laureates of the Nobel Prize, Albert Einstein World Award of Science, Right Livelihood Award, Shaw Prize, National Medal of Science, and Ramon Magsaysay will also be eligible. That is to say, potential members of the screening committee will be people who have made an outstanding contribution in improving living conditions and science in general. Other eligible members for the screening committee may be retired high-ranking generals and other people who have had experience orchestrating the actions of many thousands or millions of people. In addition to these, eligible candidates for the screening committee will be the foremost thinkers and visionaries of our time. In accordance, the grassroots organization will look at those who have been TED speakers, those who made it to the FP Top 100 Global Thinkers, winners or finalists at the Bucky Fuller Challenge, Pulitzer Prize winners for non-fiction books and the Miguel de Cervantes Prize for lifetime achievement of outstanding writers. In addition, the grassroots organization will consider those who are held in the highest regard in the realm of speculative fiction, the winners or finalists of any of the most prestigious science fiction literary awards: Nebula, Hugo, Robert A. Heinlein, and BSFA. As stated, the screening committee will be made up of nine individuals. Given the number of nominees required for the twelve ministries, the pool of screening committee members may end up in the hundreds and broken in turn into multitudes of screening teams, to expedite the process.

If the screening committee devoted many hours for grilling each and every aspiring ministry candidate, we would be at it for years—as the committee may end up going through hundreds if not thousands of hearings. This is where local chapters of the grassroots organization will prove invaluable. Those volunteers would set up open days in numerous locales throughout the world, at university campuses, high schools, churches, and auditoriums. Anyone and everyone can show up for a brief videotaped interview accompanied by a questionnaire. The screening committee would coach and instruct the local volunteer chapters on the desired level of screening and the nature of the interview The material on the more promising aspirants would be forwarded to the screening committee. After reviewing the videos and questionnaires, the screening committee would decide who to invite for a second interview. This time, the interview is to be with the committee itself. Members of the committee would tour the population centers of the world, stopping a few days in each location. In the first few days of each stopover, the committee would go through all those who made it to the second interview. In the last few days of each stopover, the committee would invite for a third round the most promising aspirants, who would go through the in-depth hearing described earlier.

It is too much to expect that the public will go in-depth through the portfolios and consider in earnest more than 30 to 40 candidates for each of the ministries. In accordance, a cap would be set at 35 possible eligible candidates for each ministry. This means that if the pool of nominees for a given ministry is already 35 strong, the screening committee can only allow an additional candidate by ejecting one who is on the list. This is not a bad thing. If the slate is filled and more applicants make it through the hearing stage, the screening committee could bump one or more people currently on the roster for the benefit of those deemed better suited. As the election will take place over a protracted period of time, if a given candidate has already garnered more than a minimal amount of votes, say 5,000, their place is secure within the pool of 35 candidates for the given ministry.

Each of the possible thirty-five nominees for each of the ministries would be given the same format and level of exposure. Each would have a dedicated web page containing a standardized curriculum vitae and links to their works. Their web pages would also feature movies that capture their lives, pertinent views, and work. Finally, the web page would also include clips of highlights from their hearings. From the moment a person becomes a candidate, he would not be allowed to publicly talk or write about his candidacy or platform. The only source of information would be his official web page. This measure is set in place to lessen, if not prevent, people from skewing the election results via better public relation efforts.

And then it is voting time—as much for the ministers as for the makeover of our society as a whole. There would be two election phases. The first will narrow the pool from the possible 35 candidates vetted by the screening committee for each ministry down to 12. The second phase will bring it from 12 down to the final 3-minister team for each of the 12 ministries.

Anyone who is eighteen years of age or older would be eligible to vote. With due authentication, people will cast their votes online. One person, one vote, or rather, twelve votes—one for each of the twelve ministries' candidates. Nothing much would happen until at least half a million people have cast their votes and thus in effect pledged their support to broadly pursue a world government and in principle the blueprint presented in this book. The time lapse between the first voter to the 500,000th voter could very well end up being measured in months. Once the first threshold is reached and half a million voters cast their votes, the votes will be tallied, and one official candidate will make it through to the next round for each of the twelve ministries. This would comprise the first election cycle. (See table 13.7.)

After that, the system waits for an additional 1 million people to cast their votes. Once this takes place, each ministry would get another official candidate. That is, by the end of this second cycle we would have a total of 2 official candidates for each ministry (24 total official candidates). At this rate, we would need 12 cycles to bring in a total of 12 official candidates for each ministry. Not to prolong the process more than needed for gaining popular support, things pick up speed a bit once we hit the fifth cycle. For the fifth election cycle to take effect, 30 million people will have to cast their vote, this time bringing on board 2 additional candidates for a total of 6 official candidates for each ministry (72 total official candidates). This will continue in such a manner until with the eighth and final election cycle, each ministry would have 12 official candidates, 144 official candidates in all.

TABLE 13.7

election cycle number	number of voters required for a given election cycle	number of elected official candidates for each ministry
1	0.5 million	1
2	1 million	2
3	3 million	3
4	10 million	4
5	30 million	6
6	50 million	8
7	100 million	10
8	(see requirements below)	12

For the results of the last cycle to be tallied, where the twelfth, final official candidate for each of the ministries will be chosen, a few criteria have to be met. In tallying the combined voting results of all 8 cycles, at least 2 percent of the eligible voters in every socio-economic region would have had to cast their votes; the total number of votes would have to be no less than 400 million; and at least one socio-economic region would have to have no less than 25 percent of its adult population cast their votes.

Subsequent elections in the future will not have to go through those stages. The twelve finalists would be chosen all in one day by the world population.

Then we get to the fun part. The twelve official candidates for each ministry will be split into four teams, each with three members. The teams will be put through their paces for an intensive month or so in a series of simulations to see how they meet the responsibilities of the office under fire. Think of a setup akin to that of a reality show. This would allow the public to get a real sense of the candidates and move past possible prejudices and sentimental reasons—such as skin color or country of origin. Throughout this month, the twelve candidates for each ministry will be placed in as many team combinations as would be deemed feasible by the screening committee.

That would be the second and last phase in the election process. From all the team combinations in the month-long test run, the public would choose the final three-person team that will head each ministry.

Under the current state of affairs, heads of state seem to organize much of their country's agenda around their ambition to get re-elected, and legislators seem to spend as much of their time fundraising for their next term as doing anything else. This is an atrocious state of affairs that the world government could ill-afford. Ministers would serve for one term only, seven years long. To prevent a situation that every seven years the entire cabinet resigns its commission, and to assure the stability and continuity of the government, subsequent ministry election would be staggered. Every two years, three ministries would be up for a vote. Naturally, the very first election cycle would not lend itself to this scheme, as the entire cabinet has to be elected concurrently. Hence, as a onetime measure, various ministries will have their team for three-, five-, seven-, and nine-year terms. This would allow future ministry elections to be staggered.

A safeguard is in order, a procedure that would allow the public to boot out of office a given minister through a vote of no-confidence, which will force the given ministry to start a new election process for a new three-minister team. What is crucial here is to set in place a procedure that from one side genuinely works, is neither frivolous nor too byzantine, and from the other side does not overshadow and dog an official's agenda.

Once every four months, each of the ministry teams would come in front of a jury one hundred members strong. Twelve ministry teams, twelve jury groups. The purpose of these day-long hearings would be to discuss matters related to a possible no-confidence. Each of the ministers would have a chance to take personal responsibility for some issues or make amends. During the hearing, there will be a significant period of time in which people from throughout the world can call or write in and ask any questions or make comments pertinent to matters at hand. At the end of the deliberations, if a few jury members feel that a no-confidence vote is in order, they can initiate one. If a majority of the jury supports the motion, the ministry team will be relieved of duty. This would be followed by a new election for that ministry.

Members of the jury would be unknown to each other until the day of the hearing. They would be chosen randomly from a pool of tens of millions of registered voters from throughout the planet who have been screened as having no apparent conflicts of interest and no criminal background. Eligible jurors will also have to pass a skill and knowledge examination that would attest to them having at least *some* understanding of the technological and logistical issues the ministry in question grapples with. In addition, eligible jurors would have to attest for possessing basic familiarity with the current operation of a given ministry.

One of the problems with laws is that once they are in the book, it is very hard to get rid of them—much as it is hard to cancel something one buys a subscription to. Thus, once a year, that is to say once every three jury hearings, instead of the regular jury proceedings, a special jury would be picked out, from a pool of professionals. This annual special hearing would be reminiscent of a house-cleaning or of a tax audit. The relevancy and merit of countless laws would be challenged and publicly debated. In the end, it would be up to the ministers to make the call of what they want to do with the discussed statutes—keep, revise, or annul them. Regardless, their offices would stand to greatly benefit from this intense outside scrutiny and debate.

———————

By definition, by the time the final election of ministers would take place, hundreds of millions of adults throughout the globe will have said yes to embarking on a new political, economic, and social paradigm. Yet, this is not enough. The tanks and tear gas are not in the hands of a social movement, not even if it constitutes the overwhelming majority. More is needed to get the existing state governments to relinquish much of their power.

The existing state governments come in two basic flavors that are relevant here: autocratic and democratic. Prominent autocratic countries include China, Indonesia, Iran, Malaysia, Russia, Saudi Arabia and Pakistan. Prominent democratic countries include Australia, India, the United States, Japan, Germany, the Philippines, and Brazil. As defined here, democracy is a system in which both the chief executive and legislature positions are filled via contested elections. In autocracy that is not the case—although oftentimes there is a charade to the contrary.

TABLE 13.8

	autocratic	democratic
number of countries	88	81
world population, percentage	51%	49%
claimed land area, percentage[a]	52%	43%

Source of data: Adapted from and largely based on Adam Przeworski et al., *Democracy and Development: Political Institutions and Well-Being in the World, 1950–1990* (Cambridge, UK: Press Syndicate of the University of Cambridge, 2000). This tally does not include the smallest of countries.
[a]Antarctica makes up the difference.

While neither autocratic nor democratic regimes are really participatory—the populace has no real say in almost anything—democratic regimes do lend themselves to reform in one respect: if in the due process of election, a vast majority of people vote for a party whose centerpiece is the establishment of a world government, it would be done. With a large enough majority, it is possible to amend the constitution of or to have due referendum in a given democratic country, as the requirement may be. In other words, the accommodation of a world government by a local democratic system can all be done by the book without a need for a single mass protest or civil resistance. The question is what it would take for people to support this blueprint. Therein lies the true challenge.

For starters, there are those with vested interests who will do all in their power to see things remain as they are.

When we average the wealth of adults across the socioeconomic spectrum of the world, we find that the 120 richest people have as much money as *170 million* average adults. In fact, these 120 modern-day pharaohs command as many resources and exert as much economic influence as all the 560 million adults of India combined—rich, poor, and otherwise.[31] The gaping disparity is not restricted to the ultra-rich. In the United States, the top 10% have 80% to 90% of stocks, bonds, trust funds, and business equity, and over 75% of real estate (excluding primary place of residence). Since financial wealth is what counts as far as the control of income-producing assets, we can say that just 10% of the people own the United States of America. The net worth of the bottom 40% is but 0.2%.[32] Thomas Dye identified 7,314 individuals who controlled almost three quarters of US industrial assets as of 2002. Those individuals also commanded over half of all assets of private foundations and universities and controlled the television networks, the national press, and the majority of newspaper chains.[33] These moguls may not necessarily be any more happy than anyone else, but they are *very* comfortable with the status quo. And so are numerous hopefuls. For every person who has gained entry to the platinum club, there are a million others dreaming about striking it rich and rising above the masses to where the air is rarified and tingles with decadence and power. All of those people reckon that when one is in pig's heaven, or trying to gain entry, one ought not to make waves.

Let's not mince words. There is a snowball's chance in hell of success in gaining popular support if the people will not muster great resolve to affect a change and will not work up the courage to think firsthand on the matters at hand. Everything depends on these two things.

ConocoPhillips reported a 66 percent gain for the fourth quarter of 2011, earning during that period USD 3.4 billion. Royal Dutch Shell reported a profit of USD 31 billion for 2011, up by 54% from the year before. Business is *good*. The top coal, oil, and gas companies are valued at USD 7,420 billion. They are valued at that level due to the fossil fuel reserves they are expected to extract and burn in due course.[34] This is a lot of money, and they are not just going to walk away from it.

It is likely that once and if the world unification movement gains a wide hold, corporations and other powerful bastions of the status quo would come together in a broad coalition and launch what would be the most astute and comprehensive public-relations and disinformation campaign in their illustrious histories. After all, these entities would be fighting for their very survival. To that end, they would secure the services of the best—and the worst—humans that money would still be able to buy. Top guns from throughout the Wild West are likely to flock to the cause: spin doctors, lobbyists, and experts-for-hire. As Allan Brandt stated, the best public-relations work leaves no fingerprints. And this is likely to be the finest of public-relations campaigns. Think-tanks and sincere-looking outreach programs that serve as fronts to multinational corporations would cultivate doubts, urge balanced cautious views, and be seemingly hard at work to solve the looming planetary crisis. Corporate-sponsored university seminars would extoll the virtues of the status quo. Well-funded grassroots movements and citizen advocacy groups that smell of home-baked apple pie will spring up. If the future would be but an extension of the past, there would hardly be a need to engage in any substantive discussion on the issues at hand; the upholders of the status quo would just need to win the battle of perception. And the prospect is grimmer yet. The powers to be are likely to either try to hijack any genuine grassroots organization from within or, as a last resort, give backing for the formation of an alternative citizen world movement, one with far more glitz and far less substance. Some of their talking heads may agree with the technical conclusions arrived at here but then regretfully proclaim that it is too late for any true mitigation measures. Instead, they would call for adaptation, that is, hunker down and circle the wagons rather than spend money we don't have and risk bringing down the borders and chance having the riffraff come in. Their campaign would have the home-field advantage. In the end they don't need to prove anything; they don't need to get anywhere; and they don't need to bring about a change. Come Monday morning, they just need to keep the people in the hamster wheels running. And spending.

None of this is to say that people would not initiate on their own any or all of the above in opposition to the proposed changes—notably to that of a world-government formation and global desegregation—without the support of the establishment.

In the eighty-eight autocratic countries, the prospect of change is different, in some ways harder, in some ways easier. We cannot vote our way into those places. We have to persuade the local junta to step down or otherwise relinquish power. There are three recent cases in the 21st century of just that: the 2011 Egyptian Revolution; the Orange Revolution in 2004–5, Ukraine; and the Rose Revolution in 2003, Georgia. In all of these cases, largely non-violent acts of mass civil resistance, (i.e., protests, strikes, sit ins, civil disobedience) did the trick. This could work in most but not in all autocratic countries.

It seems likely that sit-ins and other acts of civil disobedience would not matter in the least in North Korea, Sudan, Burma, Belarus, and China. The junta of these countries may likely kill their own citizens rather than relinquish power. The 2011 Libyan Civil War has proven that on occasion it is possible to oust a bona fide despot, but at a price of lives lost that no one outside a given country has the right to propose. Alternatively, it is conceivable that a series of financial and technology transfer incentives would persuade the despots to implement the gamut of sweeping measures outlined in this book—provided that it ends up costing less than the alternative.

Whether North Korea retains its isolation matters little to the planet, as North Korea's footprint is relatively insignificant. But what *is* worrisome and of consequence are China and Russia. Their governments hold under sway about 20% of the world's population and about 18% of world's land area—and account for quite a bit more than that in terms of their share of environmental impact.

At the end of the day, there are just us, people. This is our society, our world, and our children and grandchildren that will make a life here. And if we want to make a change, it will happen, as there is no one outside to stop us.

I have no patience left in my bones for the notion that many people are self-serving, short-sighted, or spineless. While this may be largely so at the present, we *will* lead meaningful, adult lives—given conducive circumstances.

Don't ask if people are too reactionary and self-absorbed for the needed measures, ask yourself if you are. You are not a race unto yourself, apart from the rest of humanity. If you've got what it takes when the chips are down, others may as well.

It all starts with the here and now: Seven billion passengers huddling in a big dilapidated warehouse that is our cultural and political institutions. We peer from the smoky windows at the meteorite sweeping down. We note its descent but are unable to let go of the pillars that have been holding up what we have come to call home for the last few thousand years. The warehouse may reek, be infested with cockroaches, and sport leaks—but this is all we have ever known.

It comes down to a choice: we either leave behind the current economic-political paradigm, with its obstructive constructs, or we go down with it. Thus, more than anything else, the blueprint as outlined in this book is predicated on you, on your willingness to act with moral courage and fortitude.

Reclaim our race and give our kids a chance for a future that you and I did not have. Let this be the century in which we stood as one and changed the course of humanity.

APPENDIX A

≍

rooftop-mounted PV panels and thin-film PV:
why they don't add up

ROOFTOP-MOUNTED PV PANELS

As a postscript note, I will touch upon something that did not quite add up, but is likely to be on your mind: PV rooftop installations. I, for one, wondered about them.

To do their share under the North America power plan, photovoltaic installations need to kick in about 3,783 TWh a year worth of electricity.

Roof-mounted units are fixed and do not track the sun, consequently they lose 60–90 minutes of sun input in the morning and another 60–90 minutes in the evening. Ground-based units with trackers have, on average, 6 productive hours during the day versus only 3 or 4 for a roof mounted, immobile unit.

A study assessed the technical potential of PV rooftop installations: had we installed PV panels on *every* viable square meter worth of rooftop in the contiguous United States, it would have provided a grand total of 536 GW capacity. This total roof area accounts for shading, building orientation, and roof structural soundness.[1]

I will assume it is a wash between the meager addition of Canadian rooftops and the added cooling load incurred due to the addition of the dark, absorbent PV panels on the roofs.

I assume that a PV panel with 1 kW capacity would average across the region 1.2 MWh a year. Consequently, 536 GW capacity of every available, eligible roof space in North America would yield 643,000 GWh annually. This is about 17% of the PV share—not of the total energy production—needed under the North America power plan.

The 17% of the PV share can be supplied by rooftops provided that we commandeer every available and viable square meter of roof in North America. Aside from the questionable price in aesthetic, some of this space would be better utilized by solar water-heaters, which present a far more direct and hence sensible way of cashing in on solar energy. Thus, if we accommodate solar water heaters on the roof, the PV share that can be provided to the plan by all the rooftops is less than 17%.

There is more to be said.

With very few silver resources relative to our solar energy needs, I, for one, do not care to see it employed it on the far less efficient, fixed units on rooftops. In addition, their installation and operational maintenance are far more intensive per square foot of PV panels than for massive, ground-based PV farms. Then there is the equipment. Every private roof needs its own inverter, albeit a small one. A PV farm can have a single inverter supporting over 1,500 PV panels. In addition, there is the considerable logistical burden of forming many tens of millions of separate purchasing agreements, subsequently servicing and interacting with owners of tens of millions of PV units and helping them with financing.

Finally yet importantly, if we were to assume that every household would install a PV unit of conventional size (1–3 kW capacity), the amount of electricity produced would fall to insignificant levels. All the talk about 17% was for units that utilize every square foot of viable rooftop area.

THIN-FILM PV TECHNOLOGY

Thin-film PV is less efficient and hence requires more land to produce the same yield. Sure, it is possible to embed thin-film PV panels into large windows of skyscrapers. I have wondered where that would leave us a few decades later, when the PV panels have ceased to function.

Would that mean we would have to replace the windows in millions of buildings? Maybe the answer to this resides with embedding thin film onto shingles, which are to be replaced periodically anyway. But then again, maybe not.

Doesn't matter. Thin-film PV panels don't scale.

As things stand, we need 20 metric tons of indium for each 1 GW installation of copper indium gallium selenide PV panels. With an estimated 6,000 metric tons worth of reserves, a 300 GW installation will wipe us clean. To do its share worldwide, PV technology would require installations with a total capacity of 6,500 GW. It would require twenty times more indium than is to be had. This does not take into account the existing needs of indium: 574 tons a year are used predominantly in the manufacture of LCD displays.

The other flavor of thin-film PV technology, cadmium telluride, does not scale any better. With 47,000 metric tons of tellurium in reserve, at 91 metric tons per 1 GW installation,[2] we will theoretically manage to manufacture only 515 GW worth of installations. Once again, this does not budget the existing annual consumption of about 460 tons used in unrelated industries, predominantly in the fabrication of various metal alloys.[3]

APPENDIX B

≳≪

carbon sequestration; two maybe's

THERE ARE TWO POSSIBLE venues for carbon sequestration with prospects interesting enough to mention in an appendix and not conclusive enough to mention in the book proper.

Each year, between 330 and 430 million hectares of land burn worldwide. This is a combined area half the size of the Australian continent. About 70 percent of it takes place in the African Savannah.[1] The Global Fire Emissions Database v3.1 notes that about 1.45 billion tons of carbon is going up in smoke every year.[2] To this volume of outgassing, we ought to add the carbon released due to the decomposition of vegetation killed by wildfire but not combusted. Well, in the case of savannah, this is not applicable as either the grass is burned away or it remains alive. There is no third option. However, in the case of woodlands and forests, the amount of carbon released due to fire-related decomposition is about double the amount of carbon outgassed due to outright combustion.[3] This brings us to a total of about 2 billion tons of carbon that go up in the air each year.

What I have in mind is to hijack this natural cycle.

The stuff is going to burn anyway, and if not this year then perhaps the next. If we can mimic a wildfire but capture the carbon, then we ecologically can have our cake and eat it too. We can choose an area ripe and ready to be invigorated by a wildfire, mow it down in a pattern emulating the aftermath of a wildfire, and run the harvested biomass through pyrolysis. This may accomplish on behalf of the grassland or forest what a fire service does. Just that in addition, we also lock in the carbon that otherwise would have gone up in smoke. The key is to use the science of fire modeling to predict natural fire patterns and emulate some of it—preempting some of the fires that would have occurred otherwise.

This is the theory, anyway. Hijacking a natural process is rarely successful or truly benign. As stated earlier, the sanctity of the remaining ecosystems is the paramount consideration. In the end, intensive research may endeavor to answer if this idea is ecologically viable. Even a cursory examination reveals daunting complexities and nuances that pose a question mark at every turn.

I suspect that the biomass that is not combusted but just decomposes in the aftermath of a wildfire cannot be put through the pyrolysis process without taking an essential link in the forest cycle away. Evidence suggests that char strewn on the ground in the aftermath of a fire is ecologically essential for the regeneration of the ecosystem. In one study, it was found that in a boreal forest fire, the charcoal that the fire created was very effective at reducing the poisonous effects of various shrubs.[4] If nothing else, this may mean that in a forest biome we need to leave the biochar on the ground and not cart it off to an abandoned coal mine. That is not all; the plot thickens.

Well-done biochar, as it is cooked under pyrolysis, can crumble very easily and disperse as black soot. This may be a problem, or we may find that at least in the case of forests, the subsequent leaf fall and rain activity will incorporate the biochar into the subsoil. The large proportions of biochar found at depth suggest a significant vertical transport.[5] Moreover, it is not only the above ground biomass that is affected in a wildfire. The soil itself undergoes significant changes.[6] In short, we may need to do one better if we wish to emulate the effects of wildfire for the benefit of the ecosystem. In conjunction with harvesting the biomass, running it through pyrolysis, and applying the resultant charcoal to the soil, we may also need to have a real fire going, applying heat to the soil.

If we are to revive the grassland biome, this means the return of the large herbivore herds (see tenth chapter "Nature Restoration"). Large herbivores in turn impede fire progress: their grazing practices maintain short grass,[7] and reduce to a minimum the accumulation of dead, dry stalks.[8] In short, the extent of present-day wildfires is higher than it would have been under a balanced ecology.[9]

Many of the present day fires or their absences are controlled by us—and for reasons that have little to do with regenerative ecology. Only a quarter of the land assessed in one study exhibited natural fire patterns, whether because fires were suppressed or ignited by people.[10] In truth, we cannot use existing fire patterns as a baseline for evaluation of what wildfires will be like once we restore large herbivores to the grasslands of the world and make ecosystem regeneration our paramount goal.

In conclusion, the true potential of emulating natural fires cannot be currently ascertained. Be that as it may, the volume of carbon we need to draw down in the first few decades is far beyond what is potentially possible through biochar and land-use practices. One or two billion tons of sequestered carbon annually is one order of magnitude less than what is called for.

———————

Olivine is the most common silicate in the world and makes up most of the Earth's mantle. It is notably found in the dunite and peridotite rocks, which typically have 60–90 percent olivine content. This mineral reacts with the CO_2 dissolved in groundwater and metamorphoses to another mineral as it sequesters CO_2 in the process. This is the predominant mechanism by which nature has been sequestering carbon from the air for eons. While the rate of the natural weathering process cannot even remotely keep up with our CO_2 emissions, Professor R. D. Schuiling at the Institute of Earth Science in The Netherlands suggests getting the process into hyperdrive. He proposes to grind olivine, thus gaining a far higher surface area, and spread the resultant fine powder over land and in shallow seas—notably in climates most favorable for rapid weathering of the olivine.[11]

When it comes to the rate of CO_2 sequestration, ambient temperature and the size of the olivine grains are everything. Hangx and Spiers on the faculty of Geosciences at Utrecht University maintain that an olivine grain the size of a sand particle would take hundreds of years to reach about 50 percent of the total possible CO_2 that it can sequester. With a silt-sized grain, thirty times smaller (10 microns), the same 50 percent CO_2 uptake would take but a few years. This is all at 25°C ambient temperature; the process is three times slower at 15°C ambient temperature.[12] Schuiling and his associate disagree with this time-frame, arguing that the uptake of CO_2 in natural settings is far more favorable than the alleged lab conditions Hangz and Spiers have based their calculation on. In the natural environment, fungi and lichen secrete acid that attack mineral grains. This is on top of the continuous grain-to-grain collision and scraping typical of the coastal zones.[13]

There are fringe benefits to applying crushed olivine. In regions that suffer from acid rain, it should prove possible to make the forest floor less acidic by spreading a thin layer of crushed olivine. It is possible to use it wherever quartz sand is currently being used: in sandblasting, restoring sand in beaches, and as one of the components in road construction.[14]

The crushing and grinding of olivine-containing rocks to fine clay is going to be very energy intensive. On top of that comes the energy required for blasting, excavating and hauling the rock to the crushing facility, and from there the added energy of transportation of the crushed olivine to suitable regions. Hangx and Spiers came up with an initial estimate, basing it on gold and iron ore mines in Canada. Grinding the rock to grains 10 microns in size would seem to take upward of 180 kWh per ton.

Köhler et al. estimated that roughly one ton of olivine would sequester one ton of CO_2.[15] This means that we will need to blast, excavate, haul, crush, grind, and transport upward of one billion tons of rocks rich in olivine to sequester one billion tons of CO_2.[16] This would require hundreds of gigawatt-hours of power—or markedly less if larger grains would rapidly sequester CO_2 in the natural environment, as Schuiling argues. With trucks capable of hauling 50-ton payloads, it comes to twenty million loads annually—or markedly less with the use of freight trains and barges in some suitable locales. There is nothing trivial about the logistics of this proposed measure. Yet, perhaps there is a better way to disperse the particles. The West African trade wind blows south from the Sahara for a period of about hundreds days each year, carrying and dispersing millions of tons of dust. It is conceivable that during sandstorms, large vertical tubes could blow olivine particles into the air, letting the sandstorm carry them afar.[17]

We will need to know a lot more before we release billions of tons of fine grain minerals throughout the world, some of which may very well be laced with asbsestos. Beyond the obvious health concern is the prospect of dangerously-elevated levels of pH in the waterways.[18]

The ecological implications are suspected but ultimately unknown.

NOTES

1: CLIMATE CHANGE

1. Alfredo Martínez-Garcia et al., "Southern Ocean Dust–Climate Coupling Over the Past Four Million Years," *Nature* 476 (2011): 312–15.
2. Lüthi, D. et al., 2008. EPICA Dome C Ice Core 800KYr Carbon Dioxide Data. IGBP PAGES/World Data Center for Paleoclimatology Data Contribution Series # 2008-055. NOAA/NCDC Paleoclimatology Program, Boulder CO, USA.
3. Name of data set: Law Dome Ice Core 2000-Year CO2, CH4, and N2O Data Last update: 7/2010 (Original receipt by WDC Paleo) Contributor: David Etheridge, CSIRO Marine and Atmospheric Research IGBP Pages/WDCA Contribution series number: 2010-070.
4. Within the land it has been predominantly the forests that picked it up. See Yude Pan et al., "A Large and Persistent Carbon Sink in the World's Forests," *Science* 333, no. 6045 (2011):988–93.
5. Paul N. Pearson and Martin R. Palmer, "Atmospheric Carbon Dioxide Concentrations Over the Past 60 Million Years," *Nature* 406 (2000):695–999; Aradhna K. Tripati, Christopher D. Roberts, and Robert A. Eagle, "Coupling of CO2 and Ice Sheet Stability Over Major Climate Transitions of the Last 20 Million Years," *Science* 326, no. 5958 (2009): 1394–97; Dieter Lüthi et al., "High-Resolution Carbon Dioxide Concentration Record 650,000–800,000 Years Before Present," *Nature* 453 (2008): 379–82.
6. The CO2 emissions make up more than 77 percent of the emission pie, yet they account for but 77 percent of the impact. Pound for pound, the prevalent carbon dioxide (CO2) gets a lot less traction than the more scarce methane, which in turn gets not as much traction as the scarcer yet halocarbon gases. The source of the percentage estimate is this chart: Tim Herzog, "World Greenhouse Gas Emissions in 2005," (Washington, DC: World Resource Institute, 2009).
7. Peter A. Stott, "Single-step Attribution of Increasing Frequencies of Very Warm Regional Temperatures to Human Influence," *Atmospheric Science Letters* 12, no. 2 (2011): 220–27; National Research Council, *Advancing the Science of Climate Change* (Washington, DC: The National Academies Press, 2010), 1–16.

8. Markus Huber and Reto Knutti, "Anthropogenic and Natural Warming Inferred From Changes in Earth's Energy Balance," *Nature Geoscience* 5, (2011): 31–36.
9. Richard Betts et al., "When Could Global Warming Reach 4C?" *Phil. Trans. R. Soc. A* 369, no. 1934 (2011): 67–84.
10. A. P. Sokolov et al., "Probabilistic Forecast for 21st Century Climate Based on Uncertainties in Emissions (Without Policy) and Climate Parameters," *Journal of Climate* 22 (2009): 5175–204.
11. Grant Foster and Stefan Rahmstorf, "Global temperature evolution 1979–2010," *Environmental Research Letters* 6, no. 4 (2011).
12. J. R. Petit et al., "Climate and Atmospheric History of the Past 420,000 Years from the Vostok Ice Core, Antarctica," *Nature* 399 (1999): 429–36.
13. Related citation and due analysis are found in the eleventh chapter of this book ("Drawing Down Carbon").
14. Richard A. Feely et al., "Impact of Anthropogenic CO2 on the CaCO3 System in the Oceans," *Science* 305, no. 5682 (2004): 362–66.

2: LAND

1. J. Tyler Faith and Todd A. Surovell, "Synchronous Extinction of North America's Pleistocene Mammals," *PNAS* 106, no. 49 (2009).
2. Michael R. Waters et al.,"The Buttermilk Creek Complex and the Origins of the Clovis at the Debra L. Friedkin Site, Texas," *Science* 331, no. 6024 (2011): 1599–603; A. Lawler, "Pre-Clovis Mastodon Hunters Make a Point," *Science* 334, no. 6054 (2011): 302.
3. F. W. M. Vera, E. S. Baker, & H. Olff, "Large Herbivores: Missing Partners of Western European Light-Demanding Tree and Shrub Species?" *Conservation Biology Series-Cambridge,* no. 11 (2006): 203–31.
4. D. H. Janzen, "Dispersal of Small Seeds by Big Herbivores: Foliage is the Fruit," *The American Naturalist* 123, no.3 (1984): 338–53.
5. Paulo R. Guimaraes Jr., Mauro Galetti, and Pedro Jordano, "Seed Dispersal Anachronisms: Rethinking the Fruits Extinct Megafauna Ate," *PLoS ONE* 3, no.3 (2008).
6. D. N Zaya and H. F. Howe, "The Anomalous Kentucky Coffeetree: Megafaunal Fruit Sinking to Extinction?" *Oecologia* 161, no. 2 (2009): 221–26.
7. Kenneth L. Cole et al., "Past and Ongoing Shifts in Joshua Tree Distribution Support Future Modeled Range Contraction," *Ecological Applications* 21, no. 1 (2011): 137–49.
8. Rene L. Beyers et al., "Resource Wars and Conflict Ivory: The Impact of Civil Conflict on Elephants in the Democratic Republic of Congo—The Case of the Okapi Reserve," *PLoS ONE* 6, no.11 (2011).
9. As reported in ETIS (the Elephant Trade Information System).
10. Peter D. Walsh et al., "Catastrophic Ape Decline in Western Equatorial Africa, *Nature* (April 2003): 1–3.
11. TRAFFIC, "What's Driving the Wildlife Trade? A Review of Expert Opinion on Economic and Social Drivers of the Wildlife Trade and Trade Control Efforts in Cambodia, Indonesia, Lao PDR and Vietnam," East Asia and

Pacific Region Sustainable Development Discussion Papers, East Asia and Pacific Region Sustainable Development Department, World Bank (Washington, DC: 2008).

12. See the first chapter ("Climate Change").
13. Chris Jones et al., "Committed Terrestrial Ecosystem Changes Due to Climate Change," *Nature Geoscience* 2 (2009): 484–87.
14. Carlos A. Peres, "Synergistic effects of subsistence hunting and habitat fragmentation on Amazonian forest vertebrates," *Conservation Biology* 15, no. 6 (2001): 1490–505.
15. Kathryn E. Stoner et al., "Hunting and Plant Community Dynamics in Tropical Forests: a Synthesis and Future Directions. *Biotropica* 39, no. 3 (2007): 385–92.
16. Kent H. Redford, "The Empty Forest," *BioScience* 42, no. 6 (1992): 412–22.
17. Gregory P. Asner et al., "Selective Logging and its Relation to Deforestation," 2009 *Amazonia and Global Change*, Geophysical Monograph series 186 (2009).
18. Julieta Benitez Malvido and Aurora Lemus-Albor, "Habitat Disturbance and the Proliferation of Plant Diseases," in *Emerging Threats to Tropical Forests,* ed. William F. Laurance and Carlos A. Peres (Chicago: University of Chicago Press, 2006), 165–74.
19. Gregory P. Asner et al., "Condition and Fate of Logged Forests in the Brazilian Amazon," *PNAS* 103, no. 34 (2006): 12947–50.
20. See note 14 above.
21. Carlos Peres and William Laurance, "Synergistic Effects of Simultaneous Environmental Changes," in *Emerging Threats to Tropical Forests*, ed. William F. Laurance and Carlos A. Peres (Chicago: University of Chicago Press, 2006): 81–6.
22. Approximately 837,000 km^2.
23. Derived from analysis of Liliana M. Davalos, "Forests and Drugs: Coca-Driven Deforestation in Tropical Biodiversity Hotspots," *Environmental Science & Technology* 45, no. 4 (2011): 1219–27.
24. Jennnifer J. Swenson et al., "Gold Mining in the Peruvian Amazon: Global Prices, Deforestation and Mercury Imports," *PloS ONE* 6, no. 4 (2011).
25. Elizabeth Barona et al., "The Role of Pasture and Soybean in Deforestation of the Brazilian Amazon," 2010 *Environmental Research Letters* 5, no. 2 (2010).
26. Kathryn R. Kirby et al., "The Future of Deforestation in the Brazilian Amazon," *Futures* 38 (2006): 432–53.
27. Jonatahn A. Foley et al., "Amazonia Revealed: Forest Degradation and Loss of Ecosystem Goods and Services in the Amazon Basin," *Frontiers in Ecology and the Environment* 5 (2007): 25–32.
28. Mark Cochrane, "Fire Science for Rainforests," *Nature* 421 (2003): 913–19.
29. Jos Barlow and Carlos A. Peres, "Fire-Mediated Dieback and Compositional Cascade in an Amazonian Forest," *Phil. Trans. R. Soc. B* 363, no. 1498 (2008): 1787–94.
30. Daniel Nepstad et al., "Interactions Among Amazon Land Use, Forests and Climate: Prospects for a Near-Term Forest Tipping Point," *Phil. Trans. R. Soc. B* 363, no. 1498 (2008): 1737–46; see also note 30 above.

31. Yadvinder Malhi et al., "Climate Change, Deforestation and the Fate of the Amazon," *Science* 319, no. 5860 (2007): 169–72.
32. Eric A. Davidson et al., "The Amazon Basin in Transition," *Nature* 481 (2012): 321–28.
33. Paulo M. Brando et al., "Drought Effects on Litterfall, Wood Production, and Belowground Carbon Cycling in an Amazon Forest: Results of a Throughfall Reduction Experiment," *Phil. Trans. R. Soc. B* 363, no. 1498 (2008): 1839–48.
34. G. Bruce Williamson et al., "Amazonian Tree Mortality During the 1997 El Niño Drought," *Conservation Biology* 14, no. 5 (2000): 1538–42.
35. Liang Xu et al., "Widespread Decline in Greenness of Amazonian Vegetation due to the 2010 drought," *Geophysical Research Letters* 38 (2011).
36. David Ellison, Martyn N. Futter, and Kevin Bishop, "On the Forest Cover-Water Yield Debate: From Demand- to Supply-Side Thinking," *Global Change Biology*, Published electronically December 6, 2011. doi: 10.1111/j.1365-2486.2011.02589.x.
37. Marina Hirota et al, "Global Resilience of Tropical Forest and Savanna to Critical Transitions," *Science* 14, (2011): 232–35.
38. Henk Visscher, Mark A. Sephton, and Cindy V. Looy, "Fungal Virulence at the Time of the End-Permian Biosphere Crisis?" *Geology* 39, no. 9 (2011): 883–86.
39. Curtis A. Deutsch et al, "Impacts of Climate Warming on Terrestrial Ectotherms Across Latitude," *PNAS* 105, no. 18 (2008): 6668–72.
40. Joshua J. Tewksbury, Raymond B. Huey, and Curtis A. Deutsch, "Putting the Heat on Tropical Animals," *Science* 320 (2008): 1296–97.
41. This is a combination of heat and humidity. See Steven C. Sherwood and Matthew Huber, "An Adaptability Limit to Climate Change Due to Heat Stress," *PNAS* 107, no. 21 (2010): 9552–55.
42. It would depend on their species-specific core body temperature and on their body mass.
43. The basis was the hourly humidity and dry bulb temperature values of March 2010, which was determined to be the hottest month of the year. Per the global, fossil-intensive, business-as-usual model of Hadley, mentioned in the first chapter ("Climate Change"), I have assumed an average increase of 9°C to dry bulb temperature for the Amazon basin. Source of weather data: www.freemeteo.com.
44. A. L. Westerling et al., "Warming and Earlier Spring Increase Western U.S. Forest Wildfire Activity," *Science* 313, no. 5789 (2006): 940–43.
45. Martin-Michel Gauthier and Douglass F. Jacobs, "Walnut (Juglans spp.) Ecophysiology in Response to Environmental Stresses and Potential Acclimation to Climate Change," *Annals of Forest Science* 68, no. 8 (2011): 1277–90.
46. Paul E. Hennon et al., "Shifting Climate, Altered Niche, and a Dynamic Conservation Strategy for Yellow-Cedar in the North Pacific Coastal Rainforest," *BioScience* 62 (2012): 147–58.
47. Barbara J. Bentz et al., "Climate Change and Bark Beetles of the Western United States and Canada: Direct and Indirect Effects," *BioScience* 60, no. 8 (2010): 602–13.

48. Board on Life Sciences, *Ecological Impacts of Climate Change* (Washington, DC.: National Academies Press, 2008), 26.

49. Anthony L. Westerling et al., "Continued Warming Could Transform Greater Yellowstone Fire Regimes by Mid-21st Century. *PNAS* 108, no. 32 (2011): 13165–70.

50. Jennifer R. Marlon et al., "Long-Term Perspective on Wildfires in the Western USA." *PNAS*, Published electronically February 14, 2012. doi: 10.1073/pnas.1112839109

51. William R. L. Anderegg et al., "The Roles of Hydraulic and Carbon Stress in a Widespread Climate-Induced Forest Die-Off," *PNAS* 109, no. 1 (2012): 233–37.

52. A. Park Williams et al., "Forest Responses to Increasing Aridity and Warmth in the Southwestern United States," *PNAS* 107, no. 50 (2010): 21289–94.

53. Jim Robbins, "What's Killing the Great Forests of the American West?" *Yale Environment360*, March 15, 2010, http://e360.yale.edu.

54. Jeremiah S. Lwanga, "Localized Tree Mortality Following the Drought of 1999 at Ngogo, Kibale National Park, Uganda," *African Journal of Ecology* 41 (2003): 194–96.

55. Clifford Tafangenyasha, "Decline of the Mountain Acacia, Brachystegia glaucescens in Gonarezhou National Park, Southeast Zimbabwe," *Journal of Environmental Management* 63, no. 1 (2001): 37–50.

56. Wendy Foden et al., "A Changing Climate is Eroding the Geographical Range of the Namib Desert Tree *Aloe* Through Population Declines and Dispersal Lags," *Diversity and Distributions* 13, no. 5 (2007): 645–53.

57. Abdallah Bentouati, "La Situation du Cèdre de l'Atlas dans les Aurès (Algérie)," *Forêt Méditerranéenne* 29, no. 2 (2007): 203–8.

58. P. Gonzalez, C.J. Tucker, and H. Sy, "Tree Density and Species Decline in the African Sahel Attributable to Climate," *Journal of Arid Environments* 78 (2012): 55–64.

59. Mark G. L. van Nieuwstadt, Douglas Sheil, "Drought, Fire and Tree Survival in a Borneo Rain Forest, East Kalimantan, Indonesia," *Journal of Ecology* 93, no. 1 (2005): 191–201; Jamal A. Khan et al., "Tree and Shrub Mortality and Debarking by Sambar *Cervus unicolor* (Kerr) in Gir After a Drought in Gujarat, India," *Biological Conservation* 68, no. 2 (1994): 149–54.

60. J. H. Lim et al., "Increased Declines of Korean Fir Forest Caused by Climate Change in Mountain Halla, Korea," (oral presentation, Forest Adaptation 2008, Umeå, Sweden, August 25–28, 2008).

61. Wang Hongbin et al., "Preliminary Deduction of Potential Distribution and Alternative Hosts of Invasive Pest, *Dendroctonus valens* (Coleoptera: Scolytidae)," *Scientia Silvae Sinicae* 43, no. 10 (2007): 71–76.

62. M. Vennetier et al., "Impact of Climate Change on Pine Forest Productivity and on the Shift of a Bioclimatic Limit in Mediterranean Area," *Options Méditerranéennes, Série A* 75, no. 8 (2007): 189–97.

63. Christian Körner, Dimitrios Sarris, and Dimitrios Christodoulakis, "Long-Term Increase in Climatic Dryness in the East-Mediterranean as Evidenced for the Island of Samos," *Regional Environmental Change* 5, no. 1 (2005): 27–36.

64. H. Kauhanen et al., "Extensive Mortality of Spruce Forests in Arkhangelsk Region: Satellite Image Analysis."Poster presented at the Forest Adaptation 2008, Umeå, Sweden, August 25–28, 2008.

65. Maria Laura Suarez, Luciana Ghermandi, and Thomas Kitzberger, "Factors Predisposing Episodic Drought-Induced Tree Mortality in *Nothofagus*: Site, Climatic Sensitivity and Growth Trends," *Journal of Ecology* 92, no. 6 (2004): 954–66.

66. R. J. Fensham and R.J. Fairfax, "Drought-Related Tree Death of Savanna Eucalypts: Species Susceptibility, Soil Conditions and Root Architecture," *Journal of Vegetation Science* 18, no. 1 (2007): 71–80.

67. Craig D. Allen et al., "A Global Overview of Drought and Heat-Induced Tree Mortality Reveals Emerging Climate Change Risks for Forests," *Forest Ecology and Management* 259, no. 4, (2010): 660–84.

68. Kai Zhu, Christopher W. Woodall, and James S. Clark, "Failure to Migrate: Lack of Tree Range Expansion in Response to Climate Change," *Global Change Biology*, Published electronically November 1, 2011. doi: 10.1111/j.1365-2486.2011.02571.x.

69. Michael J. Bodle, Amy P. Ferriter, and Daniel D. Thayer, "The Biology, Distribution, and Ecological Consequences of *Melaleuca Quinquenervia* in the Everglades," in *Everglades: The Ecosystem and its Restoration* (Delray Beach, FL: St. Lucie Press, 1997), 341–55; R. H. Hofstetter, "The Current Status of *Melaleuca quinquenervia* in southern Florida," in *Proceedings of the Symposium on Exotic Pest Plants*, eds. T. D. Center et al., (Washington, D. C.: University of Miami, United States Department of the Interior, National Park Service), 159–76.

70. Michael E. Dorcas et al., "Severe Mammal Declines Coincide with Proliferation of Invasive Burmese Pythons in Everglades National Park," PNAS, January 30, 2012, doi: 10.1073/pnas.1115226109.

71. R. N. Reed and G. H. Rodda, *Giant Constrictors: Biological and Management Profiles and an Establishment Risk Assessment for Nine Large Species of Pythons, Anacondas, and the Boa Constrictor: US Geological Survey Open-File Report 2009–1202* (Reston, VA: US Geological Survey, 2009).

72. Laurence G. Brewer, "Ecology of Survival and Recovery from Blight in American Chestnut Trees (*Castanea dentata* (Marsh.) Borkh.) in Michigan," *Bulletin of the Torrey Botanical Club* 122, no. 1 (1995): 40–57.

73. Bob Devine, *Alien Invasion: America's Battle with Non-Native Animals and Plants* (Washington, DC: National Geographic Society, 1998), 132.

74. Miguel Clavero and Emili García-Berthou, "Invasive Species are a Leading Cause of Animal Extinctions," *Trends in Ecology and Evolution* 20, no. 3 (2005): 110.

75. S. H. Mernild et al., "Increasing Mass Loss from Greenland's Mittivakkat Gletscher," *The Cryosphere* 5, (2011): 341–48.

76. Michael F. J. Pisaric et al., "Impacts of a Recent Storm Surge on an Arctic Delta Ecosystem Examined in the Context of the Last Millennium," PNAS 108, no. 22 (2011): 8960–5.

77. Maureen E. Raymo, et al., "PLIOMAX: Pliocene Maximum Sea Level Project," *PAGES News* 17, no. 2 (2009): 58–59.

78. Gary S. Dwyer and Mark A. Chandler, "Mid-Pliocene Sea Level and Continental Ice Volume Based on Coupled Benthic Mg/Ca Palaeotemperatures and Oxygen Isotopes," *Phil. Trans. Royal Soc. A* 367, (2009): 157–168; Kenneth G. Miller et al., "High tide of the warm Pliocene: Implications of Global Sea Level for Antarctic Deglaciation," *Geology*, Published electronically March 19, 2012. doi:10.1130/G32869.1

79. E. J. Rohling et al., "Antarctic Temperature and Global Sea Level Closely Coupled over the Past Five Glacial Cycles," *Nature Geoscience* 2 (2009): 500–4.

80. Martin Vermeer and Stefan Rahmstorf, "Global Sea Level Linked to Global Temperature," *PNAS* 106, no. 51 (2009): 21527–32.

81. S. Jevrejeva, J. C. Moore and A. Grinsted, "How Will Sea Level Respond to Changes in Natural and Anthropogenic Forcings by 2100?" *Geophysical Research Letters* 37, no. 5 (2010).

82. Detlef P. van Vuuren et al., "The Use of Scenarios as the Basis for Combined Assessment of Climate Change Mitigation and Adaptation," *Global Environmental Change* 21, no. 2 (2011): 575–91.

83. James E. Hansen and Makiko Sato, "Paleoclimate Implications for Human Made Climate Change," *NASA* Goddard Institute for Space Studies and Columbia University Earth Institute, 2011, last modified July 20, 2011 (version 3), http://arxiv.org/abs/1105.0968.

84. Orrin H. Pilkey and Rob Young, *The Rising Sea*, (Washington, DC: Island Press, 2009), 99–116.

85. Matthew L. Kirwan, "Limits on the Adaptability of Coastal Marshes to Rising Sea Level," *Geophysical Research Letters* 37, no. 5 (2010).

86. "Mapping Areas Potentially Impacted by Sea Level Rise," B. Strauss and J. L. Weiss, Department of Geosciences Environmental Studies Laboratory, University of Arizona, http://www.geo.arizona.edu.

87. Susmita Dasgupta et al., "The Impact of Sea Level Rise on Developing Countries: A Comparative Analysis" (February 1, 2007). World Bank Policy Research Working Paper 4136, http://ssrn.com/abstract=962790.

88. See note 84 above.

89. M. L. Parry et al., *Contribution of Working Group II to the Fourth Assessment Report of the Intergovernmental Panel on Climate Change, 2007* (New York: Cambridge University Press, 2007), 4.4.11.

90. Ilya M. D. Maclean and Robert J. Wilson, "Recent Ecological Responses to Climate Change Support Predictions of High Extinction Risk," *PNAS* 108, no. 30 (2011): 12337–42.

91. M. Bálint et al., "Cryptic Biodiversity Loss Linked to Global Climate Change," *Nature Climate Change* 1, (2011): 313–18.

92. Jennifer A. Sheridan and David Bickford, "Shrinking Body Size as an Ecological Response to Climate Change," *Nature Climate Change* 1 (2011): 401–6.

93. Anthony D. Barnosky, *Heatstroke: Nature in an Age of Global Warming*, (Washington, DC: Island Press, 2010), 17–58.

94. A. Michelle Lawing and P. David Polly, "Pleistocene Climate, Phylogeny, and Climate Envelope Models: An Integrative Approach to Better Understand Species' Response to Climate Change," *PLoS ONE* 6, no. 12 (2011).

95. Emily M. Rubidge et al., "Climate-Induced Range Contraction Drives Genetic Erosion in an Alpine Mammal." *Nature Climate Change*, Published electronically February 19, 2012. doi: 10.1038/nclimate1415

96. Eline D. Lorenzen, "Species-Specific Responses of Late Quaternary Megafauna to Climate and Humans," *Nature* 479, no. 7373 (2011).

97. David W. Johnston et al., "The Effects of Climate Change on Harp Seals (*Pagophilus groenlandicus*)," *PLoS ONE* 7, no. 1 (2012).

98. Board on Life Sciences, *Ecological Impacts of Climate Change* (Washington, DC: The National Academies Press, 2008), 17.

99. I-Ching Chen et al., "Rapid Range Shifts of Species Associated with High Levels of Climate Warming," *Science* 333, no. 6045 (2011): 1024–26.

100. Regan Early and Dov F. Sax, "Analysis of Climate Paths Reveals Potential Limitations on Species Range Shifts," *Ecology Letters* 14, no. 11 (2011): 1125–33.

101. Vincent Devictor et al., "Differences in the Climatic Debts of Birds and Butterflies at a Continental Scale," *Nature Climate Change*, January 10, 2012, http://www.nature.com/nclimate/journal/vaop/ncurrent/full /nclimate1347.html.

102. Mark C. Urban, Josh J. Tewksbury, Kimberly S. Sheldon. On a collision course: competition and dispersal differences create no-analogue communities and cause extinctions during climate change. *Proceedings of the Royal Society B: Biological Sciences* (2012).

103. Jon C. Bergengren, Duane E. Waliser, and Yuk L. Yung, "Ecological Sensitivity: A Biospheric View of Climate Change," *Climatic Change* 107, no. 3-4 (2011): 433–57.

104. Hamish McGallum et al., "Distribution and Impacts of Tasmanian Devil Facial Tumor Disease," *EcoHealth* 4, no. 3 (2007): 318–325; Nick Beeton and Hamish McCallum, "Models Predict that Culling is Not a Feasible Strategy to Prevent Extinction of Tasmanian Devils From Facial Tumour Disease," *Journal of Applied Ecology* 48, no. 6 (2011): 1315–23.

105. Jeffrey M. Lorch et al., "Experimental Infection of bats with *Geomyces Destructans* Causes White-Nose Syndrome," *Nature* 480 (2011): 376–8.

106. Paul M Cryan et al., "Wing Pathology of White-Nose Syndrome in Bats Suggests Life-Threatening Disruption of Physiology," BMC *Biology* 8, no. 135 (2010).

107. Matthew C. Allender et al., "Chrysosporiumsp. Infection in Eastern Massasauga Rattlesnakes." *Emerging Infectious Diseases*, Published electronically December 2011. doi: 10.3201/eid1712.110240

108. Émilie Clair et al., "A Glyphosate-Based Herbicide Induces Necrosis and Apoptosis in Mature Rat Testicular Cells *in Vitro*, and Testosterone Decrease at Lower Levels." *Toxicology in Vitro* 26, no. 2. (2012):269–79.

109. Rick A. Relyea, "New Effects of Roundup on Amphibians: Predators Reduce Herbicide Mortality; Herbicides Induce Antipredator Morphology," Ecological Applications 22, no. 2 (2012): 634. doi: 10.1890/11-0189.1.

110. Tyrone B. Hayes et al., "Demasculinization and Feminization of Male Gonads by Atrazine: Consistent Effects Across Vertebrate Classes. *The Journal of Steroid Biochemistry and Molecular Biology* 127, no. 1–2 (2011): 64–73.

111. Andrew R. Blaustein and Pieter T. J Johnson, "The Complexity of Deformed Amphibians," *Frontiers in Ecology and the Environment* 1, no. 2 (2003): 87–94.

112. Tyrone B. Hayes, "Hermaphroditic Demasculinized Frogs after Exposure to the Herbicide Atrazine at Low Ecologically Relevant Doses," *PNAS* 99, no. 8 (2002):5476–80.

113. Martin Gilbert et al, "Breeding and Mortality of Oriental White-Backed Vulture *Gyps bengalensis* in Punjab Province, Pakistan," *Bird Conservation International* 12, no. 4 (2002): 311–26.

114. Erik Kristiansson et al., "Pyrosequencing of Antibiotic-Contaminated River Sediments Reveals High Level of Resistance and Gene Transfer Elements," *PloS ONE* 6, no. 2 (2011); Timothy R. Walsh et al., "Dissemination of NDM-1 Positive Bacteria in the New Delhi Environment and its Implications for Human Health: an Environmental Point Prevalence Study," *The Lancet Infectious Diseases* 11, no. 5 (2011): 355–62.

115. Timothy M. LaPara et al., "Tertiary-Treated Municipal Wastewater is a Significant Point Source of Antibiotic Resistance Genes into Duluth-Superior Harbor," *Environmental Science & Technology* 45, no. 22 (2011): 9543–49.

116. Bonnie M. Marshall and Stuart B. Levy, "Food Animals and Antimicrobials: Impacts on Human Health," *Clinical Microbiology Reviews* 24, no. 4 (2011): 718–33.

117. Andrew E. Waters et al, "Multidrug-Resistant *Staphylococcus aureus* in US Meat and Poultry," *Clinical Infections Diseases* 52, no. 10 (2011): 1227–30.

118. Noni E. MacDonald et al., "Gonorrhea: What Goes Around Comes Around," *Canadian Medical Association Journal* 183, no. 14 (2011): 1567.

119. Peter Norberg et al., "The IncP-1 Plasmid Backbone Adapts to Different Host Bacterial Species and Evolves through Homologous Recombination," *Nature Communications* 2, no. 268 (2011).

120. Andrew F. Read, Troy Day, and Silvie Huijben, "The Evolution of Drug Resistance and the Curious Orthodoxy of Aggressive Chemotherapy," *PNAS* 108, no. Supplement 2 (2011): 10871–77.

121. Christian H. Krupke et al., "Multiple Routes of Pesticide Exposure for Honey Bees Living Near Agricultural Fields.," PLoS ONE 7, no. 1 (2012); Penelope R. Whitehorn et al., "Neonicotinoid Pesticide Reduces Bumble Bee Colony Growth and Queen Production," Science (2012).doi: 10.1126/science.1215025; Mickaël Henry et al., "A Common Pesticide Decreases Foraging Success and Survival in Honey Bees," Science (2012). doi: 10.1126/science.1215039.

122. Cyril Vidau et al., "Exposure to Sublethal Doses of Fipronil and Thiacloprid Highly Increases Mortality of Honeybees Previously Infected by *Nosema ceranae*," *PLoS ONE* 6, no. 6 (2011).

123. Angelika Hilbeck et al., "A Controversy Re-Visited: is the Coccinellid Adalia Bipunctata Adversely Affected by Bt Toxins?" *Environmental Sciences Europe*, Published electronically 15 February 2012. doi: 10.1186/2190-4715-24-10

124. Mohan Manikkam et al., "Transgenerational Actions of Environmental Compounds on Reproductive Disease and Identification of Epigenetic Biomarkers of Ancestral Exposures." *PLoS ONE*, Published electronically 28 February 2012. doi: 10.1371/journal.pone.0031901

125. Catherine L. Searle et al., "A Dilution Effect in the Emerging Amphibian Pathogen *Batrachochytrium dendrobatidis*," PNAS 108, no. 39 (2011): 16322–26.

126. Andrea Swei et al., "Is Chytridiomycosis an Emerging Infectious Disease in Asia?" *PLoS ONE* 6, no. 8 (2011).

127. Christian Hof et al., "Additive Threats from Pathogens, Climate and Land-Use Change for Global Amphibian Diversity," *Nature* 480, no. 7378 (2011): 516–19.

128. David B. Wake and Vance T. Vredenburg, "Are We in the Midst of the Sixth Mass Extinction? A View From the World of Amphibians," PNAS 105, no. Supplement 1 (2008): 11466–73.

129. Gerardo Ceballos, Andrés García, and Paul R. Ehrlich, "The Sixth Extinction Crisis Loss of Animal Populations and Species," *Journal of Cosmology* 8 (2010): 1821–31.

130. Stuart Butchart et al., "Global Biodiversity: Indicators of Recent Declines," *Science* 328, no. 5982 (2010): 1164–68.

131. Michael R. W. Rands, et al., "Biodiversity Conservation: Challenges Beyond 2010," *Science* 329, no. 5997 (2010): 1298–303.

132. The Center for Macroecology, Evolution and Climate at University of Copenhagen, "The Biodiversity Crisis: Worse than Climate Change," news release, January 19, 2012, http://news.ku.dk/all_news.

133. Stuart L Pimm and Peter Raven, "Biodiversity: Extinction by Numbers," *Nature* 403 (2000): 843–45.

3: SEA

1. David K. A. Barnes et al., "Accumulation and Fragmentation of Plastic Debris in Global Environments," *Philosophical Transactions of the Royal Society B* 364, no. 1526 (2009): 1985–98.

2. Mark Anthony Browne et al., "Accumulation of Microplastic on Shorelines Worldwide: Sources and Sinks," *Environmental Science & Technology* 45, no. 21 (2011): 9175–9.

3. G. Kukulka et al., "The Effect of Wind Mixing on the Vertical Distribution of Buoyant Plastic Debris," *Geophysical Research Letters* 39, no. 7 (2012). doi: 10.1029/2012GL051116.

4. Courtney Arthur, Joel Baker, and Holly Bamford, eds., "Proceedings of the International Research Workshop on the Occurrence, Effects, and Fate of Microplastic Marine Debris," NOAA Technical Memorandum NOS-OR&R-30 (2009).

5. Peter Davison and Rebecca G. Asch, "Plastic Ingestion by Mesopelagic Fishes in the North Pacific Subtropical Gyre," *Marine Ecology Progress Series* 432 (2011): 173–80.

6. Fiona Murray and Phillip Rhys Cowie, "Plastic Contamination in the Decapod Crustacean *Nephrops norvegicus* (Linnaeus, 1758)," *Marine Pollution Bulletin* 62, no. 6 (2011): 1207–17.

7. Mary L. Moser and David S. Lee, "A Fourteen-Year Survey of Plastic Ingestion by Western North Atlantic Seabirds," *Colonial Waterbirds* 15, no. 1 (1992): 83–94.

8. Marie Y. Azzarello and Edward S. Van-Vlee, "Marine Birds and Plastic Pollution," *Marine Ecology Progress Series* 37 (1987): 295–303.

9. V. Zitko and M. Hanlon, "Another Source of Pollution by Plastics: Skin Cleaners with Plastic Scrubbers," *Marine Pollution Bulletin* 22, no. 1 (1991): 41–42.

10. Peter G. Connors and Kimberly G. Smith, "Oceanic Plastic Particle Pollution: Suspected Effect on Fat Deposition in Red Phalaropes," *Marine Pollution Bulletin* 13, no. 1 (1982): 18–20.

11. Robin P. Angliss and Douglass P. DeMaster, "Differentiating Serious and Non-Serious Injury of Marine Mammals Taken Incidental to Commercial Fishing Operations: Report of the Serious Injury Workshop 1–2 April 1997, Silver Spring, Maryland," NOAA Technical Memorandum NMFS-OPR-13 (1998): 48; Kimberly L. Raum-Suryan, Lauri A. Jemison, and Kenneth W. Pitcher, "Entanglement of Steller Sea Lions (*Eumetopias jubatus*) in Marine Debris: Identifying Causes and Finding Solutions," *Marine Pollution Bulletin* 58, no. 10 (2009): 1487–95.

12. Katsuhiko Saido et al., "New Contamination Derived from Marine Debris Plastics," (presented at the 238th ACS National Meeting, Washington, DC, Environmental Division, August 22–26, 2009).

13. David Malakoff, "A Push For Quieter Ships," *Science* 328, no. 5985 (2010): 1502–3.

14. Rosalind M. Rolland et al., "Evidence That Ship Noise Increases Stress In Right Whales." *Proceedings of the Royal Society B*, Published electronically February 8, 2012. doi: 10.1098/rspb.2011.2429.

15. Julia Purser and Andrew N Radford, "Acoustic Noise Induces Attention Shifts and Reduces Foraging Performance in Three-Spined Sticklebacks (Gasterosteus aculeatus)," *PLoS One* 6, no. 2 (2011).

16. Michel André et al., "Low-Frequency Sounds Induce Acoustic Trauma in Cephalopods," *Frontiers in Ecology and the Environment* 9, no. 9 (2011): 498–93.

17. V. J. Fabry et al., "Present and Future Impacts of Ocean Acidification on Marine Ecosystems and Biogeochemical Cycles" (Report of the Ocean Carbon and Biogeochemistry Scoping Workshop on Ocean Acidification Research, La Jolla, CA, October 9-11, 2007).

18. Sarah Nienhuis, A. Richard Palmer, and Christopher D. G. Harley, "Elevated CO_2 Affects Shell Dissolution Rate but not Calcification Rate in a Marine Snail," *Proceedings of the Royal Society B* 277, no. 1693 (2010): 2553–58; Maoz Fine and Dan Tchernov, "Scleractinian Coral Species Survive and Recover from Decalcification," *Science* 315, no. 5820 (2007): 1811; Chris

Langdon et al., "Effect of Calcium Carbonate Saturation State on the Calcification Rate of an Experimental Coral Reef," *Global Biochem Cycles* 14, no. 2 (2000): 639–54.

19. Scott C. Doney et al., "Ocean Acidification: The Other CO_2 Problem," *Annual Review of Marine Science* 1 (2009): 169–92.

20. Bärbel Hönisch et al., "The Geological Record of Ocean Acidification." *Science* 335, no. 6072 (2012): 1058–63.

21. Jelle Bijma, "IPSO Preliminary Report on Ocean Stresses and Impacts: Case study 1," (2011).

22. Simulation in the Andy Ridgwell paper, cited below, has shown that future oceans are likely to have less carbonate in the deep ocean than that experienced during the PETM extinction event, fifty-five million years ago. See Andy Ridgwell et al., "Past Constraints on the Vulnerability of Marine Calcifiers to Massive Carbon Dioxide Release," *Nature Geoscience* 3 (2010): 196–200.

23. "Ocean Acidification Due to Increasing Atmospheric Carbon Dioxide," *The Royal Society*, June 30, 2005, http://royalsociety.org/policy/publications/2005/ocean-acidification.

24. Göran E. Nilsson et al., "Near-Future Carbon Dioxide Levels Alter Fish Behaviour by Interfering with Neurotransmitter Function," *Nature Climate Change* 2, no. 1 (2012).

25. G. W. Steller, *The Beasts of the Sea* in an appendix to David Starr Jordan, ed., *The Fur Seals and Fur-Seal Islands of the North Pacific Ocean, Part 3* (Washington, DC, 1899), 179–218.

26. Jeremy B. C. Jackson et al., "Historical Overfishing and the Recent Collapse of Coastal Ecosystems," *Science* 293, no. 5530 (2001): 629–37; Mizue Hisano, Sean R. Connolly, and William D. Robbins, "Population Growth Rates of Reef Sharks with and without Fishing on the Great Barrier Reef: Robust Estimation with Multiple Models. *PLoS ONE* 6, no. 9 (2011).

27. Christopher Costello, Steven Gaines, and Leah R. Gerber, "Conservation Science: A Market Approach to Saving the Whales," *Nature* 481, no. 7380 (2012): 139–40.

28. Marc O. Nadon et al., "Re-Creating Missing Population Baselines for Pacific Reef Sharks," *Conservation Biology* (2012). doi: 10.1111/j.1523-1739.2012.01835.x.

29. Ransom A. Myers and Boris Worm, "Rapid Worldwide Depletion of Predatory Fish Communities," *Nature* 423 (2003): 280–83.

30. Ransom A. Myers et al., "Cascading Effects of the Loss of Apex Predatory Sharks from a Coastal Ocean," *Science* 315, no. 5820 (2007): 1846–50.

31. James A. Estes, and David O. Duggins, "Sea Otters and Kelp Forests in Alaska: Generality and Variation in a Community Ecological Paradigm," *Ecological Monographs* 65, no. 1 (1995): 75–100; Robert S. Steneck et al., "Kelp Forest Ecosystems: Biodiversity, Stability, Resilience and Future," *Environmental Conservation* 29, no. 4 (2002): 436–59.

32. Laura Tremblay-Boyer et al., "Modelling the Effects of Fishing on the Biomass of the World's Oceans from 1950 to 2006," *Marine Ecology Progress Series* 442 (2011): 169–85.

33. James A. Estes et al., "Trophic Downgrading of Planet Earth. *Science* 333, no. 6040 (2011): 301–6.
34. Jeremy B. C. Jackson, "What was Natural in the Coastal Oceans?" *PNAS* 98, no. 10 (2001): 5411–18.
35. "World Review of Fisheries and Aquaculture," in *The State of World Fisheries and Aquaculture* (Rome: Food and Agriculture Organization of the United Nations, 2009), 6.
36. FAO data for 2008, as noted in S. J. Hall et al., *Blue Frontiers: Managing the Environmental Costs of Aquaculture* (Penang, Malaysia: The WorldFish Center, 2011).
37. Brad E. Erisman et al., "The Illusion of Plenty: Hyperstability Masks Collapses in Two Recreational Fisheries that Target Fish Spawning Aggregations," *Canadian Journal of Fisheries and Aquatic Sciences* 68, no. 10 (2011): 1705–16.
38. Maria José Juan-Jorda et al., "Global Population Trajectories of Tunas and Their Relatives," *PNAS* 108, no. 51 (2011): 20650–55.
39. Yvonne Sadovy de Mitcheson et al. "Fishing Groupers Towards Extinction: A Global Assessment of Threats and Extinction Risks in a Billion Dollar Fishery," *Fish and Fisheries* (2012). doi: 10.1111/j.1467-2979.2011.00455.x.
40. Christian Mullon, Pierre Fréon, and Philippe Cury, "The Dynamics of Collapse in World Fisheries," *Fish and Fisheries* 6 (2005): 111–20.
41. Malin L. Pinsky et al., "Unexpected Patterns of Fisheries Collapse in the World's Oceans," *PNAS* 108, no. 20 (2011): 8317–22.
42. Daniel Pauly, "Beyond Duplicity and Ignorance in Global Fisheries," *Scientia Marina* 73, no. 2 (2009): 215–24.
43. Food and Agriculture Organization of the United Nations, "World Agriculture: Towards 2015/30 Summary Report," (Rome: FAO, 2002), 73(2): 215–23.
44. Philippe M. Cury et al., "Global Seabird Response to Forage Fish Depletion--One-Third for the Birds," *Science*, 334, no. 6063 (2011): 1703–6.
45. Elliott A. Norse et al., "Sustainability of Deep-Sea Fisheries," *Marine Policy* 36, no. 2 (2012): 307–20.
46. J. Frederick Grassle and Nancy J. Maciolek, "Deep-Sea Species Richness: Regional and Local Diversity Estimates From Quantitative Bottom Samples," *American Naturalist* 139, no. 2 (1992): 313–41.
47. Matthew Gianni, "High Seas Bottom Trawl Fisheries and their Impacts on the Biodiversity of Vulnerable Deep-Sea Ecosystems" (Report prepared for IUCN/the World Conservation Union, Natural Resources Defense Council, WWF International and Conservation International, June 2004); Andrew G. Bauman et al., "Tropical Harmful Algal Blooms: An Emerging Threat to Coral Reef Communities?" *Marine Pollution Bulletin* 60, no. 11 (2010): 2117–22.
48. Jeremy B. C. Jackson, "Ecological Extinction and Evolution in the Brave New Ocean," *PNAS* 105, no. Supplement 1 (2008): 11458–65.
49. F. Althaus et al., "Impacts of Bottom Trawling on Deep-Coral Ecosystems of Seamounts are Long-Lasting," *Marine Ecology Progress Series* 397 (2009): 279–94.

50. Hjalmar Thiel, "Anthropogenic Impacts on the Deep Sea," in *Ecosystems of the Deep Ocean (Ecosystems of the World) Volume 28*, ed. P. A. Tyler (Amsterdam: Elsevier Science B. V., 2003), 427–72.

51. Ove Hoegh-Guldberg and John F. Bruno, "The Impact of Climate Change on the World's Marine Ecosystems," *Science* 328, no. 5985 (2010): 1523–28.

52. Sally Hall and Sven Thatje, "Temperature-Driven Biogeography of the Deep-Sea Family Lithodidae (Crustacea: Decapoda: Anomura) in the Southern Ocean," *Polar Biology* 34, no. 3 (2010): 363–70.

53. Sven Thatje et al., "Encounter of Lithodid Crab *Paralomis birsteini* on the Continental Slope off Antarctica, Sampled by ROV," *Polar Biology* 31, no. 9 (2008): 1143–48.

54. M. T. Burrows et al., "The Pace of Shifting Climate in Marine and Terrestrial Ecosystems," *Science* 334, no. 6056 2011): 652–55.

55. Christopher D. G. Harley, "Climate Change, Keystone Predation, and Biodiversity Loss," *Science* 334, no. 6059 (2011): 1124–27.

56. J. E. N. Veron et al, "The Coral Reef Crisis: The Critical Importance of <350 ppm CO_2," *Marine Pollution Bulletin* 58, no. 10 (2009): 1428–36.

57. O. Hoegh Guldberg et al., "Coral Reefs Under Rapid Climate Change and Ocean Acidification," *Science* 318, no. 5857 (2007): 1737–42.

58. Malcolm McCulloch et al., "Coral Resilience to Ocean Acidification and Global Warming Through pH Up-Regulation," *Nature Climate Change* (2012). doi: 10.1038/nclimate1473.

59. Katie L. Cramer et al., "Anthropogenic Mortality on Coral Reefs in Caribbean Panama Predates Coral Disease and Bleaching," *Ecology Letters* (2012). doi: 10.1111/j.1461-0248.2012.01768.x.

60. Douglas B. Rasher et al., "Macroalgal terpenes function as allelopathic agents against reef corals," *PNAS* 108, no. 43 (2011): 17726–31; Jeremy B. C. Jackson, "The Future of the Oceans Past," *Philosophical Transactions of the Royal Society B* 365, no. 1558 (2010): 3765–78.

61. Charles R. C. Sheppard, Simon K. Davy, and Graham M. Pilling, *The Biology of Coral Reefs: Biology of Habitats*, (Oxford, U.K.: Oxford University Press, 2009), 278.

62. Arnaud Brayard et al., "Transient Metazoan Reefs in the Aftermath of the End-Permian Mass Extinction," *Nature Geoscience* 4 (2011): 693–97.

63. J. J. Stachowicz et al., "Linking Climate Change and Biological Invasions: Ocean Warming Facilitates Nonindigenous Species Invasions," *PNAS* 99, no. 24 (2002): 15497–500.

64. Ronald E. Thresher and Armand M. Kuris, "Options for Managing Invasive Marine Species," *Biological Invasions* 6, no. 3 (2004): 295–300.

65. Nikos Streftaris, Argyro Zenetos, and Evangelos Papathanassiou, "Globalisation in Marine Ecosystems: The Story of Non-Indigenous Marine Species Across European Seas," *Oceanography and Marine Biology: An Annual Review* 43 (2005): 419–53.

66. Mark A. Albins and Mark A. Hixon, "Invasive Indo-Pacific Lionfish *Pterois volitans* Reduce Recruitment of Atlantic Coral-Reef Fishes," *Marine Ecology Progress Series* 367 (2008): 233–38.

67. Jennifer L. Molnar et al., "Assessing the Global Threat of Invasive Species to Marine Biodiversity," *Frontiers in Ecology and the Environment* 6, no. 9 (2008): 485–92.

68. Alycia L. Stigall, "Invasive Species and Biodiversity Crises: Testing the Link in the Late Devonian," *PloS ONE* 5, no. 12 (2010). Published electronically. doi:10.1371/journal.pone.0015584

69. Nancy N. Rabalais et al., "Global Change and Eutrophication of Coastal Waters," *ICES Journal of Marine Science* 66, no. 7 (2009): 1528–37.

70. Wynne Wright and Stephen L. Muzzati, "Not in My Port: The "Death Ship" of Sheep and Crimes of Agri-Food Globalization," *Agriculture and Human Values* 24, no. 2 (2007): 133–45.

71. B. Morton, "Slaughter at Sea," *Marine Pollution Bulletin* 46, no. 4 (2003): 379–80.

72. Eva Ramirez-Llodra et al., "Man and the Last Great Wilderness: Human Impact on the Deep Sea," *PLoS ONE* 6, no. 8 (2011).

73. Kathryn Patterson Sutherland et al., "Human Pathogen Shown to Cause Disease in the Threatened Eklhorn Coral Acropora palmata," *PLoS ONE* 6, no. 8 (2011).

74. Jerker Fick et al., "Therapeutic Levels of Levonorgestrel Detected in Blood Plasma of Fish: Results from Screening Rainbow Trout Exposed to Treated Sewage Effluents," *Environmental Science & Technology* 44, no. 7 (2010): 2661–66.

75. Mike E. Grigg, Stephen A. Raverty, and Melissa A. Miller, "Swimming in Sick Seas," (presented at the annual meeting of AAAS, Vancouver, Canada, February 16–20, 2012).

76. Jeremy B. C. Jackson et al., "Historical Overfishing and the Recent Collapse of Coastal Ecosystems," *Science* 293, no. 5530 (2001): 629–37.

77. Wei-Jun Cai et al., "Acidification of Subsurface Coastal Waters Enhanced by Eutrophication," *Nature Geoscience* 4 (2011): 766–70.

78. Daniel J. Conley et al., "Ecosystem Thresholds with Hypoxia," *Developments in Hydrobiology* 207 (2009): 21–29.

79. Gary Shaffer, Steffen Malskær Olsen, and Jens Olaf Pepke Pedersen, "Long-Term Ocean Oxygen Depletion in Response to Carbon Dioxide Emissions from Fossil Fuels," *Nature Geoscience* 2 (2009): 105–9.

80. A. D. Cembella, "The Toxigenic Marine Dinoflagellate *Alexandrium tamarense* as the Probable Cause of Mortality of Caged Salmon in Nova Scotia," *Harmful Algae* 1, no. 3 (2002): 313–25.

81. Leanne J. Flewelling, "Vectors of Brevetoxins to Marine Mammals," PhD. Diss., University of South Florida, 2008. http://scholarcommons.usf.edu/etd/243.

82. Nicholas J. T. Osborne, Penny M. Webb, and Glen R. Shaw, "The Toxins of Lyngbya majuscula and their Human and Ecological Health Effects," *Environment International* 27, no. 5 (2001): 381–92.

83. Georgi M. Daskalov et al., "Trophic Cascades Triggered by Overfishing Reveal Possible Mechanisms of Ecosystem Regime Shifts," *PNAS* 104, no. 25 (2007): 10518–23; Steve Hay, "Marine Ecology: Gelatinous Bells May Ring Change in Marine Ecosystems," *Current Biology* 16, no. 17 (2006): R679–82.

84. Shin-Ichi Uye, "Blooms of the Giant Jellyfish *Nemopilema nomurai*: A Threat to the Fisheries Sustainability of the East Asian Marginal Seas," *Plankton and Benthos Research* 3 (2008): Supplement 125–31.
85. Lucas Brotz et al., "Increasing jellyfish populations: trends in Large Marine Ecosystems," *Hydrobiologia* (2012). doi: 10.1007/s10750-012-1039-7.
86. Lucas Brotz,"Changing Jellyfish Populations: Trends in Large Marine Ecosystems," (master's thesis, University of British Columbia, 2011), https://circle.ubc.ca/bitstream/handle/2429/38193/ubc_2011 _fall_brotz_lucas.pdf?sequence=3; Jennifer E. Purcell "Jellyfish and Ctenophore Blooms Coincide with Human Proliferations and Environmental Perturbations" *Annual Review of Marine Science* 4 (2012): 209–35.
87. Anthony J. Richardson et al., "The Jellyfish Joyride: Causes, Consequences and Management Responses to a More Gelatinous Future," *Trends in Ecology & Evolution* (2009); See note 86 above.
88. Jeremy B. C. Jackson, "The Future of the Oceans Past," *Philosophical Transactions of the Royal Society B* 365, no. 1558 (2010): 3765–78.
89. A. D. Rogers and D. d'A. Laffoley, "International Earth System Expert Workshop on Ocean Stresses and Impacts. Summary Report," IPSO Oxford,

4: HUMAN HABITATS

1. Noah Diffenbaugh and Martin Scherer, "Observational and Model Evidence of Global Emergence of Permanent, Unprecedented Heat in the 20th and 21st Centuries," *Climatic Change* 107, no. 3–4 (2011): 615–41.
2. Freemeteo.com, Weather History.
3. Markus Huber and Reto Knutti, "Anthropogenic and Natural Warming Inferred From Changes in Earth's Energy Balance," *Nature Geoscience* 5, (2011): 31–36.
4. Steven Sherwood and Matthew Huber, "An Adaptability Limit to Climate Change Due to Heat Stress," *PNAS* 107, no. 21 (2010): 9552–55.
5. David Medvigy and Claudie Beaulieu, "Trends in Daily Solar Radiation and Precipitation Coefficients of Variation Since 1984," *Journal of Climate* 25, no. 4 (2012): 1330–39.
6. Sally Brown, Abiy S. Kebede, and Robert J. Nicholls, "Sea Level Rise and Impacts in Africa, 2000 to 2100," (University of Southampton School of Civil Engineering and the Environment, Southampton SO17, 1BJ, UK, November 4, 2011).
7. Susmita Dasgupta et al., "The Impact of Sea Level Rise on Developing Countries: A Comparative Analysis, 2007," World Bank Policy Research Working Paper No. 4136, February 1, 2007, http://ssrn.com/abstract=962790.
8. Ibid.
9. Matthew Heberger et al., "The Impact of Sea Level Rise on the California Coast," (paper from California Climate Change Center, May 2009).
10. "2009 Energy Balance for World," International Energy Agency, http://www.iea.org/stats.

11. James Galloway et al., "Transformation of the Nitrogen cycle: Recent Trends, Questions, and Potential Solutions," *Science* 320, no. 5878 (2008): 889–92.
12. Nick A. Owen, Oliver R. Inderwildi, and David A. King, "The Status of Conventional World Oil Reserves—Hype or Cause for Concern?" *Energy Policy* 38, no. 8 (2010): 4743–49.
13. E. D. Attanasi and R. F. Meyer, "Natural Bitumen and Extra-Heavy Oil," in *2007 Survey of Energy Resources*, (London: World Energy Council, 2007).
14. David Murphy & Charles A. S. Hall, "Year in Review—EROI or Energy Return on (Energy) Invested," *Ecological Economics Reviews* 1185 (2010): 102–18.
15. See note 13 above.
16. "Proposed Oil Shale and Tar Sands Resource Management Plan Amendments to Address Land Use Allocations in Colorado, Utah, and Wyoming and Final Programmatic Environmental Impact Statement," (United States Department of the Interior, Bureau of Land Management, September 2008), http://ostseis.anl.gov /documents/fpeis /vol1/ OSTS_FPEIS_Vol1_Front.pdf.
17. Steve Mohr, "Projection of World Fossil Fuel Production with Supply and Demand Interactions," (Research Doctorate, University of Newcastle School of Engineering, 2010).
18. See note 14 above.
19. Tom M. L. Wigley, "Coal to Gas: The Influence of Methane Leakage," *Climatic Change* 108, no. 3 (2011): 601–8.
20. One barrel of oil (0.146 tons) is assumed to have the energy equivalent of 5,555 cubic feet of natural gas.
21. Vello A. Kuuskraa and Scott H. Stevens, *Worldwide Gas Shales and Unconventional Gas: A Status Report*, (Arlington, VA: Advanced Resources International, Inc., December 7, 2009).
22. Sally Entrekin et al., "Rapid Expansion of Natural Gas Development Poses a Threat to Surface Waters," *Frontiers in Ecology and the Environment* 9, no. 9 (2011): 503–11.
23. Nathan Hultman et al., "The Greenhouse Impact of Unconventional Gas for Electricity Generation," *Environmental Research Letters* 6, no. 4 (2011).
24. Robert W. Howarth, Renee Santoro, and Anthony Ingraffea, "Methane and the Greenhouse-Gas Footprint of Natural Gas from Shale Formations," *Climatic Change* 106, no. 4 (2011): 679–710.
25. Jeff Tollefson, "Air Sampling Reveals High Emissions from Gas Field," *Nature News* 482, no. 7384 (2012): 139–40.
26. US Energy Information Administration, *World Shale Gas Resources: An Initial Assessment of 14 Regions Outside the United States* (Washington, DC: US Department of Energy, 2011); Vello Kuuskraa, personal communication with the author.
27. "U.S Shale Gas: Less Abundance, Higher Cost," The Oil Drum, August 5, 2011. The Oil Drum, Arthur E. Berman and Lynn F. Pittinger, August 5, 2011, http://www.theoildrum.com/node/8212.

28. See note 17 above.

29. *2010 Survey of Energy Resources*, (London: World Energy Council, 2010).

30. It is assumed that 1 ton of bituminous coal contains the energy of 0.66 Toe (tons of oil equivalent), that 1 ton of sub-bituminous coal contains the energy of 0.5 Toe, and that 1 ton of lignite coal contains the energy of 0.33 Toe.

31. US Energy Information Administration, *Natural Gas Gross Withdrawals and Production* (table), http://www.eia.gov/dnav/ng/ng_prod_sum_dcu_nus_m.htm; Nick A. Owen, Oliver R. Inderwildi, and David A. King, "The Status of Conventional World Oil Reserves—Hype or Cause for Concern?" *Energy Policy* 38, no. 8 (2010): 4743–49.

32. The total oil production at present includes not only conventional but some unconventional sources. Thus, the production of conventional oil is somewhat less than 56% of the production during its peak year, 1972.

33. Richard A. Kerr, "Peak Oil Production May Already be Here," *Science* 331, no. 6024 (2011): 1510–11.

34. Production increased by 11% for 2010 over 2008 levels. U.S. Energy Information Administration "International Energy Statistics" Database (accessed January 18, 2012), http://www.eia.gov/cfapps/ipdbproject / IEDIndex3.cfm.

35. Mikael Höök, Robert Hirsch, and Kjell Aleklett, "Giant Oil Field Decline Rates and Their Influence on World Oil Production, *Energy Policy* 37, no. 6 (2009): 2262–72.

36. The shale gas figures are extrapolated for the years 2007-2002 from "Study Analyzes Nine US, Canada Shale Gas Plays," *Oil & Gas Journal*, November 10, 2008, http://www.ogj.com/articles/print/volume-106/issue-42/drilling-production/study-analyzes-nine-us-canada-shale-gas-plays.html; Tight gas figures are extrapolated from Vello A. Kuuskraa, *Challenges Facing Increased Production and Use of Domestic Natural Gas*, (Arlington, VA: Advanced Resources International, 2009); The rest of the data is from U.S. Energy Information Administration, *Natural Gas Gross Withdrawals and Production* (table), http://www.eia.gov/dnav/ng/ng_prod_sum_dcu_nus_m .htm.

37. Mikael Höök et al., "Global Coal Production Outlooks Based on a Logistic Model," *Fuel* 89, no. 11 (2010): 3546–58.

38. Ibid.

39. Once fully operational, the Pearl GTL would generate 140,000 barrels per day worth of oil—made from a local natural gas field.

40. I assume 0.3% efficiency gain per year. This is based on US Energy Information Administration, *Annual Outlook 2011*, (Washington, DC: US Department of Energy, 2011), in which appliances and vehicles are projected to have this gain in the US market. The European market is projected to have fewer gains, due to already existing better efficiency, while developing nations are projected to have higher gains for the same reason. I take 0.3% gain as a middle-of-the-road figure.

41. *World Population Prospects: The 2010 Revision: Highlights and Advance Tables* (New York: United Nations, 2011), 2.

42. See note 17 above. Specifically, the chosen model is a dynamic projection, case 2—which is a middle-of-the-road projection. I had to harmonize some of the figures in the model that do not correspond to existing production rates.
43. 25% of shale oil energy is used for its own extraction, 33% for tar sands, 33% for extra heavy oil. See note 14 above.
44. U.S. Energy Information Administration, *Annual Energy Outlook 2010: With Projections to 2035*, (Washington, DC: U.S. Department of Energy, 2011). I keep this demand trajectory but adjust it to reflect population growth, which is based on *World Population Prospects*. See note 41 above.
45. The US Energy Information Administration projects annual global demands for coal, natural gas, and petroleum until the year 2035. From 2036 till 2100, I used the 2035 demand figures adjusted to anticipated global population for each of the subsequent years, derived from *World Population Prospects*. See note 41 above.
46. 24 million Ktoe is the equivalent of 952 quads or 1,004 exajoules or 279,120 terawatt-hours.
47. David Molden, ed., *Water for Food, Water for Life: A Comprehensive Assessment of Water Management in Agriculture* (London: Earthscan, and Colombo: International Water Management Institute, 2007), 69–75.
48. Fred Pearce, *When the Rivers Run Dry: Water–The Defining Crisis of the Twenty-First Century* (Boston: Beacon Press, 2006), 9–17.
49. See note 53 below.
50. *Charting our Water Future: Economic Frameworks to Inform Decision-Making* (2030 Water Resources Group, 2009).
51. Yoshihide Wada et al., "Global Depletion of Groundwater Resources," *Geophysical Research Letters* 37, no. 20 (2010).
52. R. K. Mall et al., "Water Resources and Climate Change: An Indian Perspective," *Current Science* 90, no. 12 (2006): 1610–26.
53. World Economic Forum, *The Bubble is Close to Bursting: A Forecast of the Main Economic and Geopolitical Water Issues Likely to Arise in the World during the Next Two Decades* (World Economic Forum, 2009), 51.
54. See note 50 above.
55. J. Wang et al., "China's Water-Energy Nexus: Greenhouse-Gas Emissions from Groundwater Use for Agriculture," *Environmental Research Letters* (2012). doi:10.1088/1748-9326/7/1/014035.
56. Michael Kugelman and Robert Hathaway, eds., Running on Empty: Pakistan's Water Crisis, (Washington, DC: Woodrow Wilson International Center for Scholars, 2009); *Pakistan—Country Water Resources Assistance Strategy: Water Economy; Running Dry* (World Bank, 2005) Report No. 34081-PK.
57. John Briscoe and Usman Qamar, *Pakistan's Water Economy: Running Dry* (New York: Oxford University Press, 2009).
58. Walter W. Immerzeel, Ludovicus P. H. van Beek, and Marc F. P. Bierkens, "Climate Change Will Affect the Asian Water Towers," *Science* 328, no. 5984 (2010): 1382–85.
59. Ibid.

60. Thomas Jacob et al. "Recent Contributions of Glaciers and Ice Caps to Sea Level Rise." *Nature* Published online 08 February 2012. doi:10.1038/nature10847; T. Bolch et al., "The State and Fate of Himalayan Glaciers," *Science* 336, no. 6079 (2012): 310. doi: 10.1126/science.1215828.

61. M. L. Parry et al., *Contribution of Working Group II to the Fourth Assessment Report of the Intergovernmental Panel on Climate Change, 2007* (New York: Cambridge University Press, 2007), 13.4.3.

62. See note 53 above.

63. Z. W. Kundzewicz et al., "The Implications of Projected Climate Change for Freshwater Resources and Their Management," *Hydrological Science Journal* 53, no. 1 (2008).

64. Richard Seager et al., "Model Projections of an Imminent Transition to a More Arid Climate in Southwestern North America," *Science* 316, no. 5828 (2007): 1181–84.

65. Seth M. Munson, Jayne Belnap, and Gregory S. Okin, "Responses of Wind Erosion to Climate-Induced Vegetation Changes on the Colorado Plateau," *PNAS* 108, no. 10 (2011): 3854–59.

66. Sujoy B. Roy et al., "Projecting Water Withdrawal and Supply for Future Decades in the U.S. under Climate Change Scenarios." *Environmental Science & Technology,* Published electronically January 12, 2012. doi: 10.1021/es2030774.

67. Susan Solomon et al., "Irreversible Climate Change Due to Carbon Dioxide Emissions," *PNAS* 106, no. 6 (2009): 1704–9.

68. Aiguo Dai, "Drought Under Global Warming: A Review," *Wiley Interdisciplinary Reviews: Climate Change* 2, no. 1 (2011): 45–65.

69. Maps are based on maps produced for the paper by Dai (see note 68), produced by UCAR, http://www2.ucar.edu/news/2904/climate-change-drought-may-threaten-much-globe-within-decades.

70. Justin Sheffield and Eric F. Wood, "Projected Changes in Drought Occurrence under Future Global Warming from Multi-Model, Multi-Scenario, IPCC AR4 Simulations," *Climate Dynamics* 31, no. 1 (2008): 79–105.

71. David Tilman et al., "Global Food Demand and the Sustainable Intensification of Agriculture," *PNAS* 108, no. 50 (2011): 20260–64.

72. William H. Kötke, *The Final Empire* (Bloomington, IN: AuthorHouse, 2007), chapter 3.

73. David R. Mongomery, *Dirt: The Erosion of Civilizations* (Berkeley and Los Angeles, CA: University of California Press, 2007), 3.

74. David Pimentel, "Soil Erosion: A Food and Environmental Threat," *Environment, Development and Sustainability* 8, no. 1 (2006): 119–37.

75. Z. G. Bai et al., "Proxy Global Assessment of Land Degradation," *Soil Use and Management* 24, no. 3 (2008): 223–34.

76. After accounting for the dying, every year 78 million are added to the world population. The widely accepted figure within FAO and otherwise, is that we have 1,500 million hectares under cultivation. Everything else being equal, it comes to the equivalent of 16.7 million hectares per 78 million people.

77. David Pimentel et al., "Will Limited Land, Water, and Energy Control Human Population Numbers in the Future?" *Human Ecology* 38, no. 5 (2010): 599–611.

78. See note 72 above.

79. I. Steen, "Phosphorus Availability in the 21st Century: Management of a Non-Renewable Resource," *Phosphorus and Potassium* 217 (1998): 25–31.

80. Dana Cordell, Jan-Olof Drangert, Stuart White, "The Story of Phosphorus: Global Food Security and Food for Thought," *Global Environmental Change* 19, no. 2 (2009): 292–305.

81. A. L. Smit et al., *Phosphorus in Agriculture: Global Resources, Trends and Developments*. (Wageningen, Netherlands: Plant Research International, 2009).

82. J. Donald Hughes, "Ancient Deforestation Revisited," *Journal of the History of Biology* 44, no. 1 (2011):43–57

83. Ren Mei-e & Zhu Xianmo, "Anthropogenic Influences on Changes in the Sediment Load of the Yellow River, China, During the Holocene," *The Holocene* 4, no. 3 (1994): 314–20.

84. Jocelyn Kaiser, "Wounding Earth's Fragile Skin," *Science* 304, no. 5677 (2004): 1616–18.

85. Germain Bayon et al., "Intensifying Weathering and Land Use in Iron Age Central Africa," *Science,* Published electronically February 9, 2012. doi: 10.1126/science.1215400

86. Philip Shearman, Jane Bryan, and William F. Laurance "Are we approaching "peak timber" in the tropics?" *Biological Conservation*, Published electronically 20 January, 2012. doi.org/10.1016/j.biocon.2011.10.036

87. Eric Katovai, Alana L. Burley, and Margaret M. Mayfield, "Understory Plant Species and Functional Diversity in the Degraded Wet Tropical Forests of Kolombangara Island, Solomon Islands," *Biological Conservation* 145, no. 1 (2012): 214–24.

88. Curtis N. Runnels, "Environmental Degradation in Ancient Greece," *Scientific American*, March 1995: 96–99.

89. J. Russell Smith, *Tree Crops: A Permanent Agriculture* (Greenwich, CT: The Devin-Adair Company, 1950), 3–4.

90. Benjamin I. Cook, Ron L. Miller, and Richard Seager, "Amplification of the North American 'Dust Bowl' Drought through Human-Induced Land Degradation," *PNAS* 106, no. 13 (2009): 4997–5001.

91. The article of Erisman et al. is updating on the earlier analysis made by Smil. Jan Willem Erisman et al., "How a Century of Ammonia Synthesis has Changed the World," *Nature Geoscience* 1 (2008): 636–39; Vaclav Smil, *Enriching the Earth: Fritz Haber, Carl Bosch, and the Transformation of World Food Production* (Massachusetts Institute of Technology, 2001), 155–98.

92. Vaclav Smil, "Nitrogen and Food Production: Proteins for Human Diets," *AMBIO: A Journal of the Human Environment* 31, no. 2 (2002): 126–31.

93. Fitzgerald Booker et al., "The Ozone Component of Global Change: Potential Effects on Agricultural and Horticultural Plant Yield, Product Quality and Interactions with Invasive Species," *Journal of Integrative Plant Biology* 51, no. 4 (2009): 337–51.

94. M. L. de Bauer, "Air Pollution Impacts on Vegetation in Mexico," in *Air Pollution Impacts on Crops and Forests – A Global Assessment,* eds. Lisa Emberson, Mike Ashmore, and Frank Murray (London: Imperial College Press, 2003), 263–86; I. A. Hassan, M. R. Ashmore, and J. N. B. Bell, "Effect of Ozone on Radish and Turnip Under Egyptian Field Conditions," *Environmental Pollution* 89, no. 1(1995): 107–14.

95. Sagar Krupa et al., "Ambient Ozone and Plant Health," *Plant Disease* 85, no. 1 (2001): 4–12.

96. Ivano Fumigalli et al., "Evidence of Ozone-Induced Adverse Effects on Crops in the Mediterranean Region," *Atmospheric Environment* 35, no. 14 (2001): 2583–87.

97. D. Velissariou, "Toxic Effects and Losses of Commercial Value of Lettuce and Other Vegetables Due to Photochemical Air Pollution in Agricultural Areas of Attica, Greece" in *Critical Levels for Ozone – Level II,* eds J. Fuhrer & B. Achermann (Bern, Switzerland: Swiss Agency for Environment, Forest and Landscape,1999), 253–56.

98. A. S. Heagle, "Ozone and Crop Yield," *Annual Review of Phytopathology* 27 (1989): 397–423.

99. Xiaoping Wang and Denise L. Mauzerall, "Characterizing Distributions of Surface Ozone and its Impact on Grain Production in China, Japan, and South Korea: 1990 and 2020," *Atmospheric Environment* 38, no. 26 (2004): 4383–402.

100. M. R. Ashmore, "Assessing the Future Global Impacts of Ozone on Vegetation," *Plant, Cell & Environment* 28, no. 8 (2005): 949–64.

5: TRANSPORTATION

1. This driving pattern is the LA4 test cycle

2. There are about 148 million vehicles registered. An undetermined, somewhat smaller number of vehicles would be on active duty, on the road.

3. Hansan Liu et al., "Mesoporous TiO2-B Microspheres with Superior Rate Performance for Lithium Ion Batteries," *Advanced Materials* 23, no. 30 (2011): 3450–54.

4. Xin Zhao et al., "In-Plane Vacancy-Enabled High-Power Si-Graphene Composite Electrode for Lithium-Ion Batteries," *Advanced Energy Materials* 1, no. 6 (2011): 1079–84.

5. I am assuming $3 per 1 gallon ($3 per 3.78 liters) and 12,000 miles a year (19,300 kilometers).

6. Stephen M. Jasinski, *Phosphate Rock* (U.S. Geological Survey, Mineral Commodity Summaries, January 2011). USGS 2011 report on phosphate rock estimates the phosphate reserves to be 65 billion tons. In the context of car batteries, it might as well be an infinite amount. In the context of fertilizing our crops, this is an entirely different ball of wax, discussed in the ninth chapter of this book ("Land Use").

7. Data is based on Syouhei Nishihama, Kenta Onishi, and Kazuharu Yoshizuka, "Selective Recovery Process of Lithium from Seawater Using Integrated Ion Exchange Methods," *Solvent Extraction and Ion Exchange* 29, no. 3 (2011): 421–31.

8. Brian Jaskula, USGS lithium mineral specialist, personal communication with the author.

9. I have stated that conventional electrical batteries are not quite up to the task of driving the heavy trucks. This is largely true. However, I do want to briefly mention an application where one can go either way: short distance, low speed driving in inner-city transportation or around the port areas. The Nautilus XE30 truck made by Balygon is doing just that. The speed is low; the range is low. At an estimated 45 mph top speed and perhaps a 30 to 50 mile range, these are adequate in the context of transporting containers around port terminals. Charge time is about six to eight hours or one hour for a fast, partial charge, but this point is moot if we allow the batteries to be swapped and set up switch stations at the ports.

10. Daniel Merki et al., "Amorphous Molybdenum Sulfide Films as Catalysts for Electrochemical Hydrogen Production in Water," *Chemical Science* 7, no. 2 (2011): 1262–67.

11. Hemamala I. Karunadasa, Christopher J. Chang, and Jeffrey R. Long, "A Molecular Molybdenum-oxo Catalyst for Generating Hydrogen from Water," *Nature* 464, no. 7293 (2010): 1329–33.

12. G. Wu et al., "High Performance Electrocatalysts for Oxygen Reduction Derived from Polyaniline, Iron, and Cobalt," *Science* 332, no. 6028 (2011): 443–47.

13. To produce all of its hydrogen needs would require 8,660,000 liters of water a month (about 2.3 million gallons). This is about the same amount of water consumed by about 200–250 households.

14. One kg of hydrogen powers 6 miles of highway driving of a fully loaded class 8 truck. Thus, a supply truck with a payload of 800 kg of hydrogen would provide the equivalent of 4,800 miles. One gallon of gasoline powers 4.8 miles of highway driving of a fully loaded class 8 truck. Thus, a tanker truck carrying 8,000 US gallons of fuel would provide the equivalent of 38,400 miles. In other words, a tanker truck can carry 8 times the amount of fuel.

15. Stephanie Flamberg, Susan Rose, and Denny Stephens – Battelle Memorial Institute, *Analysis of Published Hydrogen Vehicle Safety Research*, (National Highway Traffic Safety Administration, February 2010) Report number DOT HS 811 267.

16. See also R. R. Stephenson, *Fire Safety of Hydrogen-Fueled Vehicles; System-Level Bonfire Test*, (La Canada, CA: Motor Vehicle Fire Research Institute, 2005).

17. J. T. Ringland, *Safety Issues for Hydrogen-Powered Vehicles*, (Albuquerque, NM and Livermore, CA: Sandia National Laboratories, 1994).

18. Vivek P. Utgikar and Todd Thisen, "Safety of Compressed Hydrogen Fuel Tanks: Leakage from Stationary Vehicles," *Technology in Society* 27, no. 3 (2005): 315–20.

19. Brian L. Conley, Denver Guess, and Travis J. Williams, "A Robust, Air-Stable, Reusable Ruthenium Catalyst for Dehydrogenation of Ammonia Borane," *Journal of the American Chemical Society* 133, no. 36 (2011): 14212–15.

20. Zhuopeng Tan et al., "Thermodynamics, Kinetics and Microstructural Evolution During Hydrogenation of Iron-Doped Magnesium Thin Films," *International Journal of Hydrogen Energy* 36, no. 16 (2011): 9702–13.
21. Michael A. Tunnell and Rebecca M. Brewster, "Energy and Emissions Impacts of Operating Higher Productivity Vehicles," *Transportation Research Record: Journal of the Transportation Research Board* 1941 (2005): 107–14.
22. Michael Ogburn, Laurie Ramroth, and Amory B. Lovins, *Transformational Trucks: Determining the Energy Efficiency Limits of a Class-8 Tractor-Trailer* (Rocky Mountain Institute, July 2008).
23. Jose Antonio Carrero et al., "Traffic Related Distribution Profiles and Their Impact on Urban Soils," in *Urban Environment Proceedings of the 10th Urban Environment Symposium,* eds. Sébastien Rauch and Gregory M. Morrison (Springer Netherlands, 2012).
24. Transit Cooperative Research Program, *Report 108: Car Sharing; Where and How it Succeeds* (Transportation Research Board, 2005).
25. "Accident Protection Device for Small Cars," *Science Daily,* July 25, 2011, http://www.sciencedaily.com.
26. "Relief from 'Parking Wars': Computer Software to Revamp City Parking," *Science Daily,* October 31, 2011, http://www.sciencedaily.com/releases/2011/10/111031121219.htm.
27. International Energy Agency, "World: Statistics," http://iea.org/stats.
28. Herman Mannstein, Peter Spichtinger, and Klaus Gierens, "How to Avoid Contrail Cirrus," *Transportation Research Part D: Transport and the Environment* (2005).
29. See note 27 above.
30. A passenger is counted twice if they make additional roundtrips in a given year, but not on a return flight.
31. "Current Market Outlook 2011-2030," Boeing, http://www.boeing.com/commercial/cmo/index.html.
32. This does not account for the summer season being a lot more busy than other seasons. But then again, it does not account for the probable drastic reduction in the number of people who choose to make the trip once aviation is out of the picture.
33. J. J. Winebrake et al., "Mitigating the Health impacts of Pollution from Oceangoing Shipping: An Assessment of Low-Sulfur Fuel Mandates," *Environmental Science & Technology* 43, no. 13 (2009): 4776–82.
34. Daniel A. Lack et al., "Particulate Emissions from Commercial Shipping: Chemical, Physical, and Optical Properties," *Journal of Geophysical Research* 114 (2009).
35. Alan H. Lockwood et al., *Coal's Assault on Human Health: Executive Summary* (Physicians for Social Responsibility, November 2009).
36. The annual rate as of 2008 seems to be around 850 tons of mercury. David G. Streets et al., "All-Time Releases of Mercury to the Atmosphere from Human Activities," *Environmental Science & Technology* 45, no. 24 (2011): 10485–91.
37. "Indoor Air Pollution and Health," World Health Organization, http://www.who.int/mediacentre/factsheets /fs292/en.

38. Calculation made by Rod Adams, founder of Adams Atomic Engines, Inc., personal communication with the author.

39. The ZemShip was a hydrogen-powered tourist ship that could carry up to 100 passengers with a 72-ton displacement. It had two 50 kW fuel cells driving a 100 kW motor, and it cruised at 15 kph. Its hydrogen supply was provided by a few canisters that had a combined volume of 50 kg of hydrogen. This supply was good for 24 hours of operation. Refueling was done in 12 minutes. (http://www.zemships.eu)

40. We can compress the hydrogen in the tanks to 10,000 psi (about 690 bars), which merely gets us from having 6.8 times more volume to 5.44 times more volume to get the same mileage diesel gives us. However, the savings in volume are not that impressive, and it is cost-prohibitive to produce large storage tanks capable of storing hydrogen at this ultra-high pressure.

41. R. K. Ahluwalia et al., "Technical Assessment of Cryo-Compressed Hydrogen Storage Tank Systems for Automotive Applications," *International Journal of Hydrogen Energy* 35, no. 9 (2010): 4171–84.

42. 4,500 km journey (2,800 miles), assuming 10–11 hours of driving each day.

43. The stop time is based on Amtrak's: 2 minutes actual stop and 6 to 8 minutes for deceleration and acceleration as the train reaches and departs a station.

44. On July 23, 2011, at least 40 were killed after one high-speed train rear-ended another in China.

45. Daryl Oster, Masayuki Kumada, and Yaopig Zhang, "Evacuated Tube Transport Technologies (ET3) tm: A Maximum Value Global Transportation Network for Passengers and Cargo," *Journal of Modern Transportation* 19, no. 1 (2011): 42–50.

6: BUILDINGS

1. Robert H. Crawford and Graham J. Treloar, "Net Energy Analysis of Solar and Conventional Domestic Hot Water Systems in Melbourne, Australia," *Solar Energy* 76, no. 1–3 (2004): 159–63.

2. P. Denholm, *The Technical Potential of Solar Water Heating to Reduce Fossil Fuel Use and Greenhouse Gas Emissions in the United States* (Golden, CO: National Renewable Energy Laboratory Technical Report, 2007).

3. *Annual Energy Review 2009* (Washington, DC: U.S. Energy Information Administration, 2010).

4. Hashem Akbari, Surabi Menon, and Arthur Rosenfeld, "Global Cooling: Increasing World-Wide Urban Albedos to Offset CO_2," *Climatic Change* 94, no. 3–4 (2008): 275–86.

5. There is a range of estimates. The upper end is 3,506,565 sq km (Global Rural-Urban Mapping Project, Version 1 (GRUMPv1): Population Density Grid. Palisades, NY: Socioeconomic Data and Applications Center (SEDAC), Columbia University, 2004.); The lower limit is 400,000 sq km (Shlomo Angel, Stephen C. Sheppard, and Daniel L. Civco, *The Dynamics of Global Urban Expansion* (Washington DC: Transport and Urban Development Department, The World Bank, 2005)).

6. See note 4 above.

7. Hashem Akbari, H. Damon Matthews, and Donny Seto, "The Long-Term Effect of Increasing the Albedo of Urban Areas," *Environmental Research Letters* 7, no. 2 (2012). doi: 10.1088/1748-9326/7/2/024004.
8. Mark Z. Jacobson and John E. Ten Hoeve, *Effects of Urban Surfaces and White Roofs on Global and Regional Climate* (Stanford, CA: Stanford University Dept. of Civil and Environmental Engineering, 2011).
9. Ronnen Levinson and Hashem Akbari, "Potential Benefits of Cool Roofs on Commercial Buildings: Conserving Energy, Saving Money, and Reducing Emission of Greenhouse Gases and Air Pollutants," *Energy Efficiency* 3, no. 1 (2010), 53–109.
10. Deduced from data in Levinson and Akbari. See note 9 above.
11. Less than 2 feet rise for 12 feet of roof.
12. All reflective values here describe roofs of a minimum of three years old (versus newly applied or constructed roofs), which reflect better the reflection value through the years. Whenever age value was not available, I applied the formula used in California building code title 24 to devise aged value from new value.
13. P. Jaboyedoff et al., "Energy in Air-Handling Units: Results of the AIRLESS European Project," *Energy and Buildings* 36, no. 4 (2004): 391–99.
14. Tyler Hoyt et al., "Energy Savings from Extended Air Temperature Setpoints and Reductions in Room Air Mixing," (International Conference on Environmental Ergonomics 2009, Boston, August 2–7, 2009).
15. The building code of this plan will mandate that at higher outside temperatures buildings have a minimum of 0.3 to 0.4 meters per second air flow, which retains thermal comfort at a temperature of 1°C higher indoors than otherwise. Personal communication with Dr. Hui Zhang, research specialist focusing on human thermal comfort, Center for the Built Environment, UC Berkeley.
16. Chang Hwan Lee et al., "Counterion-Induced Reversibly Switchable Transparency in Smart Windows," *ACS Nano* 5, no. 9 (2011): 7397–403.
17. ThermalCore is an example of phase change material used in building construction. The ThermalCore is a dry wall that absorbs heat during the day and releases it at night. This drywall incorporates tiny capsules of paraffin wax that are invisible to the eye. These capsules keep a room cool in much the same way that ice cubes chill a drink: by absorbing heat as they melt. Each polymer capsule contains paraffin waxes that melt at around room temperature, enabling them to keep the temperature of a room constant throughout the day. At night, when the temperature drop the materials inside the capsules solidify and release the heat they've stored during the day.
18. A good candidate for these flexible tubes is PEX, which is made of high density polyethylene.
19. Kwang Ho Lee and Richard K. Strand, "The Cooling and Heating Potential of an Earth Tube System in Buildings," *Energy and Buildings* 40, no. 4 (2008): 486–94.
20. N. M. Thanu et al., "An Experimental Study of the Thermal Performance of an Earth Air Pipe System in a Single Pass Model," *Solar Energy* 71, no. 6 (2001): 353–64.

21. Maria Kolokotroni, "Night Ventilation Cooling of Office Buildings: Parametric Analyses of Conceptual Energy Impacts," *ASHRAE Transactions* (2001): 1–11.

22. G. Carrilho da Graça et al., "Simulation of Wind Driven Ventilative Cooling Systems for an Apartment Building in Beijing and Shanghai," *Energy and Buildings* 34, no. 1 (2002): 1–11.

23. Or 4.7 kBtu/sq.ft./ per year.

24. Measurement results collected by CEPHEUS (Cost Efficient Passive Houses as European Standards) from more than 100 dwelling units in passive houses.

25. Less common at the moment, but offering higher performance than krypton-filled, are windows with a vacuum between the panes.

26. For both the frame and the glazing insulation rating of 0.8W/m2k or less (European U standard), or 0.14 Btu/hr-ft2 F or less (American U standard).

27. Or 0.1–0.15W/m²/k.

28. May be combinations of EPS (expanded polystyrene), XPS (extruded polystyrene), and Foamglas (made of crushed glass combined with carbon).

29. "Thinner Thermal Insulation," *Science Daily,* December 2, 2011, http://www.sciencedaily.com /releases/2011/12/111202155230.htm.

30. n50 of 0.6 h-1 @ 50 Pa or less.

31. Tim Delhey Eian, passivhaus architect, Te Studio, personal communication with the author.

32. "2010 Conference–Passive House Break Even," Randy Foster, Passive House Institute U.S., http://www.passivehouse.us/passiveHouse/2010_Passive_House_Conference_Presentations,_November_5.html.

33. Jürgen Schnieders, *Passive Houses in South West Europe: A Quantitative Investigation of Some Passive and Active Space Conditioning Techniques for Highly Energy Efficient Dwellings in the South West European Region* (Passiv Haus Institut, 2009); Brian Ford, Rosa Schiano-Pham, and Duan Zhongcheng, eds., "The *Passivhaus* Standard in European Warm Climates: Design Guidelines for Comfortable Low Energy Homes," (Passive-On Project, IEEA, School of the Built Environment, University of Nottingham, 2007).

34. Jürgen Schnieders, see note 33 above.

35. The added heat required comes to 17w/m². Tim Eian, architect on record, personal communication with the author.

36. Robert Hastings and Maria Wall, eds., *Sustainable Solar Housing: Volume 2–Exemplary Buildings and Technologies* (Sterling, VA: Earthscan, 2007).

37. Francis Rubinstein et al., "Preliminary Results from an Advanced Lighting Controls Testbed," (Lawrence Berkeley National Laboratory, 1998).

38. Charles Kibert, *Sustainable Construction: Green Building Design and Delivery* (Hoboken, NJ: John Wiley & Sons, Inc., 2008).

39. As of 2010, Cree came out with a 1 W LED chip that achieved 208 lumen/watt. The assessment of such LED efficiency available commercially was provided by Theeradetch Detchprohm, Future Chips Constellation at Rensselaer Polytechnic Institute, NY, personal communication with the author.

40. Z. B. Wang et al., "Unlocking the Full Potential of Organic Light-Emitting Diodes on Flexible Plastic," *Nature Photonics* 5 (2011): 753–57.
41. L. D. Danny Harvey, *Energy and the New Reality 1: Energy Efficiency and the Demand for Energy Services* (Washington, DC: Earthscan, 2010).
42. International Energy Agency, "World: Statistics: Energy Balance for World," http://www.iea.org/stats /balancetable.asp?COUNTRY_CODE=29.
43. See note 41 above.
44. Raymond J. Cole and Paul C. Kernan, "Life-Cycle Energy Use in Office Buildings," *Building and Environment* 31, no. 4 (1996): 307–17.
45. The ratio of the area of the home's envelope to the interior floor area is less for a large home than for a small home, so large homes have less heat loss per unit of floor area than small homes.

7: ENERGY

1. At least that is the case in the United States.
2. This is the equivalent of –273.15°C or 0 K. This is the point at which no more heat can be removed from a system; the atoms are completely at rest.
3. Actually, it can be as warm as 90 Kelvin (–183°C) and still retain superconductivity.
4. In 2011, Tel Aviv University, Israel, came out with superconducting fibers using sapphire crystals, which may prove promising.
5. *Program on Technology Innovation: A Superconducting DC Cable* (Palo Alto, CA: EPRI, 2009).
6. Ibid.
7. Proceedings of Workshop on Superconducting Direct Current Electricity Transmission (January 21–22, 2010, Houston, Texas).
8. On an 800Kv overhead line, for every 1,000 km distance, we would lose 160 MWh out of 6,400 MWh transmitted total. A superconducting cable will have 20 MWh loss over the same distance: over six times less, or 0.4% loss versus 2.5% loss. Lars Weimers, Chief Engineer Marketing HVDC Systems ABB, personal communication with the author.
9. Steven Eckroad, senior technical manager and acting program manager for the underground transmission program at EPRI (Electrical Power Research Institute), personal communication with the author. It is conceivable that, combined, each pair of bipolar cables would have a reduced cooling needs, hence, reduced electricity needs—a possibility I did not account for here.
10. Mark Stemmle, Project Manager, Superconducting Cable Systems at Nexans Deutschland GmbH, Germany, personal communication with the author.
11. Specs largely provided by superconducting and research division at Nexans, manufacturer of superconducting cables. This overall cable configuration recommended by Stephen P. Ashworth, superconducting team leader at Los Alamos National Laboratory.
12. "Highway Statistics 2008," Federal Highway Administration, http://www.fhwa.dot.gov/policyinformation /statistics/2008.

13. Angela T. Bednarek, "Undamming Rivers: A Review of the Ecological Impacts of Dam Removal," *Environmental Management* 27, no. 6 (2001): 803–14.

14. Mike Staggs, John Lyons, and Kris Visser, "Habitat Restoration Following Dam Removal on the Milwaukee River at West Bend," in *Wisconsin's Biodiversity as a Management Issue: A Report to the Department of Natural Resources Managers* (Madison, WI: Department of Natural Resources, 2005), 202–3.

15. Michael J. Hill, Eric A. Long, and Scott Hardin, *Effects of Dam Removal on Dead lake, Chipola River, Florida* (Southeastern Association of Fish and Wildlife Agencies, 1993).

16. Ibid.

17. J. T. Pawloski and L. A. Cook, "Salling Dam Drawdown and Removal," Unpublished manuscript presented at The Midwest Region Technical Seminar on Removal of Dams, Association of State Dam Safety Officials, Kansas City, Missouri, September 30–October 1, 1993.

18. American Fisheries Society, North Central Division, *The Restoration of Midwestern Stream Habitat: A Symposium Held at the 52nd Midwest Fish and Wildlife Conference, Minneapolis, MN, December 4–5, 1990*, 57–65.

19. If we assume that under this plan the world would require 40,000 TWh annually, existing known inventories would run the world for 6.3 years.

20. The average in the field is 2.2 turbines per one square kilometer. An average turbine was 1.5 to 2 MW, with an average rotor diameter of 80 m. If we assume 2.5 MW with 50 m blades, the density would be reduced to 1.85 turbines per sq km.

21. D. B. Barrie and D. B. Kirk-Davidoff, "Weather Response to Management of a Large Wind Turbine Array," *Atmospheric Chemistry and Physics Discussions* 9 (2009): 2917–31; Liming Zhou et al., "Impacts of Wind Farms on Land Surface Temperature," *Nature Climate Change* (2012) doi 10.1038/nclimate1505

22. Paul M. Cryan and Robert M. R. Barclay, "Causes of Bat Fatalities at Wind Turbines: Hypotheses and Predictions," Journal of *Mammalogy* 90, no. 6 (2009): 1330–40.

23. Justin G. Boyles et al., "Economic Importance of Bats in Agriculture," *Science* 332, no. 6025 (2011): 41–42.

24. Jeffrey M. Ellenbogen et al,. *Wind Turbine Health Impact Study: Report of Independent Expert Panel* (Massachusetts Department of Environmental Protection and Massachusetts Department of Public Health, January 2012).

25. Unlike linear Fresnel and the established parabolic trough, the heliostat can track the sun; thus it has significantly higher collection efficiency. Moreover, with a single central receiver tower, less radiative losses are incurred. Finally, it achieves higher temperatures, which means the standard steam turbine used in coal and natural gas plants can be used.

26. Incidentally, the steam is re-condensed to water again by dry air-cooling fans so it can be reused.

27. Modeled after the Sandia's ATS Fourth Generation heliostat.

28. In considering both concrete towers and steel lattice-towers, it became clear that the lattice will require far fewer raw materials. Steel is needed in both tower types; however, the demand for concrete is far more modest in a lattice tower.

29. I used System Advisor Model (SAM) version 2010, made by National Renewable Energy Laboratory (NREL). For weather data of Typical Meteorological Year (TMY3), I used the National Solar Radiation Data Base developed by NREL.

30. "Joint Staff Assessment/Draft Environmental Impact Statement Released for Rice Solar Energy Project in Riverside County," (California Energy Commission, 2010).

31. Corey J. Noone, Manuel Torrilhon, and Alexander Mitsos, "Heliostat Field Optimization: A New Computationally Efficient Model and Biomimetic Layout," *Solar Energy* 86, no. 2 (2012): 792–803.

32. In addition to the silicon, traditional photovoltaic cells, thin film PVs have entered the market in recent years. Although cheaper (requiring fewer manufacturing resources), they require more footprint to produce the same electrical yield and they are based on rare-earth minerals, such as cadmium telluride and Copper Indium gallium selenide—which at truly high volume may prove to be an issue. See Appendix A.

33. Aspen Environmental Group, *California Valley Solar Ranch Conditional Use Permit, and Twisselman Reclamation Plan and Conditional Use Permit Final Environmental Impact Report* (San Francisco, January 2011).

34. Wai-Lun Chan et al., "Observing the Multiexciton State in Singlet Fission and Ensuing Ultrafast Multielectron Transfer," *Science* 334, no. 6062 (2011): 1541–45.

35. Marco Bernardi et al., "Solar Energy Generation in Three Dimensions," *Energy and Environmental Science* (2012) doi: 10.1039/c2ee21170j

36. "The Basics of Underground Natural Gas Storage," U.S. Energy Information Administration, August 2004, http://www.eia.gov/pub/oil_gas/natural_gas/analysis_publications/storagebasics/storagebasics.html. Some of the underground caverns that had very slow inflow or outflow rates were excluded as they are not viable to this plan—notably depleted aquifers, which the plan calls upon to recharge with groundwater, anyway.

37. Actually, it is more like 12.5 million tons, but we ought to budget for the possibility that at any given moment a few facilities may not be able to provide hydrogen due to some malfunction or otherwise unforeseen circumstances.

38. A maximum of 41,616 tons of hydrogen can be pumped out at any given hour.

39. Heavy equipment, such as tractors, also are to run on hydrogen. While no one knows the total fuel consumption of all heavy equipment, there is little doubt that the underground storage can easily provide for all trucks and the heavy equipment fuel needs, as well.

40. I assume 1 ton of hydrogen yields 15.9 MWh and that while to generate and compress that 1 ton of hydrogen takes about 50 MWh.

41. Willett Kempton et al., "Vehicle-to-Grid Power: Battery, Hybrid, and Fuel Cell Vehicles as Resource for Distributed Electric Power in California," (Institute of Transportation Studies, UC Davis, 2001); Jasna Tomić and Willet Kempton, "Using Fleets of Electric-Drive Vehicles for Grid Support," *Journal of Power Sources* 168, no. 2 (2007): 459–68; Willett Kempton and Steven E. Letendre, "Electric Vehicles as a New Power Source for Electric Utilities," *Transportation Research Part D: Transport and Environment* 2, no. 3 (1997): 157–75.

42. "Vehicle-to-Grid Power: Battery, Hybrid, and Fuel Cell Vehicles as Resource for Distributed Electric Power in California." See note 41 above.

43. I chose residential consumption as a basis. In contrast, total consumption includes the industrial sector, whose energy may or may not serve the local market. Rather, it may go towards manufacturing goods to be exported. Hence, residential consumption is a better representation of consumption levels of the local populace.

44. Adapted from "Energy Consumption: Residential Energy Consumption per Capita," World Resource Institute, http://earthtrends.wri.org/search-able_db/index.php?action=select_countries&theme=6&variable _ID=634.

45. *World Population Prospects: The 2010 Revision: Highlights and Advance Tables* (New York: United Nations, 2011), 2; US Population Projections, Table 6, US Census Bureau, Population Division, http://www.census.gov/population/www/projections/projectionsagesex.html

46. Derived from specs from "Joint Staff Assessment/Draft Environmental Impact Statement Released for Rice Solar Energy Project in Riverside County," (California Energy Commission, 2010).

47. Based on calculation made in Aspen Environmental Group, *California Valley Solar Ranch Conditional Use Permit, and Twisselman Reclamation Plan and Conditional Use Permit Final Environmental Impact Report* (San Francisco, January 2011).

48. A 1 GW capacity array of thermal power would emit between 3 and 37 million tons of CO_2 during construction. In comparison, the construction and first year emissions from a coal plant average about 7 million tons with subsequent annual emissions of about 6 million tons. Derived from data in table S1 in N. P. Myhrvold and K. Caldeira, "Supporting Information for Greenhouse Gases, Climate Change, and the Transition from Coal to Low-Carbon Electricity," *Environmental Research Letters*, available at stacks. iop.org/ERL/7/014019/mmedia. Having said that, the cited study assumes existing energy and cement generation practices. Under the plan on this book, the related GHG emissions will be drastically less.

49. As of 2009, the United States had 2.35 GW geothermal-capacity and a corresponding annual output of 15,209 GWh (i.e., 74% capacity factor). The US Geological Survey estimates that in the lower 48 states, there are 8.38 GW of identified resources and an additional 28.2 GW of undiscovered resources, which together total 36.58 GW. With the same ratio of capacity-to-generation that exists now, it translates to 237 TWh annually, or 27 GW hourly. With the crop-rotation style of energy harvesting described shortly thereafter for the Enhanced Geothermal Systems, it would come to 59 TWh annually and 6.76 GWh hourly.

50. Colin F. Williams et al., "Assessment of Moderate-and High Temperature Geothermal Resources of the United States," US Geological Survey Fact Sheet 3082 (2008).

51. The study was conducted by Google in collaboration with SMU Geothermal Laboratory. More information can be found at http://www.google.org/egs.

52. Enhanced Geothermal Systems resources in southern Canada have not been identified. Of course, we would need to add whatever those may turn out to be.

53. *The Future of Geothermal Energy: Impact of Enhanced Geothermal Systems (EGS) on the United States in the 21st Century* (Massachusetts Institute of Technology, 2006).

54. Massimo Canale et al., "High Altitude Wind Energy Generation Using Controlled Power Kites," *IEEE Transactions on Control Systems Technology* 18, no. 2 (2010): 279–93.

55. Cristina L. Archer and Ken Caldeira, "Global Assessment of High Altitude Wind Power," *Energies* 2, no. 2 (2009): 307–19.

56. Cristina L. Archer and Ken Caldeira, *Atlas of high altitude wind power* (Stanford, CA: Department of Global Energy, Carnegie Institute for Science, 2008), 3-19.

57. Bryan W. Roberts et al., "Harnessing High Altitude Wind Power," *IEEE Transactions on Energy Conversion* 22, no. 1 (2007): 136–44.

58. United Nations Scientific Committee on the Effects of Atomic Radiation, *Effects of Ionizing Radiation: Report to the General Assembly, with Scientific Annexes* (United Nations, 2008).

59. Alas, there are still about a dozen operational nuclear plants in Russia that are of the same design as the one in Chernobyl.

60. Satoru Monzen et al., (2011) "Individual Radiation Exposure Dose Due to Support Activities at Safe Shelters in Fukushima Prefecture," *PLoS ONE* 6, no. 11 (2011).

61. Teppei J. Yasunari et al., "Cesium-137 Deposition and Contamination of Japanese Soils Due to the Fukushima Nuclear Accident," *PNAS* 108, no. 49 (2011): 19530–34.

62. Tom Blees et al., "Advanced Nuclear Power Systems to Mitigate Climate Change," (presented at the 91st American Meteorology Society Annual Meeting, Seattle, WA, January 23–27, 2011).

63. I assume an annual household consumption of 11 MWh.

64. Based on Professor Barry Brook's calculation that 1 ton of fuel within IFR will yield about 8,760 GWh. Personal communication with the author.

65. William H. Hannum, Gerald E. Marsh, and George S. Stanford, "Smarter Use of Nuclear Waste," *Scientific American* 293, (2005): 84–93.

66. Leuren Moret, "Japan's Deadly Game of Nuclear Roulette," *Japan Times*, May 23, 2004.

67. This was calculated and deduced from the country information found in *Human Development Report 2010: The Real Wealth of Nations; Pathways to Human Development* (UNDP, 2010).

454   NOTES

8: NUTRIENT AND WATER RECYCLING

1. Estimates vary, but they seem to average around 200 grams per day per person.
2. Peter Morgan, "Experiments Using Urine and Humus Derived from Ecological Toilets as a Source of Nutrients for Growing Crops," (paper presented at 3rd World Water Forum; Kyoto, Shiga, and Osaka, Japan; March 16–23, 2003).
3. Y. Avnimelech et al., "The Use of Compost for the Reclamation of Saline and Alkaline Soils," *Compost Science & Utilization* 2, no. 3 (1994): 6–11.
4. Nyle C. Brady, 1974. *The Nature and Properties of Soils, 8th Edition* (MacMillan Publishing Co., Inc., 1974), 385–87.
5. D. B. McConnel, A. Shiralipour, and W. H. Smith, "Compost Application Improves Soil Properties," *Biocycle* 34, no. 4 (1993): 61–63.
6. M. A. Cole, Liu Zhang, and Xianzhong Liu, "Remediation of Pesticides Contaminated Soil by Planting and Compost Addition," *Compost Science and Utilization* 3, no. 4 (1995): 20–30.
7. See note 21 below.
8. See note 21 below.
9. Surendra K. Pradhan et al., "Use of Human Urine Fertilizer in Cultivation of Cabbage (*Brassica oleracea*)—Impacts on Chemical, Microbial, and Flavor Quality," *Journal of Agricultural and Food Chemistry* 55, no. 21 (2007): 8657–63.
10. Surandra K. Pradhan et al., "Human Urine and Wood Ash as Plant Nutrients for Red Beet Cultivation: Impacts on Yield Quality," *Journal of Agricultural and Food Chemistry* 58, no. 3 (2010): 2034–39.
11. G. Sridevi et al., "Evaluation of Source Separated Human Urine (ALW) as a Source of Nutrients for Banana Cultivation and Impact on Quality Parameter," *ARPN Journal of Agricultural and Biological Science* 4, no. 5 (2009): 44–48.
12. Jürgen Simons, Gitta Schirmer, and Joachim Clemens, "The Use of Separated Urine as Mineral Fertilizer," (in «Ecosan – Closing the Loop" – Proceedings of the 2nd International Symposium on Ecological Sanitation, Lübeck, Germany, April 7–11, 2003).
13. Surendra K. Pradhan, Jarmo K. Holopainen, and Helvi Henonen-Tanski, "Stored Human Urine Supplemented with Wood Ash as Fertilizer in Tomato (*Solanum lycopersicum*) Cultivation and Its Impacts on Fruit Yield and Quality," *Journal of Agricultural and Food Chemistry* 57, no. 16 (2009): 7612–17.
14. See note 17 below.
15. Surendra K. Pradhan, *Yield and Quality of Vegetables Fertilized with Human Urine and Wood Ash*, 2010. (Dissertations in Forestry and Natural Sciences, University of Eastern Finland, 2010).
16 .M. Winker, "Pharmaceutical Residues in Urine and Potential Risks Related to Usage as Fertiliser in Agriculture" (Doctoral Thesis, Hamburg University of Technology, 2009), http://doku.b.tu-harburg.de/volltexte/2009/557/pdf/PhD_Thesis_Winker.pdf.

17. Helvi Heinonen-Tanski, Surentra K. Pradhan, and Päivi Karinen, "Sustainable Sanitation—A Cost Effective Tool to Improve Plant Yields and the Environment," *Sustainability* 2, no. 1 (2010): 341–53.
18. Ryan Shaw, "The Use of Human Urine as Crop Fertilizer in Mali, West Africa," (Michigan Technological University).
19. Judit Lienert and Tove A. Larsen, "High Acceptance of Urine Source Separation in Seven European Countries: A Review," *Environmental Science & Technology* 44, no. 2 (2010): 556–66.
20. Beate I. Escher, "Monitoring the Removal Efficiency of Pharmaceuticals and Hormones in Different Treatment Processes of Source-Separated Urine with Bioassays," *Environmental Science Technology* 40, no. 16 (2006): 5095–101.
21. Amount of nitrogen before losses due to conversion to ammonia. Based on calculations from B. Vinneras et al., "The Characteristics of Household Wastewater and Biodegradable Solid Waste—A Proposal for New Swedish Design Values," *Urban Water Journal* 3, no. 1 (2006): 3–11.
22. Håkan Jönsson et al., *Guidelines on the Use of Urine and Faeces in Crop Production* (EcoSanRes Programme and Stockholm Environment Institute, 2004).
23. FAOstat Database, under Resources, Resource STAT, Land category, "Arable Land and Permanent Crops," 2008, http://faostat.fao.org.
24. Data synthesized from surveys presented in *Progress on Sanitation and Drinking Water: 2010 Update* (Geneva, Switzerland: World Health Organization and UNICEF, 2010).
25. From a design developed and described in Peter Morgan, *Toilets That Make Compost: Low-Cost, Sanitary Toilets That Produce Valuable Compost for Crops in an African Context* (Stockholm, Sweden: Stockholm Environment Institute, 2007), http://www.ecosanres.org/pdf_files/ ToiletsThatMakeCompost.pdf.
26. Anna Richert et al., "Practical Guidance on the Use of Urine in Crop Production," (Stockholm, Sweden: Stockholm Environmental Institute, EcoSanRes Series 2010- 1), http://www.ecosanres.org/pdf_files/ESR2010-1-PracticalGuidanceOnTheUseOfUrineInCropProduction.pdf.
27. *Results of a Medium-Scale Trial of Single-Use, Self-Sanitizing Toilet Bags in Poor Urban Settlements in Bangladesh* (Dhaka, Bangladesh: GIZ, 2009), http://www.gtz.de/en/dokumente/gtz2010-peepoo-bags-bangladesh-trial.pdf.
28. We produce 1.4 million tons of human manure daily. One kg yields 60 liters of biogas, and 1,000 liters yield 6 kWh worth of thermal energy, which in turn, yield about 2 kWh worth of electricity (based on the conversion rate achieved by Electrigas).
29. Deducing from data in *Municipal Solid Waste in the United States: 2009 Facts and Figures* (EPA, 2009), Americans produce 0.675 kg of organic waste for each person each day. The UK produces 0.465 Kg per person per day (The Guardian, "Tread Lightly: Compost Organic Waste," Carolyne Fry, April 17, 2008, http://www.guardian .co.uk/environment/ethicallivingblog/2008/ apr/18/compostorganicwaste). I estimate the average in developing countries to be 0.5 kg.

30. Deduced from data provided in Henning Hahn and Uwe Hoffstede, *Assessment Report on Operational Experience* (Biogasmax, 2010), http://www.biogasmax.eu/media/d2_11_biogasmax_iwes_vfinal_nov2010 _095398400_1109_10022011.pdf.

31. A 1% loss of methane and 0.28 kWh electricity is required per cubic meter of incoming biogas. Personal communication with Leif Nilsoon, technical director of water treatment at Malmberg Water.

32. Private communication with David S. Ortiz, Ph.D. Engineer, RAND Corporation. Process efficiency is assumed to be between 60% and 65%.

33. Assuming that one barrel contains 42 US gallons, one barrel of jet fuel has the heat content of 5.67 million Btu, one barrel of naphtha has the heat content of 4.62 million Btu, and 1 cubic meter of methane has the heat value of 37,080 Btu. Related information can be found in James I. Hileman et al., *Near-Term Feasibility of Alternative Jet Fuels* (Santa Monica, CA: The Rand Corporation and MIT, 2009), http://stuff.mit.edu:8001/afs /athena. mit.edu/dept/aeroastro/partner/reports/proj17/altfuelfeasrpt.pdf.

34. "Table 2.1 Nonfuel (Feedstock) Use of Combustible Energy, 2006," MECS, October 2009, http://205.254 .135.7/emeu/mecs/mecs2006/pdf/Table2_1. pdf.

35. The only exception is the nuclear waste, for which adequate storage technology has been discussed in the fifth chapter ("Transportation").

36. Tom Blees, *Prescription for the Planet: The Painless Remedy for Our Energy & Environmental Crises* (2008), 141–54.

37. The figure for the paved area is derived from some analysis shown in the sixth chapter ("Buildings").

38. Source of information about the system: http://eomega.org/omega/about/ocsl/eco-machine/.

39. Tyrone B. Hayes et al., "Demasculinization and Feminization of Male Gonads by Atrazine: Consistent Effects Across Vertebrate Classes," *The Journal of Steroid Biochemistry and Molecular Biology* 127, no. 1–2 (2011): 64–73.

40. Pengpeng Grimshaw, Joseph M. Calo, and George Hradil, "Cyclic Electrowinning/Precipitation (CEP) System for the Removal of Heavy Metal Mixtures from Aqueous Solutions," *Chemical Engineering Journal* 175 (2011): 103–9.

9: LAND USE

1. This figure includes aquaculture, which for the most part is situated at sea.

2. Sources of data and the basis for the calculations: *The State of Food Insecurity in the World* (Rome: Food and Agriculture Organization of the United Nations, 2011), 44–47; *World Health Statistics 2011* (World Health Organization), 105–13; "Dietary Energy Consumption" from the Excel file "Food Consumption Nutrients" available at http://www. fao.org/fileadmin/templates/ess/documents/food_security_statistics/ FoodConsumptionNutrients _en.xls, from the FAO Statistics Division; *World Population Prospects: The 2010 Revision: Highlights and Advance Tables* (New York: United Nations, 2011), 2.

3. Linda Scott Kantor et al., "Estimating and Addressing America's Food Losses," *Food Review* (USDA, 1997), http://www.calrecycle.ca.gov/ReduceWaste/Food/FoodLosses.pdf.
4. Lorrayne Ventour, *The Food We Waste* (WRAP, 2008), 4–6, http://www.ns.is/ns/upload/files/pdf-skrar/matarskyrsla1.pdf.
5. Jonathan Bloom, *American Wasteland: How America Throws Away Nearly Half of Its Food (and What We Can Do About It)* (Cambridge, MA: Da Capo Press, 2010).
6. Paul Hepperly et al., "Compost, Manure And Synthetic Fertilizer Influences Crop Yields, Soil Properties, Nitrate Leaching And Crop Nutrient Content," *Compost Science & Utilization* 17, no. 2 (2009): 117–26.
7. A possible exception is biochar to be discussed at length in the eleventh chapter ("Drawing Down Carbon").
8. Tara W. Hudiburg et al., "Regional Carbon Dioxide Implications Of Forest Bioenergy Production," *Nature Climate Change* 1 (2011): 419–23.
9. Estimated 86 tons of CO_2 per hectare per year (annualized over 50 years). See S. E. Page et al., "Review Of Peat Surface Greenhouse Gas Emissions from Oil Palm Plantations in Southeast Asia," International Council on Clean Transportation, White Paper Number 15 (2011).
10. D. W. Letter, R. Seidel, and W. Liebhardt, "The Performance of Organic and Conventional Cropping Systems in an Extreme Climate Year," *American Journal of Alternative Agriculture* 18, no. 3 (2003): 146–54.
11. James J. Elser, "A World Awash with Nitrogen," *Science* 334, no. 6062 (2011): 1504–5.
12. G. Philip Robertson and Peter M. Vitousek, "Nitrogen in Agriculture: Balancing the Cost of an Essential Resource," *Annual Review of Environment and Resources* 34 (2009): 97–125.
13. S. Park et al., "Trends and Seasonal Cycles in the Isotopic Composition of Nitrous Oxide Since 1940," *Nature Geoscience* 5, no. 4 (2012):261. doi:10.1038/ngeo1421.
14. See note 6 above.
15. Cover crops are crops planted expressly for building up and holding soil. They are also planted to suppress weeds.
16. David Pimentel et al., "Environmental, Energetic, and Economic Comparisons of Organic and Conventional Farming Systems," *BioScience* 55, no. 7 (2005): 573–82.
17. Oluyede Clifford Ajayi et al., "Agricultural Success from Africa: The Case of Fertilizer Tree Systems in Southern Africa (Malawi, Tanzania, Mozambique, Zambia And Zimbabwe)," *International Journal of Agricultural Sustainability* 9, no. 1 (2011): 129–36.
18. Catherine Badgley et al., "Organic Agriculture and the Global Food Supply," *Renewable Agriculture and Food Systems* 22, no. 2 (2007): 86–108.
19. Jan Willem Erisman et al., "How a Century of Ammonia Synthesis Changed the World," *Nature Geoscience* 1 (2008): 636–39.
20. Holger Kirchmann et al., "Can Organic Crop Production Feed The World?" in *Organic Crop Production - Ambitions and Limitations*, eds. Holger Kirchmann and Lars Bergström (Dordrecht, Netherlands: Springer, 2008), 39–72.

21. Verena Seufert, Navin Ramankutty, and Jonathan A. Foley, "Comparing the Yields of Organic and Conventional Agriculture," *Nature* (2012). doi: 10.1038/nature11069.

22. Sources of data and basis for the calculations: FAOstat database, under Production, Live Animals, categories "World" and "Live Animals," http://faostat.fao.org; Economic Research Service, US Department of Agriculture, "Manure Use for Fertilizer and for Energy: Report to Congress," (United States Department of Agriculture, 2009); I assume that half of the world inventory is dairy cows and half is fed cattle.

23. T. S. Cox et al., "Progress in Breeding Perennial Grains," *Crop and Pasture Science* 61, no. 7 (2010).

24. Lindsay W. Bell, Len J. Wade, and Mike A. Ewing, "Perennial Wheat: A Review of Environmental and Agronomic Prospects for Developing in Australia," *Crop and Pasture Science* 61, no. 9 (2010): 679–90.

25. C. J. Gantzer et al., "Estimating Soil Erosion after 100 Years of Cropping on Sanborn Field," *Journal of Soil and Water Conservation* 45, no. 6 (1990): 641–44.

26. G. W. Randall et al., "Nitrate Losses Through Subsurface Tile Drainage in Conservation Reserve Program, Alfalfa, and Row Crop Systems," 1997 *Journal of Environmental Quality* 26, no. 5 (1996): 1240–47.

27. Lindsay W. Bell et al., "A Preliminary Whole-Farm Economic Analysis of Perennial Wheat in an Australian System," *Agricultural Systems* 96, no. 1–3 (2008): 166–74.

28. Frank G. Dohleman and Stephen P. Long, "More Productive Than Maize in The Midwest: How Does Miscanthus Do It?" *Plant Physiology* 150, no. 4 (2009): 2104–15.

29. S. E. Bruce et al., "Pasture Cropping: Effect on Biomass, Total Cover, Soil Water & Nitrogen," CSIRO Sustainable Ecosystems, http://www.pasturecropping.com/index.php?option=com_content&view=article&id=50:pasture-cropping-effect-on-biomass-total-cover-soil-water-a-nitrogen&catid=40:research-findings&Itemid=63.

30. Ibid.

31. Christine Jones, *Adapting Farming To Climate Variability*, available at http://www.amazingcarbon.com/PDF /JONES-AdaptingFarming(April09).pdf.

32. G. D. Millar and W. B. Badgery, "Pasture Cropping: A New Approach to Integrate Crop and Livestock Farming Systems," *Animal Production Science* 49, no. 10 (2009): 777–87; see note 29 above.

33. To facilitate this irrigation and flooding regimen, there is a one-time need to dig drainage channels, 12 to 15 feet apart, during the first year.

34. To see how this estimate has been derived, see the discussion on cool roofs in the sixth chapter ("Buildings").

35. It is likely that the number is higher once we account for the patios. Naturally, it is not that each household has enough land area to feed its occupants; it is just that once we tally all the portions—small or otherwise—that are available in each backyard we come to the total, capable of feeding 20 percent of the world's adults.

36. Dickson Despommier, "Growing Skyscrapers: The Rise of Vertical Farms," *Scientific American* 301 (2009): 80–87.

37. There is a simple explanation for it. Warm-blooded animals need to use a significantly greater amount of energy just to keep their body warm, whereas cold-blooded species do not. For more information, see note 38 below.

38. Gretchen Vogel, "For More Protein, Filet Of Cricket," *Science* 327, no. 5967 (2010): 811.

39. Definitions vary and estimates range from 18% to 80% of the world's total land area. Source: H. Gyde Lund, "Accounting For The World's Rangelands," *Rangelands* 29, no. 1 (2007): 3–10.

40. "Livestock Grazing Systems & the Environment," in *Livestock & the Environment: Finding a Balance,* ed. Cees de Haan, Henning Steinfeld, and Harvey Blackburn, FAO (Suffolk, UK: WREN*media*).

41. The field diet of the hens is being supplemented with outside sources of grain, bone meal, and other forms of chicken feed.

42. Amy Rebecca Sapkota et al., "Lower Prevalence of Antibiotic-resistant Enterococci on U.S. Conventional Poultry Farms that Transitioned to Organic Practices," *Environmental Health Perspectives* 119, no. 11 (2011): 1622–28.

43. *The State Of The World Fisheries And Aquaculture 2010* (Rome: FAO, 2010). These figures exclude aquatic plants.

44. Subhas K. Venayagamoorthy et al., "Numerical Modeling Of Aquaculture Dissolved Waste Transport In A Coastal Embayment." *Environmental Fluid Mechanics* 11, no. 4 (2011): 329–52.

45. Holten C. Lützhøft, B. Halling-Sørenson, and S. E. Jørgensen, "Algal Toxicity of Antibacterial Agents Applied in Danish Fish Farming," *Archives of Environmental Contamination and Toxicology* 36, no. 1 (1999): 1–6.

46. J. F. Burka et al., "Drugs In Salmonid Aquaculture," *Journal of Veterinary Pharmacology and Therapeutics* 20, no. 5 (1997): 333–49.

47. Natasha M. Franklin, Jennifer L. Stauber, and Richard P. Lim, "Development of Flow Cytometry-Based Algal Bioassays for Assessing Toxicity of Copper in Natural Waters," *Environmental Toxicology and Chemistry* 20, no. 1 (2001): 160–70.

48. Nicole S. Webster et al., "The Effects of Copper on the Microbial Community of a Coral Reef Sponge," *Environmental Microbiology* 3, no. 1 (2001): 19–31.

49. L. E. Burridge, K. Haya, and S. L. Waddy, "The Effect of Repeated Exposure to Azamethiphos on Survival and Spawning in the American Lobster (*Homarus americanus*)," *Ecotoxicology and Environmental Safety* 69, no. 3 (2008): 411–15.

50. Hans Ackefors and Magnus Enell, "Discharge of Nutrients from Swedish Fish Farming to Adjacent Sea Areas," *Ambio* 19, no. 1 (1990): 28–35.

51. R. S. S. Wu, "The Environmental Impact of Marine Fish Culture: Towards a Sustainable Future." *Marine Pollution Bulletin* 31, no. 4–12 (1995): 159–66; The compilation and synthesis of all the environmental problem has been conducted by Global Aquaculture Performance Index. For more information, see http://www.seaaroundus .org/sponsor/gapi.aspx.

52. About 7.57 billion tons worth of crops and about 1.15 billion tons of meat and eggs. FAOstat database, under Production, category "Crops" or "Livestock Primary," http://faostat.fao.org.

53. Frederic T. Barrows, Hagerman Fish Culture Experiment Station, USDA, research physiologist (fish), personal communication with the author.

54. Andrew L. Rhyne et al., "Revealing the Appetite of the Marine Aquarium Fish Trade: The Volume and Biodiversity of Fish Imported into the United States." *PLoS ONE*, 2012; 7 (5): e35808 DOI: 10.1371/journal.pone.0035808

55. Ike Olivotto et al., "Advances in Breeding and Rearing Marine Ornamentals," *Journal of the World Aquaculture Society* 42, no. 2 (2011): 135–66.

56. FAOstat database, under Forestry, category "ForesSTAT," http://faostat. fao.org.

57. Bruce Lippke et al., "Life Cycle Impacts of Forest Management and Wood Utilization on Carbon Mitigation: Knowns and Unknowns," *Carbon Management* 2, no. 3 (2011): 303–33.

58. It comes in a brilliant white color (brightness around 96), but if a more mellow look is desired for a given book, it can be done by lightly coating the sheets with yellow (10%–15% screen).

59. It includes both semi-natural forests and forest plantations. It includes only productive plantations and forests, not those used to hold back erosion or those that are part of restoration efforts. Source of the data: A. Del Lungo, J. Ball, and J. Carle, *Global Planted Forests Thematic Study: Results and Analysis* (Planted Forests and Trees Working Paper 38, Rome: FAO, 2006). The data is for 2005, but the figure I am using is my extrapolation for estimated area for 2011.

60. Extrapolated from Jim Carle and Peter Holmgren, "Wood from Planted Forests: A Global Outlook 2005–2030," *Forest Products Journal* 58, no. 12 (2008): 6–18. There is a gap between potential and actual yield at present, as many of the productive plantations are not being harvested.

61. Ibid.

62. A. Y. Hoekstra et al., "2012 Global Monthly Water Scarcity: Blue Water Footprints versus Blue Water Availability." *PLoS ONE*, Published electronically February 29, 2012. e32688.doi:10.1371/journal.pone.0032688

63. A. Antunez et al., "Impact of Surface and Subsurface Drip Irrigation on Yield and Quality of 'Honey Dew' Melon," ISHS Acta Horticulturae 889 (2011): 417–22.

64. T. L. Thompson, Huan-cheng Pang, and Yu-yi Li, "The Potential Contribution of Subsurface Drip Irrigation to Water-Saving Agriculture in the Western USA," *Agricultural Sciences in China* 8, no. 7 (2009): 850–54.

65. V. O. Sadras, "Does Partial Root-Zone Drying Improve Irrigation Water Productivity In The Field? A Meta-Analysis," *Irrigation Science* 27, no. 3 (2009): 183–90.

66. In total, 2,700 billion cubic meters are estimated to be used annually for irrigation. Source: David Molden, ed., *Water for Food, Water for Life: A Comprehensive Assessment of Water Management in Agriculture* (London: Earthscan, 2007), 5.

67. Sandra Postel and Brian Richter, *Rivers for Life: Managing Water for People and Nature* (Washington, DC: Island Press, 2003), 1–41.
68. Pamela Chelme-Ayala, Daniel W. Smith, and Mohamed Gamal El-Din, "Membrane Concentrate Management Options: A Comprehensive Critical Review," *Canadian Journal of Civil Engineering* 36, no. 6 (2009): 1107–19.
69. Kay Paulsen and Frank Hensel, "Design of an Autarkic Water and Energy Supply Driven by Renewable Energy Using Commercially Available Components," *Desalination* 203, no. 1–3 (2007): 455–62.
70. The Canadian portion of the North America region, with its myriad lakes, does not seem to require the services of desalination and is therefore not accounted for.
71. For water requirements of photovoltaic and solar power towers see the seventh chapter ("Energy"). Water requirements of nuclear plants are assumed to be 1.8 liters per kWh. This is based on P. Torcellini, N. Long, and R. Judkoff, *Consumptive Water Use for U.S. Power Production* (Golden, CO: NREL, 2003).
72. In actuality, it is virtually certain that some of the livestock water needs would go away, as we would eliminate the current water used for power washes in the central feeding lots.
73. *2007 Census of Agriculture: Farm and Ranch Irrigation (2008)* 3 (USDA, 2009), 4–12.
74. The basis for this assumption is the existing 4.4% evaporation rate through a 336-mile canal through the desert, the Arizona Central Project, and its added 0.6% loss due to seepage. While some of the routes may be longer than 336 miles, I assume that it would still average 5% water loss as for the most part they won't traverse arid, hot terrain.

10: NATURE RESTORATION

1. Camilo Mora and Peter F. Sale, "Ongoing Global Biodiversity Loss and the Need to Move Beyond Protected Areas: a Review of the Technical and Practical Shortcomings of Protected Areas on Land and Sea," *Marine Ecology Progress Series* 434 (2011): 251–66.
2. William D. Newmark, "Legal and Biotic Boundaries of Western North American National Parks: A Problem of Congruence," *Biological Conservation* 33, no. 3 (1985): 197–208.
3. Andrew F. Bennett, *Linkages in the Landscape: The Role of Corridors and Connectivity in Wildlife Conservation* (Gland, Switzerland and Cambridge, UK: International Union for Conservation of Nature and Natural Resources, 1999), 24–29.
4. Mark L. Shaffer, "Minimum Population Sizes for Species Conservation," *BioScience* 31, no. 2 (1981): 131–34.
5. Ellen I. Damschen et al., "Corridors Increase Plant Species Richness at Large Scales." *Science* 313, no. 5791 (2006): 1284–86.
6. "Wildways: Creating Landscapes for Life," Wildlands Network, http://www.twp.org/wildways.

7. Daniel J. Simberloff et al., "Regional and Continental Restoration," in *Continental Conservation: Scientific Foundations of Regional Reserve Networks*, eds. Michael E. Soulé and John Terborgh (Washington, D. C.: Island Press, 1999), 65–98.

8. Ayn Shlisky et al., *Fire, Ecosystems and People: Threats and Strategies for Global Biodiversity Conservation* (GFI Technical Report 2007–2, Arlington, VA: The Nature Conservancy, 2007).

9. Sam Bingham, *Grassroots Restoration: Holistic Management For Villages* (Albuquerque, NM: The Savory Center).

10. Montserrat Núñez et al., "Assessing Potential Desertification Environmental Impact in Life Cycle Assessment," *International Journal Life Cycle Assessment* 15, no. 1 (2010): 67–78.

11. Boris A. Yurtsev, "The Pleistocene 'Tundra-Steppe' and the Productivity Paradox: The Landscape Approach," *Quaternary Science Review* 20, no. 1–3 (2001): 165–74.

12. S. A. Zimov et al., "Steppe-Tundra Transition: A Herbivore-Driven Biome Shift at the end of the Pleistocene," *The American Naturalist* 146, no. 5 (1995): 765–94.

13. F. W. M. Vera, E. S. Bakker, & H. Olff, "Large Herbivores: Missing Partners of Western European Light-Demanding Tree and Shrub Species?" *Conservation Biology Series – Cambridge-* 11 (2006): 203–31.

14. Daniel H. Janzen, "Dispersal of Small Seeds by Big Herbivores: Foliage is the Fruit," *The American Naturalist* 123, no. 3 (1984): 338–53.

15. Paulo R. Guimarães Jr., Mauro Galetti, & Pedro Jordano, "Seed Dispersal Anachronisms: Rethinking the Fruits Extinct Megafauna Ate," *PLoS ONE* 3, no. 3 (2008).

16. Daniel H. Janzen and Paul S. Martin, "Neotropical Anachronisms: The Fruits the Gomphotheres Ate," *Science* 215, no. 4528 (1982): 19–27.

17. Chris N. Johnson, *Australia's Mammal Extinctions: A 50000 Year History* (Cambridge, UK: Cambridge University Press, 2006), 116–19.

18. William J. Bond and John A. Silander, "Springs and Wire Plants: Anachronistic Defences Against Madagascar's Extinct Elephant Birds." *Proceedings of the Royal Society B* 274, no. 1621 (2007): 1985–92.

19. C. N. Johnson, "Ecological Consequences of Late Quaternary Extinctions of Megafauna," *Proceedings of the Royal Society B* 276, no. 1667 (2009): 2509–19.

20. Sergey A. Zimov, "Pleistocene Park: Return of the Mammoth's Ecosystem," *Science* 308, no. 5723 (2005): 796–98.

21. Inbar Friedrich Ben-Nun et al., "Induced Pluripotent Stem Cells from Highly Endangered Species," *Nature Methods* 8 (2011): 829–31.

22. R. Verma et al. "Inducing Pluripotency in Somatic Cells from the Snow Leopard (*Panthera uncia*), an Endangered Felid." *Theriogenology* 77, no. 1 (2012): 220–28.

23. A study has reconstructed the vegetation of the previous interglacial period at a site in northwest Siberia. They found forbs, sedges, grasses, and in general, pioneer communities composed of species typical of dry, disturbed sites. Furthermore, they found spores of dung fungi, indicative of a strong impact of large herbivores. Frank Kienast et al., "Continental

Climate in the East Siberian Arctic During the Last Interglacial: Implications From Palaeobotanical Records." *Global and Planetary Change* 60, no. 3–4 (2008): 535–62.

24. Sergei Zimov, "Mammoth Steppes and Future Climate," *Science in Russia* 5 (2007): 105–12.

25. Connie Barlow, "Deep Time Lags: Lessons from Pleistocene Ecology," in *Gaia in Turmoil: Climate Change, Biodepletion and Earth Ethics in an Age of Crisis,* eds. Eileen Crist and H. Bruce Rinker (Massachusetts Institute of Technology, 2010), 165–73.

26. A. El-Keblawy, T. Ksiksi, and H. El Alqamy, "Camel Grazing Affects Species Diversity and Community Structure in the Deserts of the UAE," *Journal of Arid Environments* 73, no. 3 (2009): 347–54.

27. C. Josh Donlan et al., "Pleistocene Rewilding: An Optimistic Agenda for Twenty First Century Conservation," *The American Naturalist* 168, no. 5 (2006): 660–81.

28. John A. Byers, *American Pronghorn: Social Adaptations and the Ghosts of Predators Past* (Chicago: University of Chicago Press, 1997), 10–14.

29. James A. Estes et al., "Trophic Downgrading of Planet Earth," *Science* 333, no. 6040 (2011): 301–6.

30. John W. Terborgh et al., "Ecological Meltdown in Predator-Free Forest Fragments," *Science* 294, no. 5548 (2001): 1923–26.

31. M. Crête, "The Distribution of Deer Biomass in North America Supports the Hypothesis of Exploitation Ecosystems," *Ecology Letters* 2, no. 4 (1999): 223–27.

32. Kevin R. Crooks and Michael E. Soulé, "Mesopredator Release and Avifaunal Extinctions in a Fragmented System," *Nature* 400 (1999): 563–66.

33. Robert L. Beschta, "Cottonwoods, Elk, and Wolves in the Lamar Valley of Yellowstone National Park." *Ecological Applications* 13, no. 5 (2003): 1295–309; William J. Ripple and Eric J. Larsen, "Historic Aspen Recruitment, Elk, and Wolves in Northern Yellowstone National Park, USA," *Biological Conservation* 95, no. 3 (2000): 361–70.

34. B. E. McLaren and R. O. Peterson, "Wolves, Moose and Tree Rings on Isle Royale," *Science* 266, no. 5 (1994): 1555–58.

35. Michael Soulé and Reed Noss, "Rewilding and Biodiversity: Complementary Goals for Continental Conservation," *Wild Earth* 22 (1998): 1–11.

36. Christopher C. Wilmers, "Resource Dispersion and Consumer Dominance: Scavenging at Wolf- and Hunter-Killed Carcasses in Greater Yellowstone, USA," *Ecology Letters* 6, no. 11 (2003): 996–1003.

37. Joel Berger, "A Mammalian Predator-Prey Imbalance: Grizzly Bear and Wolf Extinction Affect Avian Neotropical Migrants," *Ecological Applications* 11, no. 4 (2001): 947–60.

38. Sandra Tranquilli et al., "Lack of Conservation Effort Rapidly Increases African Great Ape Extinction Risk," *Conservation Letters* 5, no. 1 (2011): 48–55.

39. It is easy enough to monitor and record the game brought out from the various reserves, assuring that hunting is duly restricted.

40. Dan Dagget, *Gardeners of Eden: Rediscovering Our Importance to Nature* (Flagstaff, AZ: Thatcher Charitable Trust, 2005), 10–12.
41. For more information about this grazing method, see the ninth chapter ("Land Use").
42. Actual inventory is about 6.7 million sheep, 9.8 million horses, 110 million head of cattle, and 3 million goats. FAOstat database, under Production, category "Live Animals," http://faostat.fao.org.
43. I assume that the land there can support 300 head of cattle per acre per day, which is based on existing practices described in the ninth chapter ("Land Use"). I further assume that every piece of land can be sustainably grazed twice a year, which seems somewhat understated from the account of Polyface Farm.

11: DRAWING DOWN CARBON

1. In relation to 1950 levels.
2. P. Rampal et al., "IPCC Climate Models Do Not Capture Arctic Sea Ice Drift Acceleration: Consequences in Terms of Projected Sea Ice Thinning and Decline," *Journal of Geophysical Research* 116 (2011).
3. Katherine Richardson et al., *Synthesis Report* (from *Climate Change: Global Risks, Challenges & Decisions*, Copenhagen, March 10–12, 2009).
4. Jonathan Overpeck and Bradley Udall, "Dry Times Ahead," *Science* 328, no. 5986 (2010): 1642–43.
5. James Hansen et al., "Target Atmospheric CO_2: Where Should Humanity Aim?" *The Open Atmospheric Science Journal* 2 (2008): 217–31.
6. J. E. N. Veron et al., "The Coral Reef Crisis: The Critical Importance of <350 ppm CO_2," *Marine Pollution Bulletin* 58, no. 10 (2009): 1428–36.
7. Victoria J. Fabry et al., "Impacts of Ocean Acidification on Marine Fauna and Ecosystem Processes," ICES *Journal of Marine Science* 65 no. 3 (2008): 414–32.
8. Michelle Waycott et al., "Vulnerability of Seagrasses in the Great Barrier Reef to Climate Change," in *Climate Change and the Great Barrier Reef: A Vulnerability Assessment,* eds. J. E. Johnson and P. A. Marshall (Townsville, Australia: Great Barrier Reef Marine Park Authority, 2007), 193–235.
9. Participants of the Ocean Acidification Workshop, convened by The Nature Conservancy August 12–14, 2008, Honolulu, Hawaii, "The Honolulu Declaration on Ocean Acidification and Reef Management," *Journal of International Wildlife Law & Policy* 12, no. 1–2 (2009): 121–26.
10. Bradley C. Congdon et al., "Vulnerability of Seabirds on the Great Barrier Reef to Climate Change," in *Climate Change and the Great Barrier Reef: A Vulnerability Assessment,* eds. J. E. Johnson and P. A. Marshall (Townsville, Australia: Great Barrier Reef Marine Park Authority, 2007), 465–96.
11. Ibid.
12. Ivan R. Lawler, Guido Parra, and Michael Noad, "Vulnerability of Marine Mammals in the Great Barrier Reef to Climate Change," in *Climate Change and the Great Barrier Reef: A Vulnerability Assessment,* eds. J. E. Johnson and P. A. Marshall (Townsville, Australia: Great Barrier Reef Marine Park Authority, 2007), 497–513.

13. Michael J. Kingsford and David J. Welch, "Vulnerability of Pelagic Systems of the Great Barrier Reef to Climate Change," in *Climate Change and the Great Barrier Reef: A Vulnerability Assessment,* eds. J. E. Johnson and P. A. Marshall (Townsville, Australia: Great Barrier Reef Marine Park Authority, 2007), 555–92.

14. Marcus Sheaves et al., "Vulnerability of Coastal and Estuarine Habitats In the GBR to Climate Change," in *Climate Change and the Great Barrier Reef: A Vulnerability Assessment,* eds. J. E. Johnson and P. A. Marshall (Townsville, Australia: Great Barrier Reef Marine Park Authority, 2007), 593–620.

15. E. J. Rohling et al., "Antarctic Temperature and Global Sea Level Closely Coupled Over the Past Five Glacial Cycles," *Nature Geoscience* 2 (2009): 500–4.

16. Joel B. Smith et al., "Assessing Dangerous Climate Change Through an Update of the Intergovernmental Panel on Climate Change (IPCC) 'Reasons For Concern,'" *PNAS* 106, no. 11 (2009): 4133–37.

17. Çağan H. Şekercioğlu, Richard B. Primack, and Janice Wormworth, "The Effects of Climate Change on Tropical Birds," *Biological Conservation.* Published electronically 21 February, 2012. http://dx.doi.org/10.1016/j.biocon.2011.10.019

18. For related information and conclusions see: Nathan P. Gillett et al., "Ongoing Climate Change Following a Complete Cessation of Carbon Dioxide Emissions," *Nature Geoscience* 4 (2011): 83–87.

19. James P. Kennett et al. "Carbon Isotopic Evidence for Methane Hydrate Instability During Quaternary Interstadials," *Science* 288, no. 5463 (2000): 128–33.

20. Natalie Shakhova et al., "Extensive Methane Venting of the Atmosphere from Sediments of The East Siberian Arctic Shelf," *Science* 327, no. 5970 (2010): 1246–50.

21. See note 22 below.

22. David Archer, Bruce Buffett, and Victor Brovkin, "Ocean Methane Hydrates as a Slow Tipping Point in the Global Carbon Cycle," *PNAS* 106, no. 49 (2009): 20596–601.

23. Mark Maslin et al., "Gas Hydrates: Past and Future Geohazard?" *Philosophical Transactions of the Royal Society* 368, no. 1919 (2010): 2369–93.

24. See note 22 above.

25. See note 23 above.

26. Charles Tarnocai et al., "Soil Organic Carbon Pools in the Northern Circumpolar Permafrost Region," *Global Biogeochemical Cycles* 23 (2009).

27. Tingjun Zhang et al., "Spatial and Temporal Variability in Active Layer Thickness Over the Russian Arctic Drainage Basin," *Journal of Geophysical Research* 110 (2005).

28. P. Lemke et al., "Observations: Changes in Snow, Ice and Frozen Ground," in *Climate Change 2007: The Physical Science Basis; Summary for Policymakers, Technical Summary and Frequently Asked Questions,* eds S. Solomon et al. (contribution of Working Group I to the Fourth Assessment Report of the Intergovernmental Panel on Climate Change, Cambridge, UK and New York: Cambridge University Press, 2007).

29. T. E. Osterkamp, "Characteristics of the Recent Warming of Permafrost in Alaska," *Journal of Geophysical Research* 112 (2007).

30. M. T. Jorgenson and T. E. Osterkamp, "Response of Boreal Ecosystems to Varying Modes of Permafrost Degradation," *Canadian Journal of Forest Research* 35, no. 9 (2005): 2100–111.

31. Nebojsa Nakicenovic and Rob Swart, eds., *Special Report on Emissions Scenarios* (IPCC).

32. David M. Lawrence and Andrew G. Slater, "A Projection of Severe Near-Surface Permafrost Degradation During the 21st Century," *Geophysical Research Letters* 32 (2005).

33. Eliseev, A. V. et al., "Changes in Climatic Characteristics of Northern Hemisphere Extratropical Land in the 21st Century: Assessments with the IAP RAS Climate Model," *Izvestiya Atmospheric and Oceanic Physics* 45, no. 3 (2009): 271–83.

34. The assumption is that the carbon content is 10–14 kg/m^2, similar to that for soils in the temperate region. Charles Tarnocai, expert on peatlands and permafrost, personal communication with the author.

35. Kevin Schaeffer et al., "Amount and Timing of Permafrost Carbon Release in Response to Climate Warming," *Tellus B* 63, no. 2 (2011): 165–80.

36. The basis for the calculation is as follows: Averaged over 100 years, 1 ton of methane is equivalent in its impact to 25 tons of carbon dioxide. One ton of carbon makes 3.67 tons of carbon dioxide. Henceforth, I multiplied the methane figure by 25 and the carbon figure by 3.67.

37. I confirmed this to be the case with the lead authors of both models.

38. Without polar ice, one can assume an additional 2°C global annual warming. Professor Matthew Huber, Purdue Climate Change Research Center, personal communication with the author.

39. Galen A. McKinley et al., "Convergence of Atmospheric and North Atlantic Carbon Dioxide Trends on Multidecadal Timescales," *Nature Geoscience* 4 (2011): 606–10.

40. I assume that half would remain in the atmosphere. I converted the CO_2 to carbon (dividing it by 3.67) and then divided it further by 2.12 to arrive at a value of CO_2 ppm. This does not account for the prediction that the oceans and land sinks will be able to take smaller portion of CO_2 from the air than they take at present.

41. Those two figures represent the two extreme projection figures of both the land-based and ocean-based hydrates discussed earlier—and then converted to atmospheric concentration in a method described in note 36.

42. David J. Beerling and Dana L. Royer, "Convergent Cenozoic CO_2 History," *Nature Geoscience* 4 (2011): 418–20.

43. An estimate arrived at by Matthew Huber, a paleo-climatologist at Purdue, personal communication with the author.

44. Valier Galy andTimothy Eglinton, "Protracted Storage of Biospheric Carbon in the Ganges–Brahmaputra Basin," *Nature Geoscience* 4 (2011): 843–47.

45. M. Huber and R. Caballero, "The Early Eocene Equable Climate Problem Revisited," *Climate of the Past Discussions* 7 (2011): 241–304.

46. Steven C. Sherwood and Matthew Huber, "An Adaptability Limit to Climate Change Due to Heat Stress," PNAS 107, no. 21 (2010): 9552–55.

47. See note 35 above.

48. The model assumes an A1B scenario, with 700ppm reached by 2100 and remaining at that level henceforth. The model further assumes 6.75°C annual mean warming in the permafrost regions versus 10°C–11°C annual mean warming assumed under a fossil-intensive, business-as-usual scenario.

49. Derived from the permafrost carbon flux of Shaefer's model, and figures are scaled by a factor of 4.68 to account for estimated carbon pool of 1,466 billion tons versus the 313 billion tons that were fed into the model. The model defines the outgassing in terms of carbon being outgassed. I converted it to CO_2 equivalent, assuming the outgassing would be 93% CO_2 and 7% methane.

50. Gerald A. Meehl et al., "Model-Based Evidence of Deep-Ocean Heat Uptake During Surface-Temperature Hiatus Periods," Nature Climate Change 1 (2011): 360–64.

51. Gerald A. Meehl et al., "How Much More Global Warming And Sea Level Rise?" Science 307, no. 5716 (2005): 1769–72.

52. Derived from the conclusion of the paper of Hansen et al. in which long-term climate sensitivity is 6°C (4–8°C range). See note 5 above. [James Hansen]

53. RCP Concentration Calculation & Data Group, M. Meinshausen et al., "Annual Average, Global Mean Radiative Forcing," RCP8.5, Final Release November 26, 2009.

54. There is a high level of uncertainty as to the total combined radiative forcing of aerosols (both direct and indirect). See Johannes Quaas et al., "Aerosol Indirect Effects–General Circulation Model intercomparison and Evaluation with Satellite Data," Lawrence Berkeley National Laboratory, July 14, 2010, http://escholarship.org/uc/item/5hx3j5wd and Daniel Rosenfeld et al., "Aerosol Cloud-Mediated Radiative Forcing: Highly Uncertain and Opposite Effects From Shallow and Deep Clouds." Paper presented at the WCRP Open Science Conference, Denver, CO, October 2011.

55. I did not account here for the slow, natural decrease in GHG in the atmophere. But then again, I also did not account for the fact that so far the ocean and air are not in heat equalibrium; so far the ocean soaks up much of the heat.

56. Dean Roemmich, W. John Gould, and John Gilson, "135 Years of Global Ocean Warming Between the Challenger Expedition and the Argo Programme," Nature Climate Change (2012). doi:10.1038/nclimate1461.

57. Lüthi, D., et al.. 2008. EPICA Dome C Ice Core 800KYr Carbon Dioxide Data. IGBP PAGES/World Data Center for Paleoclimatology Data Contribution Series # 2008-055. NOAA/NCDC Paleoclimatology Program, Boulder CO, USA.

58. Based on Long Cao's gracious adaptation for this book of his model. See Long Cao and Ken Caldeira, "Atmopheric Carbon dioxide removal: long-term consequences and commitment," Environmenal Research Letters, Published electronically 30 June, 2010. doi:10.1088/1748-9326/5/2/024011

59. Whether it outgasses as CO_2 or methane, it will all convert to CO_2 in a matter of a few years. So in the end, it comes to carbon.

60. We average a total of 32 billion tons of emitted CO_2 annually from fossil fuel and cement production. In addition, we are responsible for other greenhouse gases whose combined greenhouse effect is equivalent to 12 billion tons of CO_2. Thus, in total, we average 44 billion tons of CO_2 equivalent.

61. Lauren Anthony et al., *Carbon Capture and Storage: An Assessment; A V600 Capstone Course* (The Indiana University CREE, 2010).

62. "Post-Combustion Capture Research," US Department of Energy, http://www.fe.doe.gov/programs /powersystems/pollutioncontrols/Retrofitting_Existing_Plants.html.

63. Edward S. Rubin, Alumni Professor of Environmental Engineering and Science, personal communication with the author.

64. Alain Goeppert et al., "Carbon Dioxide Capture from the Air Using a Polyamine Based Regenerable Solid Adsorbent," *Journal of the American Chemical Society* 133, no. 50 (2011): 20164–67.

65. Ryan C. Moffet and Kimberly A. Prather, "In-Situ Measurements of the Mixing State and Optical Properties of Soot with Implications for Radiative Forcing Estimates," *PNAS* 106, no. 29 (2009): 11872–7; J. M. Haywood and V. Ramaswamy, "Global Sensitivity Studies of the Direct Radiative Forcing Due to Anthropogenic Sulfate and Black Carbon Aerosols," *Journal of Geophysical Research* 103, no. D6 (1998): 6043–58.

66. Mark Z. Jacobson, "Short-Term Effects of Controlling Fossil-Fuel Soot, Biofuel Soot and Gases, and Methane on Climate, Arctic Ice, and Air Pollution Health," *Journal of Geophysical Research* 115, no. D14209 (2010).

67. T. M. L. Wigley, "A Combined Mitigation/Geoengineering Approach to Climate Stabilization," *Science* 314, no. 5798 (2006): 452–54.

68. A. Jones et al., "Geoengineering by Stratospheric SO_2 Injection: Results From the Met Office HadGEM2 Climate Model and Comparison With the Goddard Institute for Space Studies ModelE," *Atmospheric Chemistry and Physics* 10 (2010): 5999–6006.

69. Kelly E. McCusker, David S. Battisti, and Cecilia M. Bitz, "The Climate Response to Stratospheric Sulfate Injections and Implications for Addressing Climate Emergencies," *Journal of Climate*, Published electronically 2011. doi: http://dx.doi.org/10.1175/JCLI-D-11-00183.1.

70. Susan Solomon et al., "The Role of Aerosol Variations in Anthropogenic Ozone Depletion at Northern Midlatitudes," *Journal of Geophysical Research*, 101, no. D3 (1996): 6713–27; Susan Solomon, "Stratospheric Ozone Depletion: A Review of Concepts and History," *Reviews of Geophysics* 37, no. 3 (1999): 275–316.

71. Adverse effects on the hydrological cycle may also result from blocking sunlight before it reaches the Earth's surface. More energy absorbed at the surface returns to the atmosphere through evaporation than through radiation, and the latent heat released occurs elsewhere as water vapor and is transported many hundreds of kilometers before it condenses in the form of rain or snow. For more details, see Kevin E. Trenberth and Aiguo Dai,

"Effects of Mount Pinatubo Volcanic Eruption on the Hydrological Cycle as an Analog of Geoengineering," *Geophysical Research Letters* 34, no. L15702 (2007).

72. P. J. Irvine, R. L. Sriver, and K. Keller, "Tension Between Reducing Sea-Level Rise And Global Warming Through Solar-Radiation Management," *Nature Climate Change* 2 (2012): 97–100.

73. Justin McClellan et al., *Geoengineering Cost Analysis: Final Report*, (Cambridge, MA: Aurora Flight Sciences, 2010).

74. Losses estimated for the period 1850–1998, R. Lal, "Soil Carbon Sequestration Impacts on Global Climate Change and Food Security," *Science* 304, no. 5677 (2004): 1623–27.

75. Ibid.

76. Gundula Azeez, *Soil Carbon and Organic Farming* (Soil Association, 2009).

77. 1.5 billion hectares are under cultivation worldwide.

78. See note 75 above.

79. 20 to 40 kilograms per hectare (kg/ha) for wheat, 10 to 20 kg/ha for maize, and 0.5 to 1 kg/ha for cowpeas.

80. Jonathan Sanderman, Ryan Farquharson, and Jeffrey Baldock, *Soil Carbon Sequestration Potential: A Review for Australian Agriculture* (CSIRO, 2010).

81. M. A. Liebig et al., "Soil Carbon Storage by Switchgrass Grown for Bioenergy," *Bioenergy Research* 1, no. 3–4 (2008): 215–22.

82. Douglas B. Kell, "Breeding Crop Plants with Deep Roots: Their Role in Sustainable Carbon, Nutrient and Water Sequestration," *Annals of Botany* 108, no. 3 (2011): 407–18.

83. Christine Jones, *Carbon That Counts*, available at http://soilcarboncoalition.org/files/JONES-Carbon-that-counts-20Mar11.pdf.

84. Christine Jones, *Mycorrhizal Fungi—Powerhouse of the Soil*, available at http://www.amazingcarbon.com /PDF/JONES-MycorrhizalFungiEVERGREEN(Sept09).pdf.

85. Christine Jones, "Liquid Carbon Pathway Unrecongnised," *Australian Farm Journal*, July 2008, 15–17.

86. Equivalent to about 120 billion tons of carbon.

87. Adriana Downie, Alan Crosky, and Paul Munroe, "Physical Properties of Biochar," in *Biochar for Environmental Management: Science and Technology*, eds. Johannes Lehmann and Steven Joseph (London: Earthscan, 2009), 13–32.

88. Christoph Steiner et al., "Long Term Effect of Manure, Charcoal and Mineral Fertilisation on Crop Production and Fertility on a Highly Weathered Central Amazonian Upland Soil." *Plant and Soil* 291, no. 1–2 (2007): 275–90.

89. L. Van Zwieten et al., "Papermill Char: Benefits to Soil Health and Plant Production " (Conference of the International Agrichar Initiative, Terrigal, NSW, Australia, April 30–May 2, 2007); Bruno Glaser et al., "Potential of Pyrolysed Organic Matter in Soil Amelioration" (12th International Soil Conservation (ISCO) Conference, Beijing, 2002).

90. Kishimoto, S. and Sugiura, G. 1985 "Charcoal as a Soil Conditioner," in symposium on Forest Products Research, International Achievement for the Future, vol 5 pp12–23; Mikan, C. J. and Abrams, M. D. 1995 "Altered Forest Composition And Soil Properties." *Bioresources Technology*, vol 96, pp 699–706.
91. "K. Yin Chan and Zhihong Xu Biochar: Nutrient Properties and Their Enhancement" in *Biochar Environmental Management* et. Johannes Lehmann and Stephen Joseph. (2009)
92. See note 89 above.
93. K. Yin Chain and Zhihong Xu, "Biochar: Nutrient Properties and Their Enhancement," in *Biochar for Environmental Management: Science and Technology*, Johannes Lehmann and Steven Joseph, eds. (London: Earthscan, 2009), 67–84.
94. The use of heat recovery technique would have made it more efficient. K.C. Das et al., "Steam Pyrolysis and Catalytic Steam Reforming of Biomass for Hydrogen and Biochar Production," *Applied Engineering in Agriculture* 26 (2010): 137–46.
95. 44 percent of the 126.6 kg total yield of syngas.
96. Almudena Hospido et al., "Environmental Evaluation of Different Treatment Processes for Sludge From Urban Wastewater Treatments: Anaerobic Digestion Versus Thermal Processes." *The International Journal of Life Cycle Assessment* 10, no. 5 (2005): 336–45.
97. Extrapolated from Fridolin Krausmann et al., "Global Patterns of Socioeconomic Biomass Flows in the Year 2000: A Comprehensive Assessment of Supply, Consumption and Constraints," *Ecological Economics* 65, no. 3 (2008): 471–87.
98. I assume that about half of the processed residue will be used as compost and half will be applied as biochar.
99. This is the equivalent of capturing about 8.17 billion tons of carbon.
100. The source of data for the specs of the air-capture device are derived from K. S Lackner, "Capture of Carbon Dioxide from Ambient Air," *The European Physical Journal – Special Topics* 176, no. 1 (2009): 93–106.
101. Jennifer J. Roberts, Rachel A. Wood, and R. Stuart Haszeldine, "Assessing the Health Risks of Natural CO_2 Seeps in Italy," *PNAS* 108, no. 40 (2011): 16545–48.
102. IPCC, *Carbon Dioxide Capture and Storage* (New York: Cambridge University Press, 2005).
103. Y. Raza, "Uncertainty Analysis of Capacity Estimates and Leakage Potential for Geologic Storage of Carbon Dioxide in Saline Aquifers" (master's thesis, MIT, 2009), http://sequestration.mit.edu/pdf/YamamaRaza_Thesis_May09.pdf.
104. National Energy Technology Laboratory, *2009 Carbon Sequestration Atlas of the United States and Canada, 2nd ed.* (US Department of Energy).
105. Christine Ehlig-Economides and Michael J. Economides, "Sequestering Carbon Dioxide in a Closed Underground Volume," *Journal of Petroleum Science and Engineering* 70, no. 1–2 (2010): 123–30.
106. Howard J. Herzog, "Scaling Up Carbon Dioxide Capture and Storage: From Megatons to Gigatons," *Energy Economics* 33, no. 4 (2010): 597–604.

107. B. Peter McGrail et al., "Potential for Carbon Dioxide Sequestration in Flood Basalts," *Journal of Geophysical Research* 11, no. B12201 (2006).

12: CONSUMPTION

1. Marc L. Imhoff et al., "Global Patterns in Human Consumption of Net Primary Production," *Nature* 429, no. 6994 (2004): 870–73.
2. Brian Maurer, "Relating Human Population Growth to the Loss of Biodiversity," *Biodiversity Letters* 3 (1996):1–5.
3. Ibid.
4. *World Population Prospects: The 2010 Revision: Highlights and Advance Tables* (New York: United Nations, 2011). Medium Projection. Under this scenario we are projected to hit 10 billion around 2080 and 10.1 billion around 2100.
5. Paul A. Murtaugh and Michael G. Schlax, "Reproduction and the Carbon Legacies Of Individuals," *Global Environmental Change* 19, no. 1 (2009): 14–20.
6. Christopher E. Doughty and Christopher B. Field, "Agricultural Net Primary Production in Relation to That Liberated by the Extinction of Pleistocene Mega-Herbivores: An Estimate of Agricultural Carrying Capacity?" *Environmental Research Letters* 5, no. 4 (2010).
7. Kalantary, I. "Family Planning Program Will Be Established in Iran", *Kyhan International* (2 December). (1989).
8. Farnaz Vahidnia, "Case Study: Fertility Decline in Iran," *Population and Environment* 28, no. 4–5 (2007): 259–66.
9. Ibid.
10. Friedrich Schmidt-Bleek, *The Earth: Natural Resources and Human Intervention* (Haus Publishing, 2009), 75.
11. Ibid., 20.
12. "Material Intensity of Materials, Fuels, Transport Services, Food," (Wuppertal Institute for Climate, Environment and Energy, 2011).
13. Stuart Licht et al., "STEP Cement: Solar Thermal Electrochemical Production of CaO Without CO2 Emission," *Chem. Commun.* (2012). doi: 10.1039/C2CC31341C
14. C. Liedtke, C. Manstein, and T. Merten, "MIPS, Resource Management and Sustainable Development," (paper presented at the Conference on the Recycling of Metals, Amsterdam, October 1994).
15. N. Quijorna, G. San Miguel, and A. Andrés, "Incorporation of Waelz Slag into Commercial Ceramic Bricks: A Practical Example of Industrial Ecology," *Industrial & Engineering Chemistry Research* 50, no. 9 (2011): 5806–14.

13: ECONOMIC AND POLITICAL PARADIGM

1. William L. Laurence, "Lung Cancer Rise Laid to Cigarettes; Increase in Smoking is Blamed by Ochsner at Session of Surgeons in Chicago," New York Times, October 26, 1940, 17; Alton Ochsner, M. D. and Michael DeBakey, M. D., "Carcinoma of the Lung," *Archives of Surgery* 42, no. 2 (1941): 209–58.

2. William L. Laurence, "Cigarettes Linked to Cancer in Lungs: Study Of 200 Male Sufferers Shows 95.5% Were Heavy Smokers 20 Years," New York Times, February 27, 1949, 39.
3. Robert Schrek, M. D. et al., "Tobacco Smoking as an Etiologic Factor in Disease. I. Cancer," Cancer Research 10 (1950): 49–58.
4. Ernest L. Wynder and Evarts A. Graham, M. D., "Tobacco Smoking as a Possible Etiologic Factor in Bronchiogenic Carcinoma," The Journal of the American Medical Association 143, no. 4 (1950): 329–36.
5. Ernest L. Wynder, Evarts A. Graham, and Adele B. Croninger, "Experimental Production of Carcinoma with Cigarette Tar," Cancer Research 13 (1953): 855–64.
6. Levin, Goldstein, and Gerhardt, 1950; Mills and Porter, 1950; McConnel, Gorden and Jones, 1952; Koulumies, 1953; Dungal, 1950; Gsell, 1951; Sadowsky, Gilliam and Cornfield, 1953; Watson and Congten, 1954; Richard Doll and Bradford Hill, "The Mortality of Doctors in Relation to their Smoking Habits," British Medical Journal 1 (1954): 4877.
7. Allan M. Brandt, The Cigarette Century: The Rise, Fall, and Deadly Persistence of the Product That Defined America (New York: Basic Books, 2007), 148.
8. Robert K. Plumb, "Study on Smoking and Cancer is Set," New York Times, October 20, 1954, 39.
9. Hrayr S. Karagueuzian et al., "Cigarette Smoke Radioactivity and Lung Cancer Risk," Nicotine & Tobacco Research 14, no. 1 (2012): 79–90.
10. Suggestions for Research on Polonium-210 in Tobacco. 7 Feb 1964. American Tobacco Company. Bates No. 950142942/2945. http://legacy.library.ucsf.edu/tid/uhl31a00;jsessionid=E86559F5D507480F94DF3615F1B EED72 .tobacco03.
11. C. J. P. de Siqueira. Souza Cruz. Radioactivity Measurements on Tobacco Leaf & Cigarettes. 30 Aug 1988. British American Tobacco. Bates no. 400751698 /1699. http://legacy.library.ucsf.edu/tid/pwn62a99.
12. "Nuclear Radiation and Health Effects," World Nuclear Association, November 2011, http://www.world-nuclear.org/info/info5.html.
13. See note 9 above.
14. See note 7 above.
15. The Importance of Younger Adults. R.J. Reynolds. Bates No. 503418151/8156. http://legacy.library.ucsf.edu /tid/eyn18c00.
16. Paul M. Fischer, M. D. et al., "Brand Logo Recognition by Children Aged 3 to 6 Years: Mickey Mouse and Old Joe the Camel," Journal of the American Medical Association 266, no. 22 (1991): 3145–48; Patrick J. Coughlin and Frank Janacek, Jr., "A Review of R. J. Reynolds' Internal Documents Produced in Mangini vs. R. J. Reynolds Tobacco Company, Civil Number 939359: The Case that Rid California and the American Landscape of 'Joe Camel,'" (Milberg, Weiss, Bershad, Hynes & Lerach LLP, 1998).
17. Joseph R. DiFranza et al., "Tobacco Promotion and the Initiation of Tobacco Use: Assessing the Evidence for Causality," Pediatrics 117, no. 6 (2006): 1237–48.
18. Mattias Öberg et al., "Worldwide Burden Of Disease From Exposure To Second-Hand Smoke: A Retrospective Analysis Of Data From 192 Countries," The Lancet 377, no. 9760 (2011): 139–46.

19. Lauren M. Petrick, Alona Svidovsky, and Yael Dubowski, "Thirdhand Smoke: Heterogeneous Oxidation of Nicotine and Secondary Aerosol Formation in the Indoor Environment," *Environmental Science & Technology* 45, no. 1 (2011): 328–33.

20. "WMO Greenhouse Gas Bulletin: The State of Greenhouse Gases in the Atmosphere Based on Global Observations Through 2010," no. 7. (2011): 1–4.

21. Glen P. Peters et al., "Rapid Growth In CO_2 Emissions After the 2008–2009 Global Financial Crisis," *Natural Climate Change* 2 (2012): 2–4.

22. Peter Joseph, "Arriving at a Resource Based Economy," (London Z-Day 2011, London, March 13, 2011).

23. The analogy of the needed open-source economy to the existing science culture can only be pushed so far. Practices within the science culture are contaminated. Under the existing paradigm, producing research papers is the economic lifeline of the scientist. This encourages hoarding of data and findings until publication and does not lend itself to collaboration wiki style and in real time among thousands of scientists from throughout the world. Furthermore, some segments of scientific research are permeated with patents, and data sharing may be riddled with red tape.

24. Richard Wilkinson, *The Impact of Inequality: How to Make Sick Societies Healthier* (New York: The New Press, 2005); Adam Whitworth, "Inequality and Crime across England: A Multilevel Modelling Approach," *Social Policy and Society* 11, no. 1 (2012): 27–40.

25. Philip K. Howard, "Four Ways to Fix a Broken Legal System," (TED Conference, February 9–12, 2010).

26. Ibid.

27. Sabina Alkire et al., *Multidimensional Poverty Index 2011* (Oxford: OPHI, 2011).

28. John Burbidge ed., *Approaches That Work In Rural Development: Emerging Trends, Participatory Methods and Local Initiatives* (New York: K. G. Saur, 1988).

29. Some provinces, like Virginia and New York, broke off, others, like Nova Scotia and East Florida, remained.

30. In addition to ministries, there would be secondary agencies that may include a world postal service, non-utilitarian science exploration, election commission, atomic regulatory agency, public relations, public safety & security, and building council (This matter would be mostly dealt with by local government; however, some *footprint* caps and general policy guidelines in this matter would be issued by the building council.)

31. "The World's Billionaires," Forbes, http://www.forbes.com/wealth/billionaires; James B. Davies et al., "Estimating the Level and Distribution of Global Household Wealth," (United Nations University World Institute for Development Economics Research, 2007).

32. Edward N. Wolff, "Recent Trends in Household Wealth in the United States: Rising Debt and the Middle Class Squeeze—An Update to 2007," (Working paper No. 589, Levy Economics Institute of Bard College, 2010).

33. Thomas R. Dye, *Who's Running America?: The Bush Restoration* (Prentice Hall, 2002).

34. Nicholas Stern, "A Profound Contraction at the Heart of Climate Change Policy," *Financial Times*, December 8, 2011.

APPENDIX A

1. J. Paidipati et al., *Rooftop Photovoltaics Market Penetration Scenarios* (Golden, CO: NREL, 2008).
2. Ken Zweibel, "The Impact of Tellurium Supply on Cadmium Telluride Photovoltaics," *Science* 328, no. 5979 (2010): 699–701.
3. Selya Price and Robert Margolis, *2008 Solar Technologies Market Report* (US Department of Energy, 2010); Andrea Feltrin and Alex Freundlich, "Material Considerations for Terawatt Level Deployment of Photovoltaics,"*Renewable Energy* 33, no. 2 (2008): 180–85.

APPENDIX B

1. L. Giglio et al., "Assessing Variability and Long-Term Trends in Burned Area by Merging Multiple Satellite Fire Products," *Biogeosciences* 7 (2010): 1171–86.
2. The total is around 2 billion tons a year. However, some man-made fires ought to cease, such as forest clearing and the burning of agricultural waste. What will be left are grassland and open savanna fires, with 0.85 GtC emissions; woodland fires, with annual average of about 0.3 GtC; and forest fires (excluding deforestation) with about 0.3 GtC. Thus, we arrrive at a total of 1.45 GtC.
3. Professor Guido van der Werf, a leading expert on wildfire emissions, personal communication with the author.
4. The shrubs produce phenolic compounds. Olle Zackrisson, Marie-Charlotte Nilsson, and David A. Wardle, "Key Ecological Function of the Charcoal from Wildfire in the Boreal Forest," *OIKOS* 77, no. 1 (1996): 10–19.
5. Johannes Lehmann et al. "Stability of Biochar in Soil," in Johannes Lehmann and Steven Joseph eds., *Biochar for Environmental Management: Science and Technology* (London: Earthscan, 2009).
6. José A. González-Pérez et al., "The Effect of Fire on Soil Organic Matter—a Review," *Environment International* 30, no. 6 (2004): 855–70.
7. Matthew S. Waldram, William J. Bond, and William D. Stock, "Ecological Engineering by a Mega-Grazer: White Rhino Impacts on a South African Savanna," *Ecosystems* 11, no. 1 (2008): 101–112.
8. C.N. Johnson, "Ecological Consequences of Late Quaternary Extinctions of Megafauna.". *Proceedings of the Royal Society B* 276, no. 1667 (2009): 2509–19.
9. Jacquelyn L. Gill et al., "Pleistocene Megafaunal Collapse, Novel Plant Communities, and Enhanced Fire Regimes in North America," *Science* 326, no. 5956 (2009): 1100–3.
10. Ayn Shlisky et al., "Fire, Ecosystems and People: Threats and Strategies for Global Biodiversity Conservation," (Boulder, CO: The Nature Conservancy Global Fire Initiative, 2007).

11. R. D. Schuiling and P. Krijgsman, "Enhanced Weathering of Olivine as a Cheap and Sustainable CO_2 Capture Strategy," *Climate Change* 74 (2006): 349–54.
12. Suzanne J. T. Hangx & Christopher J. Spiers, "Coastal Spreading of Olivine to Control Atmospheric CO_2 Concentrations: A Critical Analysis of Viability," *International Journal of Greenhouse Gas Control* 3, no. 6 (2009): 757–67.
13. R. D. Schuiling and P.L. de Boer, "Coastal Spreading of Olivine to Control Atmospheric CO2 Concentrations: a Critical Analysis of Viability. Comment: Nature and Laboratory Models are Different," *International Journal of Greenhouse Gas Control* (2010).
14. See note 11 above.
15. Peter Köhler, Jens Hartmann, and Dieter A. Wolf-Gladrow, "Geoengineering Potential of Artificially Enhanced Silicate Weathering of Olivine," *PNAS* 107, no. 47 (2010): 20228–33.
16. This allows for the fact that rocks contain less than 100 percent olivine content.
17. R.D. Schuiling & O. Tickell, "Olivine Against Climate Change and Ocean Acidification," available at http://www.innovationconcepts.eu/res/literatu- urSchuiling/olivineagainstclimatechange23.pdf.
18. See note 15 above.

INDEX

A123 Systems, 97
advertising, 379
aerosols, 9, 308, 315
Africa, 238
Africa Center for Holistic Management, 295
Agassi, Shai. *See* Better Place
agriculture. *See also* grains, perennial; *See also* nitrogen: bio-availability
    Africa, 238
    Austria, 247–249
    chemical fertilizers, 82
    existing status quo, 236–237
    fertilizers, 237
    Japan, 243–246
    Kenya, 246
    organic, 237–240
    pasture cropping, 241–242
    rice growing, 243–246
    yields, 238–240
air transport. *See* aviation
algae, algae bloom, 45, 111
alternative energy. *see individual technologies (e.g., wind turbines)* solar power towers
Amazon rainforest, 13–21
    climate change effects, 13, 20
    deforestation, 16–18
    drought, 19–20
    fires, 17
    Harvesting, 15
    hunting within, 15–16
    insects, 20
    logging and its effects, 15–16
    Mammals, 20
    pathogens and fungus, 19–20
ammonia, 237

anaerobic digesters, 222–223
Antarctica, 5, 26, 40, 41, 423, 435
apes, 13
Apparent Temperature, 48–51
aquaculture, 262–268
    problems, 263–264
    zero-discharge tank systems, 264–267
Arizona Central Project, 275
ASHREA, 134
Asia, 289
Australia, 9, 76, 96, 289, 306
automobiles. *See* road vehicles
aviation, 111–114
    air routes to retain under this plan, 112–114
    NOx, 112, 114
    super-saturated zones, 112
axis. *See* orbit, Earth
Ballard, 185
Balsam woolly adelgids. *See* invasive species: animals
Bangladesh, 221
Barefoot Power, 213
bats, 30
battery swapping. *See* Better Place
Belarus, 412
Bernard Shaw, George, 360
Better Place, 91–94, 118
bicycle programs, 107–108
bicycles, 107–108
    electric bike, 108
biochar, 321–324
biofuel, 111, 236
biogas, 111, 222–225
biometric ID, 379
birth control. *See* overpopulation: two-child policy

bison, 11, 12
black carbon, 314
blast furnaces, 349
Brandt, Allan, 411
buildings. *See* housing
Burkina Faso, 217
Burma, 412
cadmium telluride, 416
calamity, 3
camelina, 111
Canada, 23
carbon capture and storage
    biochar, 321–324
    capture (fire) smoke, 417–419
    direct-air capture machines, 324–327
    flue-gas carbon capture, 313–315
    land, 316–320
        grassland, 318
        peatlands, 320
    olivine, using, 419–421
    sequestration, 327–329
carbon dioxide
    CO2 concentration, 8, 298
    emissions of, 6–7, 8, 301–308, 360
        permafrost, 301–303
    heightened levels at sea, 37
carbon-negative economy, 309
carbon targets, 308–313
Caribbean monk seals, 37
carnivores. *See* predators
cars. *See* road vehicles
car-sharing, 106–107
cattle. *See* livestock
cement, 348
Center for Marine Biotechnology, 264
charcoal. *See* biochar
Cheatgrass. *See* invasive species: plants
cheetah, 290, 291, 292
Chernobyl Disaster, 206
China, 72, 80, 97, 138, 306, 412
chlorofluorocarbon (CFC) gases, 309
CityCar, 108, 109
Clemens Terminal, 184, 186
climate dynamics, 5–7
    Hothouse World, 7
    Icehouse House, 7
    orbit, Earth, 6
    sun, 7
climate refugees, 55
clover, 238, 239, 243, 244, 245
Club of Rome, 99

CO2. *See* carbon dioxide
coal, 62, 67–68
    coal power plants, 167
    health hazards, 121
Columbia, 16
Common Law, 381, 382
community bicycle programs, 107–108
community gardens, 388
Community Service, 382
Compost, 215–217
    composition, 217
composting toilet, 220
concrete, 348
Congo, DR, 13
consumerism, 339–347, 353–361
cooking stoves, solar, 213
copper indium gallium selenide, 416
coral reefs, 42–43, 299
corn, 12, 193, 194, 240, 246, 248, 251, 257,
    261, 332, 349
corporation, 382
cropland, 77
crops. *See* agriculture
crowdsourcing, 401
Dai, Aiguo, 76
dams
    environmental problems, 272
Dead Sea, 280
deforestation, 79
Delhi. *See* Indo-Gangetic Plain
Deliberatorium, 401
Denmark, 96
Dennis, Neil, 256
desalination scheme, 272–278
Desmodium, 246
Despommier, Dickson, 251
Diclofenac, 31
DK Group, 124
driverless car, 110–111
drought, 75–76. *See also* water shortage
drugs and antibiotics, 30–31, 36
dugongs, 37
Dust Bowl, 75, 76, 442
Dye, Thomas, 410
ecological footprint, 333, 334, 340–346,
    340–347, 342. *See also* "footprint"
economy
    called-for ("open source"), 361–364,
    372–380
        currency, 374
        loan scheme, 372

production, 375–377
property, 388, 394–395
unemployment benefits, 373
wages, 374
existing, 351–361, 377
inequality, 409
Soviet economy, 380
Ecosystems
redundancy, 21
Ecovative, 350
Edmonton Composting Facility, 215, 225
education, 389–391
Einstein, 363
electric arc furnace, 349
electric cars. *See* road vehicles: electric
cars
electricity. *See* power, electricity
electrolyzer, 185
elephants, 13
El Niño, 5
Emerald Cities, 134
emission scenarios, 9
energy. *See* power, electricity
enhanced geothermal systems (EGS),
203–205
Eocene, 10, 304, 305, 306, 466
Eocene Epoch, 10
e-reader, 269
ethnic groups, 400
Europe, 76
Everglades, 27
excreta. *See* manure, human
extinctions, 32–33, 44, 285
family planning. *See* overpopulation: two-
child policy
FCX Clarity, 99–100
feces. *See* manure, human
fertility. *See* overpopulation
financing the blueprint, 4
Finland, 389
fires. *See* wildfires
fire suppression, 21
Fischer-Tropsch process, 224
Five Nations of the Iroquois, 13
Fluence Z.E., 93, 98
food. *See also* agriculture
calories, consumption, 232
waste, 233–234
food production, 76–83
"footprint", 342–347
allotment for the poor, 369

gift or inheritance, 370
loan scheme, 366
forests. *See also* wood
Africa, 23
Asia, 23
Beetles, 21–22
Europe, 23
fungus, 25
piñon and juniper woodlands, 22–23
tree die-off, 23–24
fossil fuel, 89, 116
cheap, 56
scarcity and depletion, 3, 56–71
freedom of movement, 385–386
frogs, 30, 32, 33
fuel cell (hydrogen), 100–101, 364–371
Fukuoka, Masanobu, 243–246
Fukushima, 207, 211, 453
fungus, 30, 32
Future Farm Industries Cooperative
Research Centre, 241
Gabon, 13
Gaviotas, 213
Gemasolar, 174
geoengineering, sulfur into the atmosphere, 315–316
geothermal power, 203
glacial. *See* ice ages
"global dimming". *See* aerosols
global warming, 7–10, 48–52, 298–299,
305–306
Google Driverless car, 110–111
government
existing, 359–361, 395
types, 408–409
local governments, future functions
of, 399
world government, 395–413
charter, 398
election process, 402–413
make up and working of, 396–397
radical transparency, 401
safeguards, 399, 407
grains, perennial, 240–241
Grand Turk, 37
grassroots organization, called for, 402
Great Depression, 351
Great Plains, 171, 196, 288, 289
Greeks, 79
Greenhouse Effect, 6–8
weathering, 7

greenhouse gases (GHG)
   emission of, 3, 8
      cattle, 8
Greenland, 26
Greenlight Planet, 213
grid. *See* power grids, transcontinental
Groundhog Day, movie, 352
Grow Biointensive, 250, 251
Grow Fish Anywhere, 264
Guatemala, 394
Gulf of Mexico, 37
habitat
   fragmentation, 29
hairy vetch, 238
Haiti, 38
Hammarby Sjöstad development, 225
Hansen, James, 27
health care, 362
Himalayas, 74
Hindenburg, 103
Holistic Management, 252–260, 294–297
Holzer, Sepp, 247–249
homes. *See* housing
honeybee, 31
horses, 235
housing. *See also* passivhaus
   building code, 147–152
   heating and cooling loads, 131–143
   insulation, 139
   orientation, 136
   phase-change materials, 136
   roofs, 131–134
   Solar Reflectance Index (SRI), 133
   thermally comfortable interiors,
      134–136
   ventilation, 136–138
Howard, Philip, 381
Hydraulic fracturing. *See* natural gas:
      shale gas
hydrogen
   source of power, boats, 126–127
   specifications, North America grid,
      199
Hyperion Power Generation, 117, 118
ice ages, 6, 9
ice melt, 74–75
India, 31, 72, 73, 137, 306
Indo-Gangetic Plain, 48–51
insecticides, 32
Institute of Cultural Affairs (ICA),
      392–393

interglacial. *See* ice ages
invasive species
   animals
      Balsam woolly adelgids, 25
   plants, 24–25
   sea, 43–44
Iran, 72, 338
Iritani, Akira, 287
iron and steel, 349
   electric arc furnace, 349
irrigation, 270–272
   effects of, 81
Israel, 96
Ivanpah Solar Power Facility, 174
Jamaica, 38, 39
Japan, 31
jatropha, 111
jellyfish, 46–47
Jones, Christine, 319
Jordan, 280
Joseph, Peter, 364
Joshua tree, 12
Judy, Greg, 256
junk mail, 268
Kentucky coffee tree, 12
Kilimanjaro Energy, 325
Kindle, 269
Köhler, 420
labor. *See* workforce
Lal, Rattan, 318
landfill, 348
Land Institute, 241
laws, effects of, 380–383
   transformation, 381
Lawton, Geoff, 280
Leaf. *See* Nissan Leaf
Leopard, 13
lighting, 144–147
   Core Sunlighting System, 144–145
   fluorescent lightbulbs, 145
   Light Emitting Diode (LED), 146–147,
      213
   light pipe, 144
Lincoln Composites, 126
lions, 294
lithium, in batteries, 98–99
litigation, 381, 382
Little Ice Age, 5
livestock
   emissions from, 8

environment, and the. *See* Holistic
  Management
inventory, 258–260
Living Machines. *See* water reclamation
  system
lumber. *See* wood
Madagascar, 286
Maliwada, 393
mammoths, 11, 287, 289
manatees, 39
Manatees, 46
manure, human, 214–217
marine transport, 114–117, 117–126,
  126–127
  fuel saving measures, 123–124
  nuclear powered, 117–126
  ocean liners, 114–116
  power source for smaller boats,
    126–127
The Matrix, the movie, 279, 360
McGrail, Pete, 329
Mediterranean, 80
"megafauna fruit", 285
melaleuca tree. *See* invasive species:
  plants
mercury, 36
methane
  emissions, 300–303
  generation, 223–224
methane hydrate, 300
Middle East, 76
Mid Miocene Climatic Optimum period, 9
Mitsubishi Heavy Industries, 313
Mohr, Steve, 59, 69, 70
moneymaking, 377–378, 410
Monteverde Cloud Forest Preserve in
  Costa Rica, 32
Mount Pinatubo, 5
mycorrhizal fungi, 319
naphtha, 224, 456
NAS battery, 212
National Wildlife Refuge, 388
natural gas, 60
  coalbed methane, 60
  shale gas, 60–61
  tight gas, 61
neonicotinoids, 31
newspapers, 268
New Zealand, 9, 37, 39
Nissan Leaf, 90
nitrogen, 240

bio-availability, 237–238
noise pollution, 36
North Africa, 80
North Korea, 412
nuclear power
  accidents. *See* Chernobyl Disaster;
    See Fukushima
  acts of terror, potential, 119–120, 210
  integral fast reactor (IFR), 206–211
  long term storage, 121–123
  nuclear power plants, existing, 168
  reserves, 125
nutrition. *See* food
ocean. *See* sea
ocean acidification, 36–37
octopuses, 36
oil (petroleum), 57–59
  extra-heavy oil, 58
  Natural bitumen, 57
  oil shale, 58–59
oil sands (tar sands). *See* oil (petroleum):
  Natural bitumen
Okiluoto, Finland, 121
Olivine, 419, 475
Open Source Ecology, 393
open-source economy. *See* economy:
  called-for ("open source")
Open Yale Courses, 389
orbit, Earth, 6
Oriental white-back vulture, 31
overpopulation, 331–339
  ecological costs, 332–333
  Iran, 338
  rates of, trends, 332
  related targets, 334–335
  two-child policy, 335–339
ozone, 315
ozone, ground level, 83
Pacific Pyrolysis, 323
Pakistan, 72, 73, 73–74
Pampas, 12
paper, 268–270
passivhaus, 138–143
  cost, 141
  energy recovery ventilator, 140–141
  envelope, insulation, 140
  Isabella Eco Home, Minnesota, 142
  Le Boise House, 138
  Louisiana, 138
  Mediterranean, 141–142
  subarctic climate, 143

windows, 139
pasture cropping, 241–242
paved area, globally, 132
"Peak Oil". *See* fossil fuel: scarcity and
     depletion
Peepoo bag, 221
perennial grains, 240–241
permaculture, 246–249, 280
     birds, 249
     livestock, 248
     urban setting, 249–252
permafrost, 301–303, 311
Peru, 16
Peterson, Chad, 253, 255
petrochemicals, 57
pets, 234–235
phosphate, 78
Phosphorus, 78
photovoltaic, 181–183
     performance, 181
     proposed location, 182, 198
     raw materials, 198
     roof-mounted, 414–415
     thin-film PV, 181, 415–416
Pied Flycatcher, 29
Mount Pinatubo, 5, 315
plantations, wood, 269
plasma arc, 225–227
plastic, debris, 35–36
plastic, production, 349–350
Pleistocene Park, 287, 288
Pliocene Epoch, 9, 26
plowing, effects of, 79–80
poaching, 13, 293
pollutants. *See* toxins
polonium, 357
Polyface Farm, 260–262
population. *See* overpopulation
poverty, 361, 391
     charity, 392
     Institute of Cultural Affairs (ICA),
        392–393
power, electricity. *see also individual renew-*
        able technologies, e.g., wind turbines
     capacity, 155
     capacity factor, 155
     coal and natural gas power plants, 167
     Dams, 167–168, 272
     energy, measurement units, 155
     HVDC, 158

land requirements, North America,
     193
off grid, 211–213
     cooking, 213
present and future energy needs,
     North America, 160–167
     boats, power requirements, 163
     industrial sector, power require-
        ments, 162
     residential and commercial sec-
        tors, power needs, 161
     road vehicles, power require-
        ments, 164
superconducting cable, 156–160
voltage source converters (VSC), 157
power grids, transcontinental, 153–160
     land requirements, North America,
        193–196
     raw materials, 195, 199
     workforce, 202
predators, 291–297
pronghorns, 11
property, 388, 394–395
PyroChar 4000, 323
Quaternary Megafaunal Extinction,
     11–14, 285–287
RadWaste, 123
rail transport. *See* trains
Rann of Kutch, 28
recycling. *See also* water reclamation
     system
     graywater, 228–230
     rainwater, 227–228
renewable energy. *see individual technolo-*
        gies (e.g., wind turbines) solar power
        towers
rewilding, nature restoration
     core areas, 281–284, 292
     corridors, 281–284
     Great Plains, North America, 289–297
     predators, role, 291–297
     Siberia, 287, 288
     via livestock, 294–297
rhinoceros, 13
Rice Solar Energy Project, 174
right of passage, residence, and work,
     385–386
roads, 134
road vehicles, 89–99, 99–105
     electric cars, 89–99. *See also* Nissan
        Leaf

battery, 95–96, 97
    buses, 97
    charging, 91–92
    electromagnetic radiation, 97
    lithium reserves for battery, 98–99
    motorcycles, 97
    vehicle-to-grid (V2G), 189–190
    heavy commercial vehicles, 99–105
        aerodynamics, 104–105
    hydrogen-powered vehicles, 99–105
        hydrogen fuel station, 102–103
Rodale Institute, 237, 238
Roman Empire, 39
Romans, 79
Russia, 412
Sasol, 111
Savory, Allan, 255
Schmidt-Bleek, 339, 341
sea
    acidification, 36–37
    dead zones, 45
    deep sea, 40–41
    dumping, 41, 44
    Mediterranean Sea, 41
    overfishing, 39–41
    Sea of Japan, 46
    sewage, 32, 45
    trawling, 41
    warming, 41–42
sea level rise, 26–27, 52–56
    Bangladesh, 55
    Egypt, 54, 55
    Mekong Delta, 55
    mitigation, 55–56
    Salt marshes and mangroves, 27
    Viet Nam, 55
seals, 36
seaweeds, 43
Second Life, 378
sequestration. See carbon capture and
        storage
Shaefer, Kevin, 307
Shanghai, 48, 52
shark rays, 42
sharks, 39
Shasta ground sloth, 12
Siberia, 285, 289
silicon, 181, 198, 200, 451
Silver, 201
sixth mass extinction, 33
SkySails, 123–124

Sky Windpower, 206
Smith Electric Vehicles, 97
Smith, Russell, 80
Smits, Willie, 281
smoking, 355–358
soil, 76–78
    earthworms, 77
    erosion, 77, 78, 79, 79–80
    formation of, 76
    lack of micro nutrients, 81–82
solar cookers. See cooking stoves, solar
Solar heaters, 131
solar photovoltaic. See photovoltaic
solar power towers, 174–180
    day and night time installations,
        configuration, 176–179
    environmental impact, 179
    optimized configuration, 180
    proposed location, 180, 196
    raw materials, 196–197
South America, 12, 289
steel. See iron and steel
Steller's sea cows, 38
Stinging Limu, 46
Sudan, 412
sulfur aerosols. See geoengineering, sulfur
        into the atmosphere
Sundarbans, 27
SunFire Solutions, 213
sunspot activity, 5
supergrid. See power grids,
        transcontinental
Sutherland, Rory, 341
switch station. See Better Place
syngas, 324
Tasmanian devil, 30
taxation, 358, 360, 376
terrorism, threats of. See nuclear power:
        acts of terror, potential
tiger, 13
Tipton, Jerrie, 294
tobacco. See smoking
Tonga, 35
toxins
    land, 30–32
    sea, 44–45
trains, 127–130
trees. See also wood
    acacia tree, 238, 286
    chestnut tree, 25
    palmer oaks, 289

Walnut tree, 22
    yellow-cedar, 22
trophic cascade. *See* predators
turtles, 39
Tyrano, 101, 102
unification, 385–386
United Nations, 395
United States
    Alaska, 22
    Deep South, 306
    Great Lakes, 12
    Great Plains, 11, 12
    Midwest, 9
    Mojave Desert, 12
    Southwest deserts, 22, 75
    western regions, 22
UNSCEAR, 207
uranium, 118, 119, 123, 125, 168, 169, 209,
    210
urine, 217–219
    composition, 218
US Constitution, 398
vactrain, 129–130
Van Gogh, 363
vehicles. *See* road vehicles
vehicle-to-grid (V2G), 189–190
Vertical farming, 251
Voisin, Andre, 255
volcanic activity, 5
water heaters, 131
water reclamation system, 228–230
water shortage, 3, 71–75, 270–271. *See
    also* drought
    Indus River, 74
weather, 9
West Antarctic Peninsula, 35, 42
Wet Bulb Globe Temperature (WBGT),
    49–51
wet-bulb temperature, 52
whales, 36, 37, 39
Wikipedia, 345, 363
wildfires, 21, 285
Wildlands Network, 283, 461
Windcatcher X-Air, 137
wind turbines, 169–174
    at sea, 170
    footprint, 170
    high-altitude wind power, 205–206
    performance, 171
    raw materials, 195–196
    related animal fatalities, 174

wind energy and hydrogen (storage),
    183–189
wood, 268–270
    paper, 268–270
    plantations, 269–270
    wood fuel, 268
workforce, 383–385
workplace environment, 380–383
world government. *See* government:
    world government
Yangtze River Delta, 48–51
Greater Yellowstone Ecosystem, 22
Yupo paper, 269
Zeitgeist Movement, 364
ZemShip, 126, 446
Zidisha, 395
Zimov, Sergey, 288
Zonda, 97

16627942R10285

Made in the USA
Charleston, SC
03 January 2013